AF352798

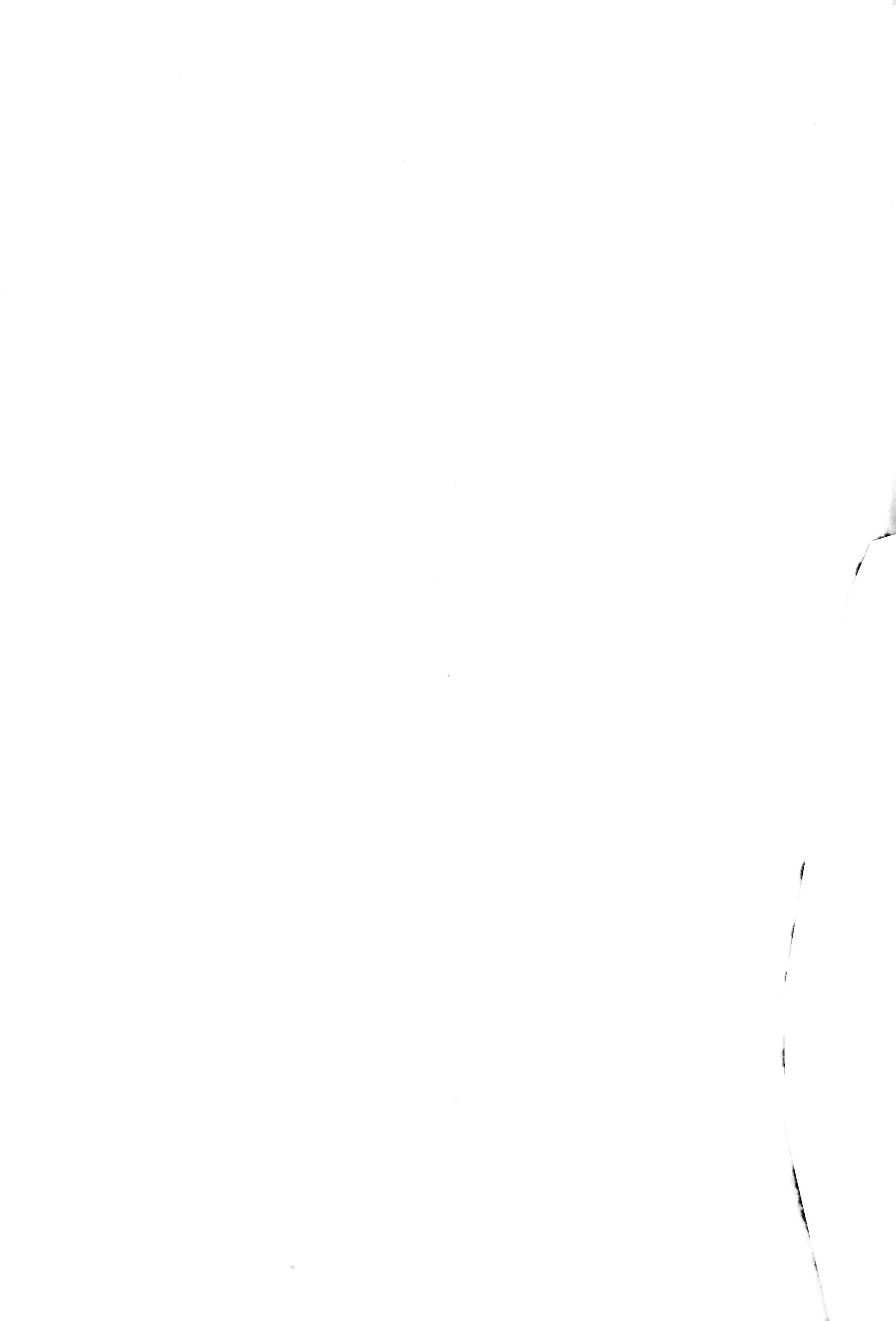

Earth Observation in Forest Biophysical/Biochemical Parameter Retrieval

Earth Observation in Forest Biophysical/Biochemical Parameter Retrieval

Editors

Prashant K Srivastava
Ramandeep Kaur M. Malhi
Mukunda Dev Behera
G. Sandhya Kiran
Prem Chandra Pandey
George P. Petropoulos

MDPI • Basel • Beijing • Wuhan • Barcelona • Belgrade • Manchester • Tokyo • Cluj • Tianjin

Editors
Prashant K Srivastava
Banaras Hindu University
India

Ramandeep Kaur M. Malhi
Banaras Hindu University
India

Mukunda Dev Behera
Indian Institute of Technology
(IIT)
India

G. Sandhya Kiran
M S University of Baroda
India

Prem Chandra Pandey
Shiv Nadar University
India

George P. Petropoulos
Harokopio University of
Athens
Greece

Editorial Office
MDPI
St. Alban-Anlage 66
4052 Basel, Switzerland

This is a reprint of articles from the Special Issue published online in the open access journal *Remote Sensing* (ISSN 2072-4292) (available at: https://www.mdpi.com/journal/remotesensing/special_issues/Forest_Biophysical_Retrieval).

For citation purposes, cite each article independently as indicated on the article page online and as indicated below:

LastName, A.A.; LastName, B.B.; LastName, C.C. Article Title. *Journal Name* **Year**, *Volume Number*, Page Range.

ISBN 978-3-0365-6525-5 (Hbk)
ISBN 978-3-0365-6526-2 (PDF)

Contents

About the Editors

Prashant K Srivastava

Prashant K Srivastava: Dr Srivastava received his Ph.D. from Department of Civil Engineering, University of Bristol, Bristol, United Kingdom, sponsored by British High Commission, United Kingdom under the Commonwealth Scholarship and Fellowship Plan (CSFP) and MHRD, GOI. He received his B.Sc. in Agriculture Sciences from the Institute of Agricultural Sciences, Banaras Hindu University (BHU), and his M.Sc. in Environmental Sciences from School of Environmental Sciences (SES), Jawaharlal Nehru University (JNU), India. At present, he is working as an Assistant Professor at the Institute of Environment and Sustainable Development, Banaras Hindu University, Varanasi, India. He is the recipient of several awards, such as NASA Fellowship, United States of America; University of Maryland Fellowship, United States of America; Commonwealth Fellowship, United Kingdom. Additionally, he received JRF-NET fellowships from Council of Scientific and Industrial Research and University Grant Commission, Government of India. He is leading many national and international projects funded by reputed agencies. He has published over 140 peer-reviewed journal papers, seven books, many book chapters, and has presented his work in several conferences. He is also serving as an associate editor of many scientific peer-reviewed journals. He is also a member of European Geosciences Union (EGU), Indian Society of Geomatics, Indian Society of Remote Sensing, Indian Association of Hydrologists (IAH), and International Society for Agro-meteorology (INSAM), International Association for Hydro-Environment Engineering and Research (IAHR), and International Association of Hydrological Sciences (IAHS).

Ramandeep Kaur M. Malhi

Ramandeep Kaur M. Malhi (Ph.D) is working as an Assistant Professor in the Department of Botany, Faculty of Science at The Maharaja Sayajirao University of Baroda in Vadodara, Gujarat, India. Prior to her current position, she was the recipient of a DST-SERB National Postdoctoral fellowship, for which she worked at Banaras Hindu University in Varanasi, Uttar Pradesh, India. Dr. Malhi received her Bachelor's, Master's and Doctoral degrees in Botany from The Maharaja Sayajirao University of Baroda, Vadodara, Gujarat, India. Her research interests include hyperspectral remote sensing, forestry and agriculture.

Mukunda Dev Behera

Mukund Dev Behera: Prof. Mukunda Dev Behera has made outstanding contributions to the fields of forest remote sensing and ecological climatology through theorizing, modeling, and conducting innovative experiments and field-based measurements. His research innovations have fundamentally transformed the study of phytogeography, paying attention to the ecological functioning of three ecosystem components: plant diversity, water, and energy. His research highlighted the spatial variations in climate with plant diversity and quantified the contribution of climate drivers in an Indian context. He has developed innovative protocols for inclusive biodiversity assessment and new methods of estimating a range of vegetation photosynthetic/structural variables using a suite of satellite data products and modeling protocols. Dr. Behera has guest edited six Special Issues in journals and served on the editorial boards of three Springer journals (*Biodiversity and Conservation, Environmental Monitoring and Assessment*, and *Tropical Ecology*). With over 20 years of research and teaching experience, he has published over 100 papers in journals and supervised 11 Ph.Ds. and 3 post-doctoral researchers.

G. Sandhya Kiran

Prof. Dr. G. Sandhya Kiran (Ph.D.) is Professor and Former Head of Department of Botany, Faculty of Science at The Maharaja Sayajirao University of Baroda in Vadodara, Gujarat, India. She earned her undergraduate, graduate and Ph.D. degrees in Botany from the same university. She conducted postdoctoral research at Department of Geography, Geology and Anthropology at Indiana State University, Terre Haute, U.S.A. She was awarded the Boys cast fellowship by the Department of Science and Technology, New Delhi, India. Her areas of interest are plant ecology, remote sensing and plant physiology.

Prem Chandra Pandey

Prem Chandra Pandey: Dr. Pandey received his PhD from University of Leicester, United Kingdom, under Commonwealth Scholarship and Fellowship Plan. He completed his post-doctoral research at Tel Aviv University Israel. Currently, he is working as an Assistant Professor at School of Natural Sciences, Shiv Nadar Institution of Eminence (deemed to be University), UP India. Previously, he has been associated with Banaras Hindu University India as SERB-NPDF fellow. He received B.Sc and M.Sc. degrees from Banaras Hindu University and M.Tech degree (Remote Sensing) from Birla Institute of Technology, India. He has worked as a Professional Research fellow on remote sensing applications in the National Urban Information System funded by NRSC Government of India. He has been a recipient of several awards, including Commonwealth Fellow United Kingdom, INSPIRE fellow GoI, MHRD-UGC fellow GoI, Malviya Gold Medal from Banaras Hindu University, SERB-NPDF from Government of India and Young Investigator Award. Dr. Pandey is working in three projects related to the monitoring of wetlands/chilika lake, mainly focusing on ramsar sites, along with other natural-resources-based research work funded by the NGP and SERB Government of India. He has published several peer-reviewed journal papers ¿50, 7 Edited Books, several book chapters and presented his work at national and international conferences. He is serving as a member of the editorial board for Geocarto International Journal, Taylor & Francis and acted as guest editor for Remote Sensing, MDPI. Additionally, he is also a member of the Indian Society of Geomatics (ISG), Indian Society of Remote Sensing (ISRS), IUCN-CEM (2017-2025), Society of Wetland Scientists (2021-2022), SPIE and AAG. Dr. Pandey focuses his research on remote sensing for natural resources, including forestry, agriculture, urban studies, environmental pollutant modeling and climate change.

George P. Petropoulos

George Petropoulos: Dr. George P. Petropoulos is an Assistant Professor in Geoinformation at Harokopio University of Athens, Greece. He received his Ph.D. (2008) from King's College London, his M.Sc. in Remote Sensing (2003) from University College London (UCL), and his B.Sc. in Natural Resources Development and Agricultural Engineering (1999) from the Agricultural University of Athens, Greece. His research focuses on the use of EO, alone or synergistically with land surface process models, in deriving key state variables of the Earth's energy balance and water budget. He also has a strong interest in the development of EO and GIS geospatial analysis techniques in geohazards (mainly floods, wildfires, and frost) and in quantifying land use/cover and its changes using technologically advanced EO-sensing systems and synergistic multisensor modeling techniques. He also contributes to the development of open source software tools in EO modeling and the benchmarking of EO operational algorithms/products and surface process models. Dr. Petropoulos currently serves as a council member of the Remote Sensing and Photogrammetric Society (RSPSoC). He is the editor of SENSED (the RSPSoc Newsletter), and an associate editor or editorial board member of several international scientific journals focusing on EO and environmental modeling. He has edited six books and has coauthored more than 80 research articles in international peer-reviewed journals.

Article

Use of Hyperion for Mangrove Forest Carbon Stock Assessment in Bhitarkanika Forest Reserve: A Contribution Towards Blue Carbon Initiative

Akash Anand [1], Prem Chandra Pandey [2,*], George P. Petropoulos [3,4], Andrew Pavlides [4], Prashant K. Srivastava [1,5], Jyoti K. Sharma [2] and Ramandeep Kaur M. Malhi [1]

[1] Institute of Environment and Sustainable Development, Banaras Hindu University, Varanasi 221005, India; anand97aakash@gmail.com (A.A.); deep_malhi56@yahoo.co.in (R.K.M.M.); prashant.iesd@bhu.ac.in (P.K.S.)

[2] Center for Environmental Sciences and Engineering, School of Natural Sciences, Shiv Nadar University, Greater Noida, Uttar Pradesh 201314, India; jyoti.sharma@snu.edu.in

[3] Department of Geography, Harokopio University of Athens, El. Venizelou St., 70, Kallithea, Athens 17671, Greece; gpetropoulos@hua.gr

[4] School of Mineral Resources Engineering, Technical University of Crete, Crete 73100, Greece; apavlidis@isc.tuc.gr

[5] DST-Mahamana Centre for Excellence in Climate Change Research Institute of Environment and Sustainable Development, Banaras Hindu University, Varanasi 221005, India

[*] Correspondence: prem.pandey@snu.edu.in or prem26bit@gmail.com; Tel.: +91-9955303852

Received: 6 December 2019; Accepted: 5 February 2020; Published: 11 February 2020

Abstract: Mangrove forest coastal ecosystems contain significant amount of carbon stocks and contribute to approximately 15% of the total carbon sequestered in ocean sediments. The present study aims at exploring the ability of Earth Observation EO-1 Hyperion hyperspectral sensor in estimating aboveground carbon stocks in mangrove forests. Bhitarkanika mangrove forest has been used as case study, where field measurements of the biomass and carbon were acquired simultaneously with the satellite data. The spatial distribution of most dominant mangrove species was identified using the Spectral Angle Mapper (SAM) classifier, which was implemented using the spectral profiles extracted from the hyperspectral data. SAM performed well, identifying the total area that each of the major species covers (overall kappa = 0.81). From the hyperspectral images, the NDVI (Normalized Difference Vegetation Index) and EVI (Enhanced Vegetation Index) were applied to assess the carbon stocks of the various species using machine learning (Linear, Polynomial, Logarithmic, Radial Basis Function (RBF), and Sigmoidal Function) models. NDVI and EVI is generated using covariance matrix based band selection algorithm. All the five machine learning models were tested between the carbon measured in the field sampling and the carbon estimated by the vegetation indices NDVI and EVI was satisfactory (Pearson correlation coefficient, R, of 86.98% for EVI and of 84.1% for NDVI), with the RBF model showing the best results in comparison to other models. As such, the aboveground carbon stocks for species-wise mangrove for the study area was estimated. Our study findings confirm that hyperspectral images such as those from Hyperion can be used to perform species-wise mangrove analysis and assess the carbon stocks with satisfactory accuracy.

Keywords: blue carbon; hyperspectral data; mangrove forest; carbon stock; Bhitarkanika Forest Reserve; regression models; machine learning

---

1. Introduction

Mangrove forest coastal ecosystems provide several beneficial functions, both to terrestrial and marine resources [1,2]. Mangrove forests contain significant amount of carbon stocks and are one of the sources of carbon emissions [3]. Coastal habitat contributes more than half of the total carbon sequestrated in ocean sediments, only 2% of the total carbon is sequestered by coastal habitat [4]. Mangroves provide essential support to the ecosystem, thus, their decline also results in socio-economic loss. Previous studies demonstrated the existence of mangrove forests in several countries (about 120 in total) including tropical as well as sub-tropical ones, with coverage of 137,760 km^2 across the earth [5]. Recently, Hamilton and Casey (2016) provided key information concerning mangrove forest distribution worldwide. The total mangrove area in India is 4921 km^2, which comprises about 3.3% of global mangroves [6]. Due to their valuable contribution in biomass, carbon sinks as well as numerous other benefits for biodiversity of mangrove forests ecosystem are considered as a valuable ecological and economic resources worldwide [7,8].

Resources are declining and continuously limiting in its spatial extent due to human induced as well as natural factors which is putting pressure with every passing time [9], thus, the rapid altering of the composition, structure, and behavior of the ecosystem and their capability to deliver ecosystem services is declining [10–12]. This decline happens at a fast rate by 0.16% to 0.39% annually at global level [13]. It is estimated that mangroves store 1.23 ± 0.06 Pg of carbon globally sequestered from coastal ecosystem is one of the integral parts of the global carbon circulation [14]. Annually, around 131–639 km^2 of mangrove forests are being destroyed; in terms of overall carbon loss, it goes up to 2.0-75 TgCYr^{-1} [13].

Valiela et al. [15] demonstrated that mangrove forests in tropical countries are the most threatened ecosystems. The major threat is the conversion of mangrove forests in other land use types and categories, such as aquaculture, coastal development, construction of channels, agriculture, urbanization, coastal landfills, and harbors, or deterioration due to indirect effects of pollution [1,16]. Allen et al. [17] described about the impact of natural threats on mangrove forest which includes sea level rise, tropical storm, insects, lightning, tsunami affected [18], and climate change. Yet, those are considered as minor threats, as the mangrove forest degradation rate is much less because of natural causes than anthropogenic factors. Several studies have provided evidence of the decline of mangroves population, which are already critically endangered [15] or approaching the state or verge of extinction in some of countries where these eco-sensitive fragile ecosystems exist (data demonstrated that approximately 26 are listed where mangrove are in grave situation out of a total 120 countries) [12,19]. It is therefore imperative to monitor mangrove forests for their biodiversity, biomass, and carbon stocks at regular time intervals to provide suitable database and help in conservation strategies. There are critical studies [20–22] the mangrove forest ecosystem and its biodiversity in India [23], where authors stressed on the importance of mangrove forests [24] and conservation priorities [21]. Some authors also demonstrated the degradation of mangrove and their impact [20,23–25]. There have been several published studies that focused on assessing the blue carbon stored in the mangroves around the world and in India; yet, a species-wise blue carbon analysis with significant accuracy is missing. Species-wise blue carbon analysis can be used to evaluate the impact of global climate change on different types of mangrove species and can also help in ecosystem services and policy makers to accurately evaluate the ecological as well as economical trade off associated with the management of mangroves ecosystem.

Blue carbon is nothing but the carbon stored and captured in coastal and marine ecosystems in different forms globally, such as biomass and sediments from mangrove forest, tidal marshes, and seagrasses. About 83% of global carbon is circulated through oceans. A major contribution is through coastal ecosystems [4] such as mangrove forests in form of biomass and carbon stocks [26]. Thus, blue carbon stock assessment of tropical regions, especially mangrove forests, is an issue for global change research [27], in order to effectively manage such ecosystems to reduce loss of biomass and carbon stock. Therefore, these ecosystems provide an exceptional candidate for research such as carbon change mitigation program such as REDD+ (Reducing Emissions from Deforestation and Forest Degradation) in third world countries or developing countries [28–30] and Blue Carbon studies around the coastal regions in the world [31,32]. The coastal line covers a large area, which can be surveyed at a high temporal resolution with a very cost-effective way through remote sensing approach and is able to generate databases for each of the mangrove forest sites. Use of technologies such as Remote

Sensing is crucial as a tool for assessing and monitoring mangrove forests, primarily because many mangrove swamps are inaccessible or difficult to field survey [33].

Previous work by the authors as well as other researchers has allowed assessing the biomass of the several mangrove plant species and has provided the biomass of species individually. Chaube et al. [34] employed AVIRIS-NG (Airborne Visible InfraRed Imaging Spectrometer Next Generation) hyperspectral data to map mangrove species using a SAM (Spectral Angle Mapper) classifier. Authors identified 15 mangrove species over Bhitarkanika mangrove forest, reporting an overall accuracy (OA) of 0.78 (R^2). They also concluded that the hyperspectral images are very useful in discriminating mangrove wetlands, and having a finer spectral and spatial resolution can be crucial in investigating fine details of ground features. Kumar et al. [35] used the five most dominant classes of mangrove species present in Bhitarkanika as training sets to classify using SAM on Hyperion hyperspectral images, and archived an OA of 0.64. Ashokkumar and Shanmugam [36] demonstrated the influence of band selection in data fusion technique; they performed classification using support vector machine and observed that factor based ranking approach shown better results (R^2 of 0.85) in discriminating mangrove species than other statistical approaches. In another study, Padma and Sanjeevi [37] used an identical algorithm by integrating Jeffries-Matusita distance and SAM to map the mangrove species within the Bhitarkanika using Hyperion Image with an OA of 0.86 (R^2 value).

Presently, the spatial distribution maps of mangroves are generated using Earth Observation (EO) Hyperion datasets [26]. Table 1 illustrates the wetland research, which employed several algorithms for the assessment using various data types. Identifying different species in a mangrove forest is a fundamental yet difficult task, as it requires a high spatial and spectral resolution satellite images. To identify different species within the study area, EO-1 Hyperion hyperspectral data is currently acquired and field-sampling points are taken to generate the endmember spectra. This study demonstrated the use of vegetation indices (in this paper NDVI (Normalized Difference Vegetation Index) and EVI (Enhanced Vegetation Index)) for estimating carbon stock within an area with a significant accuracy. Presently, the field inventory data were incorporated with the hyperspectral image to derive the carbon stock. Three different NDVI and EVI based models were used to determine the total blue carbon sequestered by each species within the study area.

In purview of the above, this study aimed at evaluating the net above ground carbon stocks present at Bhitarkanika mangrove forest ecosystem, particularly with relevant field inventory and remote sensing approaches.

Table 1. Showing the recent studies in mangrove classification and mapping using different techniques.

Technique Used	Datasets	Study Location	Ref.	Year
Maximum Likelihood Classifier (MLC)	Aerial Photographs	Texas, USA	[38]	2010
MLC and The Iterative Self-Organizing Data Analysis Technique (ISODATA) algorithm	Landsat, Radar Satellite (RADARSAT), Satellite Pour l Observation de la Terre (SPOT)	Vietnam	[39]	2011
MLC	IKONOS	Sri Lanka	[40]	2011
Unsupervised	Landsat and The Linear Imaging Self Scanning Sensor (LISS-III)	Eastern coast of India	[41]	2011
Sub-Pixel	Moderate Resolution Imaging Spectroradiometer (MODIS)	Indonesia	[42]	2013
Spectral Angle Mapper (SAM)	Hyperion	Florida	[34, 43]	2013
Neural Network	Landsat	Global	[44]	2014
Object based	Landsat	Vietnam	[45]	2014

Table 1. *Cont.*

Technique Used	Datasets	Study Location	Ref.	Year
Object based	Advanced Land Observing Satellite (ALOS) Phased Array type L-band Synthetic Aperture Radar (PALSAR)/ Japanese Earth Resources Satellite 1 (JERS-1) Synethetic Aperture Radar (SAR)	Brazil and Australia	[46]	2015
Hierarchical clustering	Hyperspectral Imager for the Coastal Ocean (HICO) and HyMap	Australia	[47]	2015
Tasseled cap transformation	Landsat	Vietnam	[48]	2016
NDVI	Landsat	Vietnam	[49]	2016
MLC	IKONOS, QuickBird, Worldview-2	Indonesia	[50]	2016
Object based Support Vector Machine	SPOT-5	Vietnam	[36, 51]	2017
Iso-cluster	Landsat	Madagascar	[52]	2017
Random Forest	Landsat	Vietnam	[53]	2017
K-means	Landsat	West Africa	[54]	2018
Decision Tree	Landsat	China	[55]	2018
Data Fusion	ALOS PALSAR & Rapid Eye	Egypt	[56]	2018
	Compact Airborne Spectrographic Imager (CASI) and Bathymetric Light Detection and Ranging (LiDAR)	Mexico	[57]	2016
Structure from Motion (SfM) Multi-View Stereo (MVS) Algorithm	Unmanned Aerial Vehicle (UAV)	Australia	[58]	2019
Hybrid decision tree/ Support Vector Machine (SVM)	Hyperspectral	Galapagos Islands	[33]	2011
Hierarchical cluster analysis	Compact Airborne Spectrographic Imager (CASI)	South Caicos, United Kingdom	[59]	1998
Feature Selection Algorithm	CASI	Galeta Island, Panama	[60]	2009
SAM	Airborne Imaging Spectrometer for Applications (AISA)	South Padre Island, Texas	[61]	2009
SVM	Earth EO-1 (Earth Observation) Hyperion	Bhit arkanika National Park, India	[35]	2013
MLC & Hierarchical neural network	CASI	Daintree river estuary, Australia	[62]	2003
Object based Classification	UAV based Hyperspectral Image	Qi'ao Island, China	[63]	2018
SAM	Airborne Visible/Infrared Imaging Spectrometer (AVIRIS)	Everglades National Park, Florida, USA	[64]	2003
SAM	EO-1 Hyperion	Talumpuk cape, Thailand	[65]	2013
Pixel based and Object based classification	CASI-2 (CASI-2)	Brisbane River, Australia	[66]	2011
SAM	Airborne Visible/Infrared Imaging Spectrometer—Next Generation (AVIRIS-NG)	Lothian Island and Bhitarkanika National Park, India	[34]	2019

2. Materials and Methods

2.1. Study Area

Our study site is located in the Kendrapara district of Odisha, India, which lies between 20°41′36.70″ and 24°45′28″ N latitude and 86°54′17.29″ and 86°92′8.96″ E longitude (as shown in Figure 1). Geographically, it covers an area of around 41.05 Km² of which mostly low-lying (10–25 m above mean sea level) covered with dense mangrove forests. The Bhitarkanika Forest Reserve is a protected forest reserve with a unique habitat and ecosystem. About two-third of the Bhitarkanika Forest Reserve is covered by the Bay of Bengal, and this estuarial region (lies within Bramhani-Baitarni) is a predominant inter tidal zone. Bhitarkanika Forest Reserve is home to a diverse types flora and fauna including some endangered species; it is the second largest mangrove forest in India formed by the estuarial formation of Brahmani-Baitarni, Dhamra, and Mahanadi rivers [67].

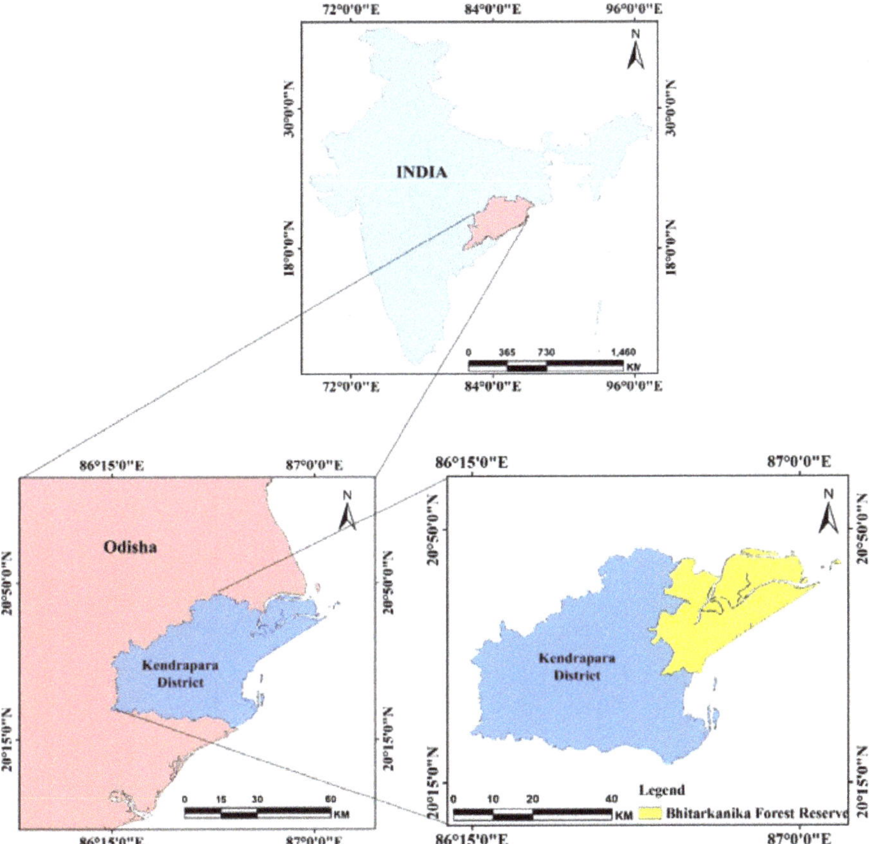

Figure 1. Location map of the Bhitarkanika Forest Reserve, Odisha India.

The study area comes under the humid sun-tropical climatic region broadly having three seasons namely, summer in which the temperature reaches up to 43 °C, winter in which the temperature goes down to as low as 10 °C, and the rainy season in which this region faces flash floods and frequent cyclones between the months of June to October. The Bhitarkanika Forest Reserve was chosen for the present study because it contains variety of heterogeneous species. In our work, the 10 most dominant mangrove species (as shown in Table 2) were identified and used for further analysis.

Table 2. In-situ measurements of different mangrove species in the Bhitarkanika forest reserve.

	Species	Tree Height (m)	Diameter at Breast Height (DBH) (cm)	No of Trees	Wood Density (g/cm^3)	Stem volume (m^3)	Biomass $(t.\ ha^1)$	Carbon stock $(t.\ C\ ha^1)$
1	*Excoecaria agallocha* L.	18.45 ± 2.11	20.14 ± 2.56	11	0.49	6.46	222.74 ± 11.17	104.68 ± 5.24
2	*Cynometra iripa* Kostel	17.23 ± 1.62	16.54 ± 4.39	10	0.81	3.70	231.43 ± 29.09	108.77 ± 13.67
3	*Aegiceras corniculatum* (L.)	15.03 ± 1.82	22.17 ± 2.81	9	0.59	5.22	262.44 ± 13.84	123.34 ± 6.50
4	*Heritiera littoralis Dryand ex Ait.*	18.17 ± 2.17	17.21 ± 2.56	10	1.06	4.22	339.13 ± 23.85	159.39 ± 11.21
5	*Heritiera fomes* Buch.-Ham.	12.35 ± 1.03	18.83 ± 2.94	12	0.88	4.13	287.66 ± 12.81	135.20 ± 6.02
6	*Xylocarpus granatum* Koenig	14.13 ± 2.01	27.52 ± 4.28	5	0.67	4.20	379.64 ± 38.10	178.43 ± 17.90
7	*Xylocarpus mekongensis* Pierre	15.38 ± 1.98	20.28 ± 3.40	8	0.73	3.97	162.13 ± 26.30	76.20 ± 12.36
8	*Intsia bijuga* (Colebr.) Kuntze	12.29 ± 1.38	26.69 ± 4.90	9	0.84	6.18	196.92 ± 32.78	92.55 ± 15.40
9	*Cerbera odollam* Gaertn.	12.24 ± 1.86	28.56 ± 5.05	6	0.33	4.70	355.36 ± 24.69	167.01 ± 11.60
10	*Sonneratia apetala* Buch.-Ham.	11.25 ± 1.67	21.85 ± 4.06	10	0.53	4.22	351.14 ± 23.14	165.03 ± 10.87
	Average						278.86 ± 23.57	131.06 ± 11.08

2.2. EO Data Acquisition

EO-Hyperion images (L1Gst) were obtained over the study area from the United States Geological Survey (USGS). The specifications of Hyperion sensor are illustrated in Table 3. Hyperion has a spatial resolution of 30 m and 242 spectral bands covering 356 nm to 2577 nm wavelengths. The Hyperion data strip passing over Bhitarkanika Forest Reserve is shown in Figure 2. Out of the 242 spectral bands, 46 bands are considered as bad bands (including 1–7, 58–78, 120–132, 165–182, 185–187, and 221–242 bands), and thus, these were not considered in further analysis. Bad bands have a high amount of noise caused by the water absorption in atmosphere, band overlaps, and lack of proper illumination. The performed image pre-processing includes noise removal and cross track illumination correction. In addition, atmospheric correction has been applied to remove atmospheric noises using the FLAASH (Fast Line-of-sight Atmospheric Analysis of Hyper Spectral-cubes) module in ENVI (v. 5.2) software [68]. After completing this step, endmember extraction was performed for each of the targeted species using the final Hyperion reflectance image and the in-situ GPS (Global Positioning System) locations.

Table 3. Hyperion Data Description

Satellite Data	EO-Hyperion
Path/Row	139/45
Spatial Resolution	30 meters
Flight Date	31 December 2015
Inclination	97.97 degree
Cloud Cover	<5%

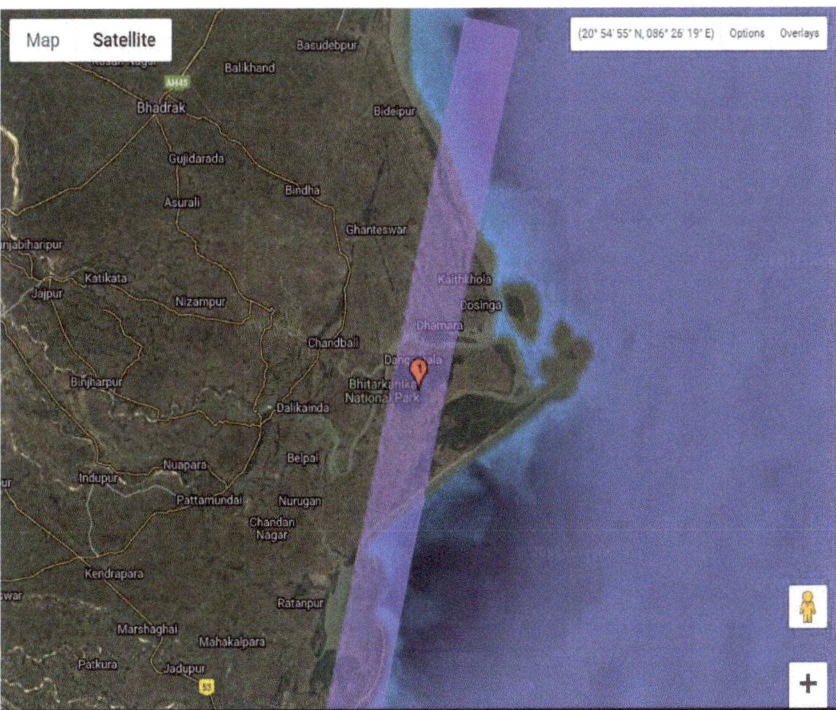

Figure 2. Footprint of Hyperion data available for the Bhitarkanika Forest reserve; it illustrates the region covered for Hyperion data for conducting the present study.

2.3. Field-Inventory Based Biomass Measurement

Field sampling was undertaken during 2015 for the study site. The foremost steps are the prior knowledge of the mangrove plant species; their location and its structure were essential for collecting the sample data for geospatial analysis. Random and the most homogenous patches within the Bhitarkanika Forest Reserve were selected for the field survey to measure tree height, number of samples (trees), Diameter at Breast Height (DBH), and total number of species within the plot.

As the study site selected is 36.42 km^2 falling within the range of Hyperion data strip (Figure 2). Hyperion image has limited coverage over the Bhitarkanika forest range, and for this reason, a region was selected that falls within the area covered by the Hyperion field of view. The samples were collected by making a 90 × 90 m^2 grid and it is further divided into nine equal 30 × 30 m^2 sub-grids, i.e., 90 sub-grids were examined. The most homogenous grid was taken into consideration. This process was then repeated to identify the 10 most homogenous mangrove plant species within the study area and samples were collected using GPS and Clinometer. The field data records the vegetation parameters using GPS in multiple directions. The number of tree species was counted within the plot in random sampling design in the Bhitarkanika Forest Reserve [69]. An overview of the methodology implemented is available in Figure 3. These major species were identified for the study site and their spectral profile was extracted using EO-1 Hyperion dataset. Total area covered by these species was 36.42 km^2 (see Figure 2). Non-vegetative regions were masked out from the study region.

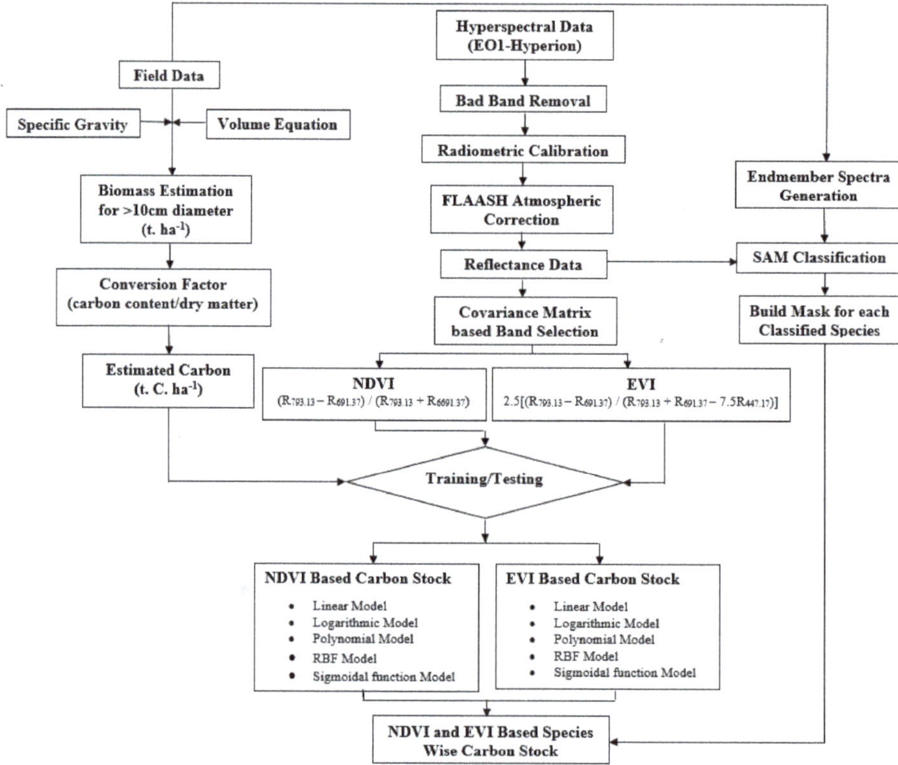

Figure 3. Flowchart providing an overview of the methodology implemented where NDVI stands for Normalized Difference Vegetation Index, EVI stands for Enhanced Vegetation Index and RBF forRadial Basis Function.

The Spectral Angle Mapper (SAM) supervised classification algorithm was used for the land use/cover classification using ENVI software [70,71]. SAM is a physically-based spectral classification algorithm, according to [72] that calculates the spectral similarity between a pixel spectrum and a reference spectrum as "the angle between their vectors in a space with dimensionality equal to the number of bands" [72]. SAM uses the calibrated reflectance data for classification and thus relatively insensitive to illumination and albedo effects. End-member reference spectra used in SAM were collected directly from acquired hyperspectral images. SAM compares the angle between reference spectrum and each pixel of an image in n-D space [72–74]. This 'spectral angle' (α) is calculated as:

$$\alpha = \cos^{-1} \frac{(\, t.r\,)}{(\, \|t\| \, \|r\| \,)} \tag{1}$$

where α is the angle between reference spectra and endmember spectra, t is the endmember spectra, and r is the reference spectra.

A thorough and detailed investigation was performed to develop a criterion to estimate different species and determine variety of communities present in that ecosystem. To perform the sampling, firstly, the area is sub-divided into homogeneous patches or units, and furthermore, the samples were taken within these homogenous patches. The total number of transect sampling units to determine the allowable error was calculated using (Chacko, 1965) as follows:

$$N = \frac{t(CV)^2}{E^2}. \tag{2}$$

where N is the total number of samples, t is the Student's (t-statistics) value at a 95% significance level, CV is the coefficient of variation (in %), and E is the confidence interval (in mean %).

While performing the field sampling, a transect of 30 m $\times$ 30 m plot was laid on the most dominant patch for each species inside the protected area of Bhitarkanika forest reserve. The collected field sampling points were further distributed, and 2/3 of the samples were used for generating the models, whereas 1/3 of the samples were used for validation purpose. Table 2 has shown the field measurements of each species, e.g., scientific name, tree height, DBH, total number of trees within the sample plot, wood density of each species, biomass, and carbon stock. The trees whose girth height was below 1.32 m and DBH < 10 cm were not taken under consideration. The geographical location (latitude and longitude) was recorded using hand-held GPS. There were several mathematical equations developed and used by researchers for biomass estimation of trees [75–81]. These equations are species specific, particularly in the tropics. The general equation has been developed in modified form. It is more general in nature ([78,82,83]) and applicable in field. It is not possible to cut all the trees to estimate their biomass. Considering the mathematical terms, the models were developed by [76,77,83,84]. The model developed by [75] (1989) to estimate above ground biomass has been used in the present investigation. The literature revealed that this method is non-destructive and is the most suitable method. The biomass for each tree is calculated using the following allometric equation [76,83,85]:

$$Y = \exp\left[-2.4090 + 0.9522 \ln \left(D^2 \times H \times S\right)\right]. \tag{3}$$

where Y is above ground biomass (t. ha^1), D is the diameter at breast height, H is the tree height, and S is the wood density. The average wood density (S) for each species is taken from the wood density database provided by the International Council for Research in Agroforestry (ICRAF). From the acquired wood density, it was found that the wood density of *Cerbera odollam* Gaertn. was lowest (0.3349 gcm^3), followed by *Excoecaria agallocha* L. (0.49 gcm^3) among all. *Heritiera littoralis* Dryland ex Ait. had the highest (0.848 gcm^3) wood density. The above ground carbon was calculated using the following formula to estimate biomass [83,85,86]:

$$Y = B * 0.47 \tag{4}$$

where Y is the above ground carbon stock (t. ha^1) and B is the above ground biomass per hectare (t. C ha^1).

The precise location of the in-situ ground control points of each species were further used to generate the spectral profile using Hyperion hyperspectral data as shown in Figure 4. The generated spectra of each species were given as an input to the SAM classifier. It is observed that *Intsia bijuga* (Colebr.) Kuntze is showing the highest reflectance among other observed species, whereas *Aegiceras corniculatum* (L.) has the lowest reflectance.

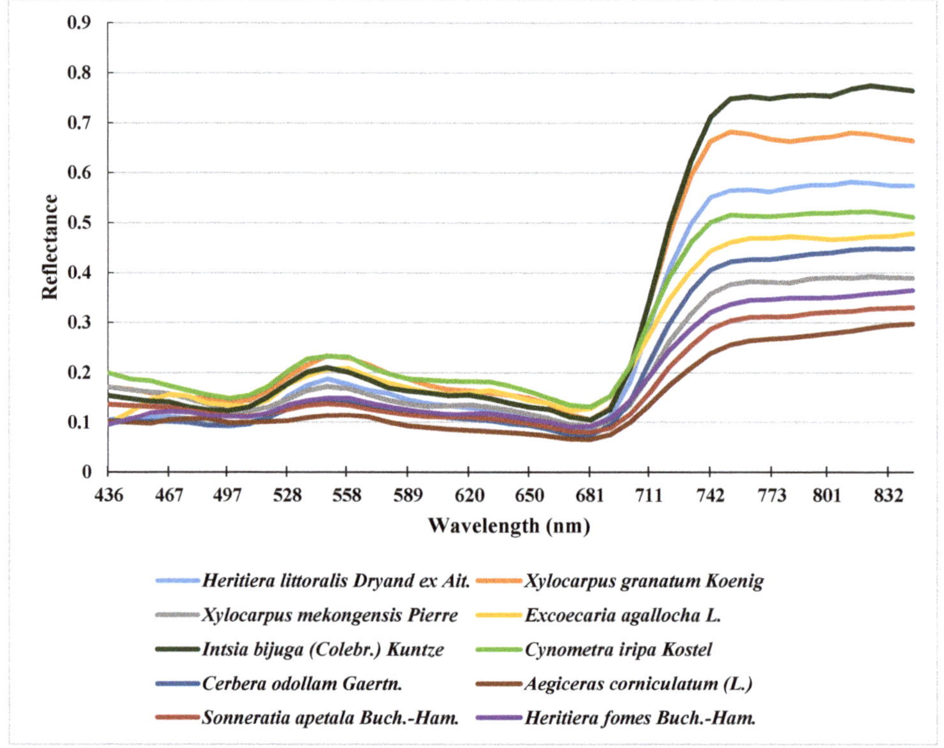

Figure 4. Spectral reflectance curve of the observed mangrove species.

2.4. Covariance Matrix Based Band Selection

Hyperspectral data are a set of hundreds of narrow bands at different wavelengths posing problems related to computational complexity, high data volume, bad bands, etc. Therefore, dimensionality reduction of hyperspectral data is considered as one of the solutions for the aforementioned issue. The dimensionality reduction technique is further classified into two groups, namely, feature extraction and feature selection. In the present study, an approach has been made to select the best band for calculation of different vegetation indices. Band selection generally involves two major steps, which are selection of criterion function and optimum band searching. The selection criterion applied in this study is the one proposed by [87], which was named Maximum ellipsoid volume criterion (MEV).

Mathematically it can be formulated as:

$$J(s) = \det\left(\frac{1}{M-1}\right)S^T S$$

where M is the number of pixels and S is the selected bands with $S = [x_1, x_2, \ldots, x_n]$ and S^T is the column vector with $S^T = [x_1, x_2, \ldots, x_m]^T$. Here, n and m are the number of bands and m is the number of number of pixels.

Additionally, for the band searching purpose, sequential forward search was implemented, which basically works on the principle of "down to top". Here, the first band is defined as the band with maximum variance and the remaining band is compared one by one. While selecting the optimum band, the constant value $\left(\frac{1}{M-1}\right)$. is neglected. Thus, Equation (4) can also be written as:

$$B_k = S_k^T S_k \tag{5}$$

where B_k is the covariance matrix and $S_k = [x_1, x_2, \ldots, x_k]$. Therefore, we have:

$$B_k = S_k^T S_k \tag{6}$$

$$= [x_1, x_2, \ldots, x_k]^T [x_1, x_2, \ldots, x_k]$$

$$= \begin{bmatrix} x_1^T x_1 & x_1^T x_2 & \cdots & x_1^T x_k \\ x_2^T x_1 & x_2^T x_2 & \cdots & x_2^T x_k \\ \cdots & \cdots & \cdots & \cdots \\ x_k^T x_1 & x_k^T x_2 & \cdots & x_k^T x_k \end{bmatrix}$$

According to the rule of determination, the relation between B_k and B_{k+1} is described as:

$$\det(B_{k+1}) = \det(B_k)\left(a_k - d_k^T B_k^{-1} d_k\right) \tag{7}$$

Equation (7) was further used for determining the optimum band; the band that maximizes the value of $\det(B_{k+1})$ was termed as the optimum band. This band selection method was applied at blue, red, and near infrared bands to further calculate the NDVI and EVI indices.

2.5. NDVI and EVI

In our study, the vegetation indices of NDVI and EVI were employed, which were computed from the Hyperion hyperspectral data to assess the total above ground carbon stock using different allometric regression models [26]. The covariance matrix based band selection algorithm as per described in Section 2.4 determines the specific band for the calculation of vegetation indices. It was observed that the optimum band in NIR (Near-Infrared) region is $R_{793.13}$ (surface reflectance at 793.13 nm), in Red region, it is $R_{691.37}$ (surface reflectance at 691.37 nm), and in Blue region the optimum band is observed at $R_{447.17}$ (surface reflectance at 447.17 nm). The NIR and Red bands were used to calculate the NDVI; as shown in Equation (5), its value ranges from -1 to $+1$. The negative NDVI values shows waterbody and bare soil, whereas positive values are the green vegetation. The higher the NDVI value, the higher will the density of forest or vegetation be because of the high NIR reflectance and low Red reflectance coming from dense vegetation [88,89]. NDVI has been widely used to monitor vegetation health, density, changes, amount and condition of vegetation:

$$NDVI = \frac{(R_{793.13} - R_{691.37})}{(R_{793.13} + R_{691.37})} \tag{8}$$

EVI (Enhanced Vegetation Index) was originally developed as an improvement over NDVI; EVI is basically an optimized vegetation index that is used to enhance the sensitivity of high biomass region and it decouples the background variables as well as the atmospheric influences [90,91]. EVI is calculated as follows:

$$EVI = 2.5* \frac{(R_{793.13} - R_{691.37})}{(R_{793.13} + 6*R_{691.37} - 7.5*R_{447.17} + L)} \tag{9}$$

where L is the adjustment factor, generally 1.

In the present study, both NDVI and EVI were employed to correlate the carbon stock of the Bhitarkanika mangrove forest. EVI is considered as more robust proxy of biomass and carbon stock estimation, as it has better resilience to saturation and resistant to atmospheric contamination and soil [90,92].

Five different models, linear, polynomial, logarithmic, Radial Basis Function (RBF), and sigmoidal function, were utilized for assessing carbon using hyperspectral data derived from NDVI and EVI indices. The relationship of field measured above ground carbon with the NDVI and EVI vegetation indices for all the five models were calculated. The field measured above ground carbon was trained with NDVI and EVI values retrieved from hyperspectral image in each of the five models. The 2/3 of the in-situ measurements were used for training the data, while 1/3 of the remaining data were used for testing the models.

3. Results

This section provides a concise and precise description of the experimental results for blue carbon for a mangrove forest.

3.1. Spatial Distribution of Species

This section demonstrates the species-wise carbon stock spatial distribution and overall carbon stock of the Bhitarkanika forest reserve and delivers a brief analysis on the overall results. SAM classification (Figure 5) achieved an OA of 84% and a kappa coefficient (k) of 0.81. These results indicate that SAM classification algorithm performed very well in determining the major plant species. These outputs were further taken into account and were used to derive the estimated carbon stock for each species using NDVI and EVI models and illustrating the species-wise carbon stock.

As per Table 4, it has been observed that the total aboveground carbon from EVI and NDVI derived aboveground carbon are 459.82 kt. C and 514.47 kt. C, respectively. The NDVI derived carbon is showing higher value than the EVI derived carbon because NDVI values can be influenced by the atmospheric contaminants, topography, soil, and dense biomass. These can lead to the increase in the irradiance of the NIR band and result in bias. It should also be noted that NDVI saturates in dense vegetation so that the accuracy of NDVI values differ by land use, topography, and atmospheric conditions [90,93–95]. Santin-Janin et al. [96] used non-linear model coupled with NDVI and EVI estimates to estimate the biomass and carbon stock. Wicaksono et al. [97] employed 13 vegetation indices to assess the above ground carbon of mangrove forest and concluded that the best fitted above ground carbon model for mangrove species derived from vegetation indices was EVI1 (R^2=0.688), whereas for below ground carbon GEMI (R^2=0.567) showed the best fit. Similarly, Adam et al. [95] utilized the narrow band vegetation indices with all possible band combinations using hyperspectral data for above ground biomass and concluded EVI is more robust for the assessment. Different band selections were used by them to enhance the predictive accuracy, the best three combinations for estimating EVI are (a) 445 nm, 682 nm, and 829 nm, (b) 497 nm, 676 nm, and 1091 nm, and (c) 495 nm, 678 nm, and 1120 nm.

Table 4. (**a**) Species-wise carbon stock derived from NDVI and (**b**) EVI for the Bhitarkanika forest reserve.

(a)	Species Name	NDVI Derived Carbon Stocks				
		Area (km^2)	Total carbon (kt. C)	Min carbon (t. C ha^{-1})	Max carbon (t. C ha^{-1})	Ave. carbon $\pm$ SD (t. C ha^{-1})
1	*Excoecaria agallocha* L.	3.80	52.25	68.14	258.23	143.48 $\pm$ 17.39
2	*Cynometra iripa* Kostel	3.77	42.20	55.28	226.90	115.88 $\pm$ 19.61
3	*Aegiceras corniculatum* (L.)	0.96	54.59	69.66	254.65	149.90 $\pm$ 5.57
4	*Heritiera littoralis Dryand ex Ait.*	2.07	53.08	83.76	225.30	145.55 $\pm$ 7.88
5	*Heritiera fomes* Buch.-Ham.	4.21	51.69	72.47	258.83	141.95 $\pm$ 10.60
6	*Xylocarpus granatum* Koenig	6.41	54.69	55.28	252.01	150.50 $\pm$ 15.51
7	*Xylocarpus mekongensis* Pierre	0.48	47.48	67.35	258.84	130.39 $\pm$ 12.70
8	*Intsia bijuga (Colebr.)* Kuntze	1.66	50.21	83.36	256.40	137.87 $\pm$ 12.57
9	*Cerbera odollam* Gaertn.	8.34	56.36	68.52	219.66	154.78 $\pm$ 18.39
10	*Sonneratia apetala* Buch.-Ham.	4.72	51.84	76.91	254.54	142.34 $\pm$22.46
	Total Area (36.42 km^2)	**36.42**	**514.47**			

(b)	Species Name	EVI Derived Carbon Stocks				
		Area (km^2)	Total carbon (kt. C)	Min carbon (t. C ha^{-1})	Max. carbon (t. C ha^{-1})	Ave. carbon $\pm$ SD (t. C ha^{-1})
1	*Excoecaria agallocha* L.	3.80	45.22	56.57	*225.45*	*124.18 $\pm$ 10.15*
2	*Cynometra iripa* Kostel	3.77	31.02	61.25	*241.22*	*85.19 $\pm$ 26.29*
3	*Aegiceras corniculatum* (L.)	0.96	44.35	63.30	*222.70*	*121.80 $\pm$ 16.38*
4	*Heritiera littoralis Dryand ex Ait.*	2.07	42.45	57.17	*190.22*	*116.57 $\pm$ 22.72*
5	*Heritiera fomes* Buch.-Ham.	4.21	47.38	55.28	*229.22*	*130.11 $\pm$ 32.21*
6	*Xylocarpus granatum* Koenig	6.41	46.90	67.66	*253.04*	*128.78 $\pm$ 15.70*
7	*Xylocarpus mekongensis* Pierre	0.48	50.60	66.66	*218.84*	*138.95 $\pm$ 20.75*
8	*Intsia bijuga (Colebr.)* Kuntze	1.66	53.10	97.24	*253.40*	*145.83 $\pm$ 18.84*
9	*Cerbera odollam* Gaertn.	8.34	48.56	61.51	*209.66*	*133.36 $\pm$ 10.19*
10	*Sonneratia apetala* Buch.-Ham.	4.72	50.19	61.05	*235.54*	*137.83 $\pm$ 15.30*
	Total Area (36.42 km^2)	**36.42**	**459.82**			

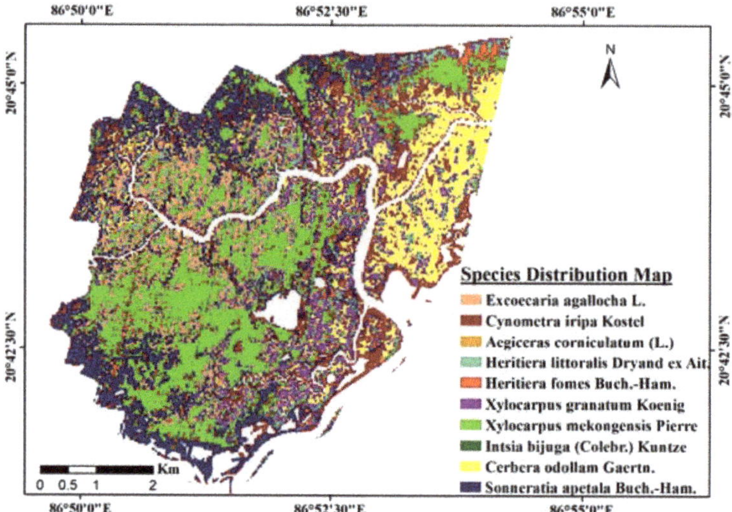

Figure 5. Distribution map of major species-wise mangrove analysis in the study site using EO-1 Hyperion.

3.2. Estimation of Carbon Stock Using Spectral Derived Indices

This section presents the carbon stock assessment for mangrove forest using different models namely, linear, logarithmic, polynomial (second degree), RBF, and sigmoidal function. All the models were trained with the EVI and NDVI generated relations with the ground measured data as well as tested with the modeled biomass and observed carbon stock as shown in Figure 6. The latter figure illustrates the performance of each model for EVI and NDVI based estimations; it can be observed that the RBF model performed better than the others.

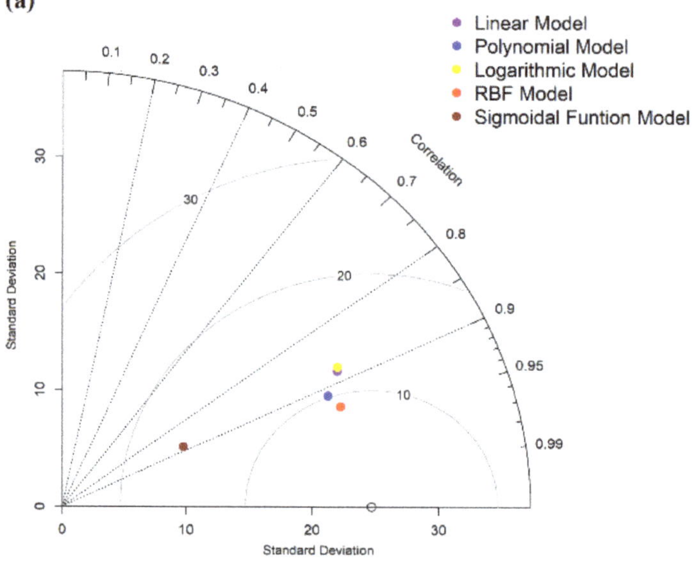

Figure 6. *Cont.*

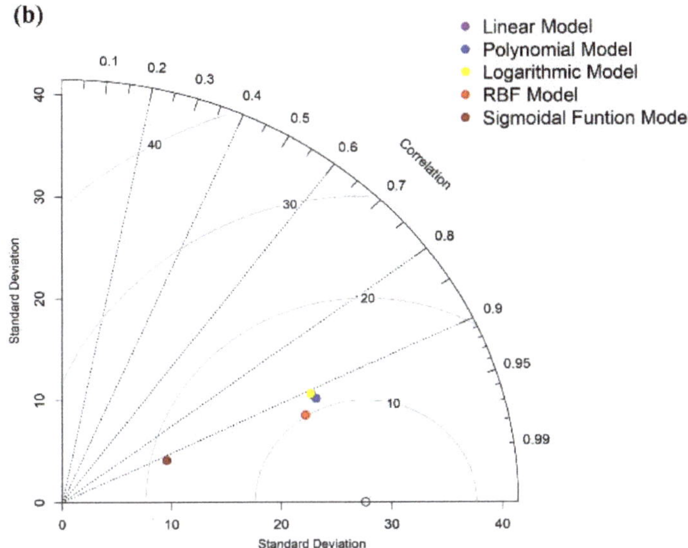

Figure 6. (a) Performance analysis of different models with EVI based carbon estimation and in-situ measurements (b) Performance analysis of different models with NDVI based carbon estimation and in-situ measurements. In both cases, the index-derived carbon estimation shows good agreement between measured and estimated carbon stock and either index could provide a good estimation. From the results EVI (R^2 = 86.98%) seems to perform slightly better than NDVI (R^2 = 84.1%). However, since the sample size is small (10 observations) the results are too close to say with statistical confidence that this hypothesis is true. However, the literature (see Section 3.1) indicates that this is indeed the case. The EVI and NDVI based carbon stock for each species (identified in the present study) is shown in Table 4.

According to the distributed EVI value, it has been concluded that a good amount of area is under dense coverage of forest species; moreover, it has shown higher estimation of carbon stock than NDVI. EVI varies from 0.35 to 6.9 and it is more sensitive to branches and other non-photosynthetic parts of the vegetation (parts different from leaves). EVI is more sensitive to plant parameters, as it avoids the atmospheric effects as well as the soil background. The results illustrate that EVI derived carbon varies from 27.22 to 215.35 t. C ha^{-1} for linear, 85.39 to 236.66 t. C ha^{-1} for log, 104.72 to 306.70 t. C ha^{-1} for polynomial, 55.281 to 253.4 t. C ha^{-1} for RBF and 54.068 to 363.7 t. C ha^{-1} for sigmoidal function models (See Figure 7A–E). NDVI derived carbon varies from 111.11 to 184.14 t. C ha^{-1} for linear, 112.53 to 187.50 t. C ha^{-1} for log, and 109.85 to 181.57 t. C ha^{-1} for polynomial, 55.281 to 258.84 t. C ha^{-1} for RBF, and 46.5 to 357.17 t. C ha^{-1} for sigmoidal function models (See Figure 7F–J). Estimated carbon is highest for EVI derived sigmoidal function model with highest carbon content up to 363.7 t. C ha^{-1} and lowest for linear regression models reaching up to only 27.22 t. C ha^{-1}. Lowest estimated carbon for NDVI derived carbon stocks comes to be 46.5 t. C ha^{-1} for the sigmoidal function model and highest values was observed as 357.17 t. C ha^{-1} for the sigmoidal function model.

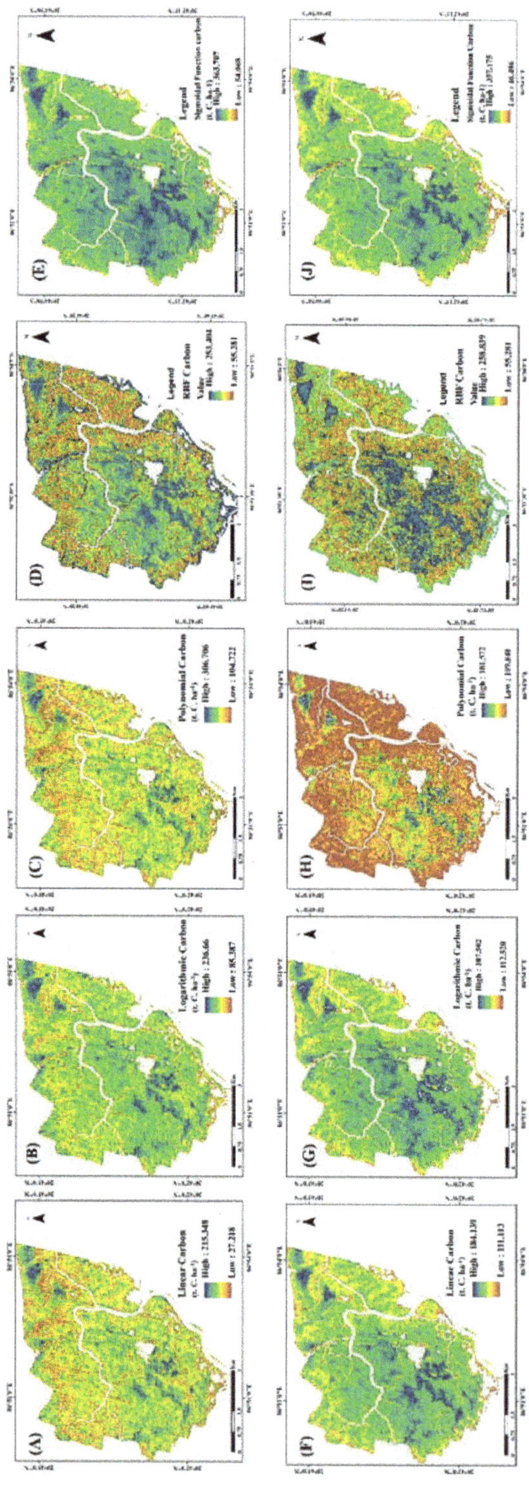

Figure 7. Estimated carbon derived for the Bhitarkanika mangrove forest reserve: from EVI and NDVI indices using different regression models. (**A–E**) EVI derived carbon maps and (**F–J**) NDVI derived carbon maps for Bhitarkanika Site for Linear, Log, Polynomial, RBF (Radial Basis Function), and Sigmoidal models, respectively.

The carbon stock values from the satellite-derived indices fall within the expected ranges for mangrove carbon stocks. NDVI values range from 0.5 to 0.65; the latter shows a healthy, dense mangrove forest in Bhitarkanika. The final interpretation result reveals that the middle northern part of the study area is showing higher biomass values (~250 t. C ha^{-1}). Thus, it is concluded that these regions are highly dense and stores an ample amount of blue carbon in it.

The polynomial regression model using EVI is found to be suitable for the estimation of carbon stock at the study site, with an R^2 of 0.87. EVI has shown high amount of estimated carbon ranges as it is more sensitive to biomass, and ultimately affecting the carbon estimation as compared to the NDVI and can be seen from Figure 7 and Table 4 whereas, NDVI has shown more consistent outcomes in the case of minimum and maximum estimated carbon stocks.

3.3. Species-Wise Carbon Stock Assessment

The classification results generated from SAM classifier and the covariance matrix based optimum band selection for generating vegetation indices were further used to extract the species-wise carbon stock as well as the area covered by each species in the Bhitarkanika forest reserve (see Figures 8 and 9). Figure 9 illustrates the NDVI derived carbon distribution map for each major species, while Figure 8 demonstrates the EVI derived carbon distribution map for each major species. It is also important to notice that the carbon stock of each species shows some variance, which is investigated and presented in Figures 10 and 11. Furthermore, the outcome of species-wise carbon stocks depends upon the species classification accuracies for species distribution classification maps.

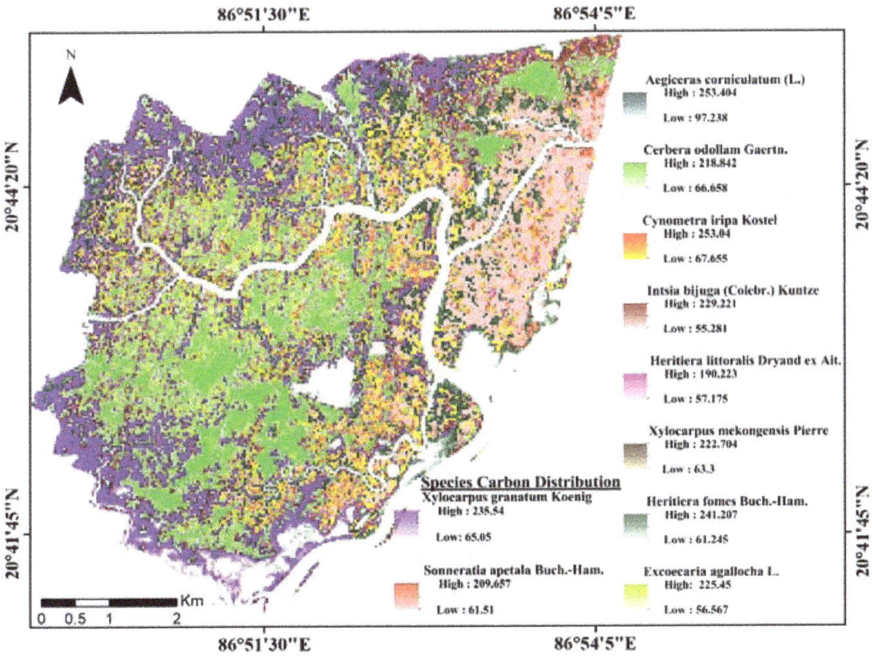

Figure 8. Species-wise estimated carbon map of the study area derived from the EVI indices.

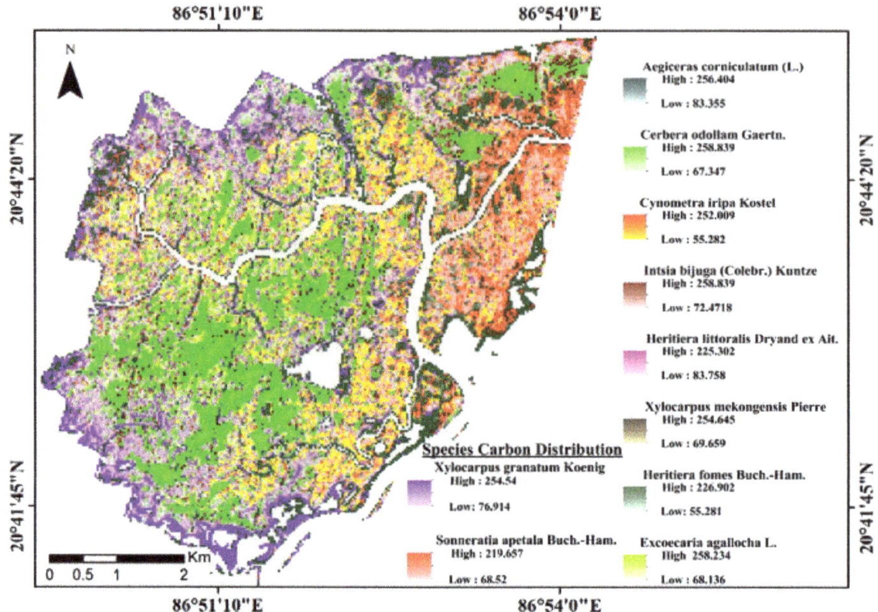

Figure 9. Species-wise estimated carbon map of the study area derived from the NDVI indices.

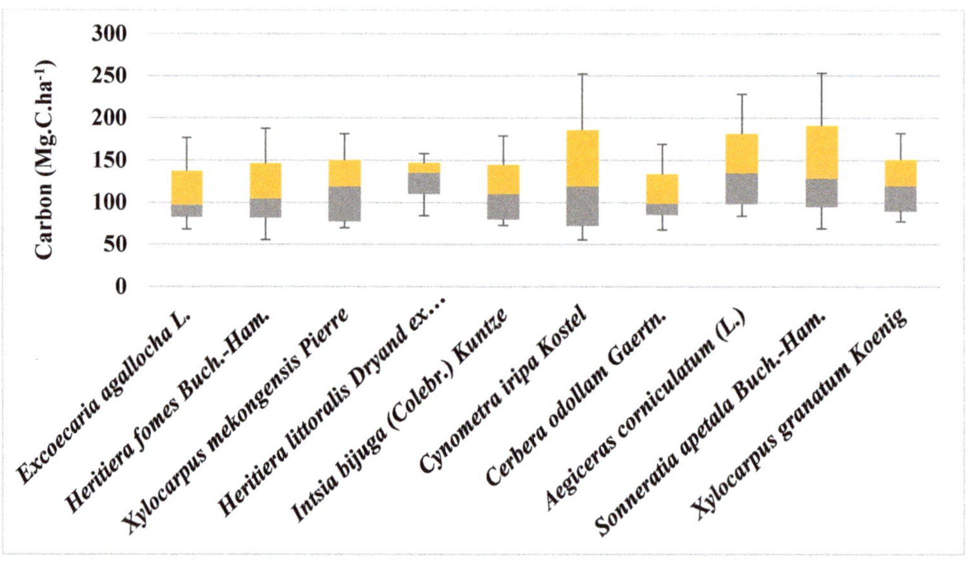

Figure 10. Box plot showing species-wise above ground carbon stock derived from NDVI.

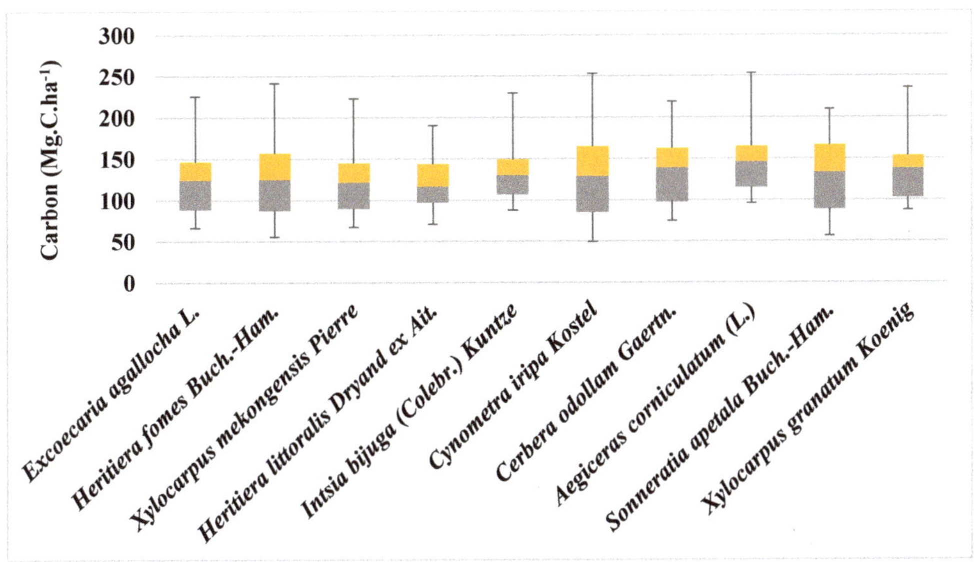

Figure 11. Box plot showing species-wise above ground carbon stock derived from EVI.

Total area covered by the major mangrove species was around 36.42 km^2. *Cerbera odollam* Gaertn covers the largest part of the forest, approximately 22.90% of the total area. Total estimated carbon for the EVI derived indices is 49.82 kt. C. and total carbon estimated for the Bhitarkanika forest derived from NDVI indices is 514.47 kt. C. Using EVI-derived carbon stocks, the highest contribution of carbon stock is the *Intsia bijuga* (Colebr.) Kuntze species with 53.10 kt. C (11.54%). From the NDVI derived carbon stocks, *Cerbera odollam* Gaertn seems to contribute the most with 56.36 kt. C (10.95%). Field measured carbon was recorded lowest for the species *Xylocarpus mekongensis* Pierre, which was 76.20 t. C ha^{-1}. Figure 8 shows the spatial distribution of carbon derived from EVI for each species. *Intsia bijuga* (Colebr.) Kuntze shows highest carbon content up to 253.4 t. C ha^{-1}. The highest carbon stocks as derived from NDVI were displayed for *Xylocarpus mekongensis* Pierre at 258.84 t. C ha^{-1}.

As such, while *Cerbera odollam* Gaertn covers most of the area (22.9%), differences in carbon per hectare (Carbon area density) promote *Intsia bijuga* (Colebr.) Kuntze as the highest contributing species in the Bhitarkanika forest with EVI-derived carbon stocks. This is due to the large difference between EVI and NDVI derived carbon area density for *Cerbera odollam* Gaertn (average 128.78 ± 15.702 t. C ha^{-1} and 150.498 ± 15.51 t. C ha^{-1}). Cross-referencing with the measured values presented in Table 2 (165.03 ± 10.87167.02 t. C ha^{-1}), leads to the conclusion that the NDVI derived carbon stocks for *Cerbera odollam* Gaertn are more accurate. This conclusion is not reflective of all the species. Out of the 10 species examined, the average Carbon area density of EVI is closer to the measured value in six of them, while NDVI derived Carbon area density is more accurate in the other four. The greatest divergence between EVI and NDVI estimated carbon area densities is for *Cerbera odollam* Gaertn. Significant differences are also shown for *Intsia bijuga* (Colebr.) Kuntze and *Xylocarpus mekongensis* Pierre.

A species-wise box-plot is generated to assess the variation in different species-wise carbon stock estimated using EVI and NDVI, which is shown in Figures 10 and 11, with the minima, maxima, median, 25% quartile, and 75% quartile. The average carbon stock measured from field sampling is 131.07 t. C ha^{-1}. Average EVI derived carbon stock ranges from 77.86 t. C ha^{-1} to 135.28 t. C ha^{-1} and for NDVI derived carbon stock 116.57 t. C ha^{-1} to 145.82 t. C ha^{-1} for the Bhitarkanika mangrove forest. As such, both EVI and NDVI estimated averages are in agreement with the average carbon stock measured from the field.

4. Conclusions

Mangrove forests store a large quantity of blue carbon in plants, both in the form of biomass and as sediment in the soil. Anthropogenic activities threaten these forests nowadays due to conversion to other land use types. Such transition of forest areas is a major source of carbon emissions to the atmosphere. As such, carbon stock assessment is essential to reduce the loss of biomass in such ecosystems. Species-wise blue carbon analysis can be used to assess the impact of global climate change on different mangrove species as well as to help policy makers to accurately evaluate the ecological and economical trade off associated with the management of mangroves ecosystem. The present study aimed at demonstrating the use of hyperspectral EO data for species identification in a highly diversified mangrove ecosystem and for calculating total carbon stored. The Bhitarkanika forest in India was chosen as a study site and Hyperion hyperspectral images were used.

There have been several studies on the blue carbon stored in mangroves, however, thus far, a species wide blue carbon analysis with significant accuracy was missing. This study attempts to mitigate that gap of knowledge by estimating the above-ground carbon stocks for each of the 10 major species that were identified and found dominant in the study area.

Hyperspectral data from EO-1 Hyperion were collected and processed to extract the biophysical parameters of interest. Near co-orbital field measurements of biomass and carbon measurements were acquired for validation. The in-situ locations of mangrove species were used to generate spectral profile. The spatial distribution of the major mangrove species was identified using the SAM classification algorithm, which performed reliably well (e.g., kappa coefficient $\kappa = 0.81$). NDVI and EVI radiometric indices were calculated from the optimum bands, obtained by covariance matrix based band selection algorithm. Several models were tested to relate NDVI and EVI with carbon stocks. The RBF model performed best ($R^2 = 86.98\%$ for EVI and $R^2 = 84.1\%$ for NDVI) and was subsequently used in this study to estimate carbon stocks for the 10 dominant species and the entire study area.

Despite the significance of mangrove ecosystem and blue carbon for local as well as global climate, the drastic transformation of mangrove forests into other land use types is directly affecting the livelihood around it, which can be seen through the shortage of firewood, regular soil erosion, and decrease in fishing zones. Therefore, there should be adequate digital information about the coverage, biomass, and carbon content of the mangrove forest for quick management and planning. The present study provides evidence that NDVI and EVI indices have a very promising potential to be applied in classifying the dominant species of mangrove forests and coastal ecosystems according to their carbon content. These indices can provide adequate estimates of maximum, minimum, and average carbon content for a large area and show the spatial distribution of carbon, and thus, biomass. The above-ground carbon stocks for each species were estimated and presented in this study. For the whole study area, the carbon stocks were estimated 459.82 kt. C. from EVI and 514.47 kt. C. from NDVI.

The only limitation faced in this study was the limited availability of Hyperion data and that too covering a part of Bhitarkanika as shown in Figure 2. Using the same methodology with spectral images from different satellites could provide better coverage, and thus carbon stock estimations of different areas. Future studies could focus on different ecosystems to assess the effectiveness for this method and estimate carbon stock for different areas and ecosystems in order to provide the tools for a better evaluation of biomass and global carbon stocks; this remains to be seen.

Author Contributions: Conceptualization, P.C.P. and P.K.S.; Data curation, A.A. and P.C.P.; Formal analysis, P.C.P., A.A.; Investigation, A.A., P.K.S. and A.P.; Methodology, P.C.P., G.P.P., P.K.S. and A.P.; Resources; A.A., P.K.S., and R.K.M.M.; Software, A.A., P.C.P., P.K.S.; Supervision, J.K.S., P.C.P., P.K.S.; Validation, P.K.S.; A.A., P.C.P.; Visualization, P.C.P. and G.P.P.; Writing—original draft, P.C.P.; Writing—review and editing; P.C.P.; P.K.S., G.P.P., AP, R.K.M.M. and J.K.S. All authors have read and agreed to the published version of the manuscript.

Funding: G.P.P.'s participation was supported financially by the European Union's Horizon 2020 Marie Skłodowska-Curie Project "ENViSIoN-EO" (grant agreement No 752094).'

Acknowledgments: The authors gratefully acknowledge the USGS for Hyperion data of the study site, free of cost. Pandey also acknowledges Shiv Nadar University, Greater Noida for support and facility. G.P.P.'s contribution was supported by the FP7- People project ENViSIoN-EO (project reference number 752094) and the author gratefully acknowledges the European Commission for the support provided. The author would like to thank NMHS, MOEF and CC, Government of India and to the reviewers for their comments that resulted to improving the manuscript.

Conflicts of Interest: The authors declare no conflict of interest.

References

1. Saenger, P.; Hegerl, E.; Davie, J.D. *Global Status of Mangrove Ecosystems*; International Union for Conservation of Nature and Natural Resources: Gland, Switzerland, 1983.
2. Barbier, E.B. The protective service of mangrove ecosystems: A review of valuation methods. *Mar. Pollut. Bull.* **2016**, *109*, 676–681. [CrossRef]
3. Houghton, R.; Hall, F.; Goetz, S.J. Importance of biomass in the global carbon cycle. *J. Geophys. Res. Biogeosci.* **2009**, *114*. [CrossRef]
4. Conservation-International. The Blue Carbon Initiatives. Available online: https://www.thebluecarboninitiative.org/ (accessed on 15 May 2019).
5. Giri, C.; Ochieng, E.; Tieszen, L.L.; Zhu, Z.; Singh, A.; Loveland, T.; Masek, J.; Duke, N. Status and distribution of mangrove forests of the world using earth observation satellite data. *Glob. Ecol. Biogeogr.* **2011**, *20*, 154–159. [CrossRef]
6. FSI. Mangrove Cover. Available online: http://fsi.nic.in/isfr2017/isfr-mangrove-cover-2017.pdf (accessed on 23 May 2019).
7. Osland, M.J.; Feher, L.C.; Griffith, K.T.; Cavanaugh, K.C.; Enwright, N.M.; Day, R.H.; Stagg, C.L.; Krauss, K.W.; Howard, R.J.; Grace, J.B. Climatic controls on the global distribution, abundance, and species richness of mangrove forests. *Ecol. Monogr.* **2017**, *87*, 341–359. [CrossRef]
8. Himes-Cornell, A.; Pendleton, L.; Atiyah, P. Valuing ecosystem services from blue forests: A systematic review of the valuation of salt marshes, sea grass beds and mangrove forests. *Ecosyst. Serv.* **2018**, *30*, 36–48. [CrossRef]
9. Gilman, E.L.; Ellison, J.; Duke, N.C.; Field, C. Threats to mangroves from climate change and adaptation options: A review. *Aquat. Bot.* **2008**, *89*, 237–250. [CrossRef]
10. Kairo, J.G.; Lang'at, J.K.; Dahdouh-Guebas, F.; Bosire, J.; Karachi, M. Structural development and productivity of replanted mangrove plantations in Kenya. *For. Ecol. Manag.* **2008**, *255*, 2670–2677. [CrossRef]
11. Bosire, J.O.; Dahdouh-Guebas, F.; Walton, M.; Crona, B.I.; Lewis, R., III; Field, C.; Kairo, J.G.; Koedam, N. Functionality of restored mangroves: A review. *Aquat. Bot.* **2008**, *89*, 251–259. [CrossRef]
12. Duke, N.C.; Meynecke, J.-O.; Dittmann, S.; Ellison, A.M.; Anger, K.; Berger, U.; Cannicci, S.; Diele, K.; Ewel, K.C.; Field, C.D. A world without mangroves? *Science* **2007**, *317*, 41–42. [CrossRef]
13. Hamilton, S.E.; Casey, D. Creation of a high spatio-temporal resolution global database of continuous mangrove forest cover for the 21st century (CGMFC-21). *Glob. Ecol. Biogeogr.* **2016**, *25*, 729–738. [CrossRef]
14. Hamilton, S.E.; Friess, D.A. Global carbon stocks and potential emissions due to mangrove deforestation from 2000 to 2012. *Nat. Clim. Chang.* **2018**, *8*, 240. [CrossRef]
15. Valiela, I.; Bowen, J.L.; York, J.K. Mangrove Forests: One of the World's Threatened Major Tropical Environments. *Bioscience* **2001**, *51*, 807–815. [CrossRef]
16. Alongi, D.M. Present state and future of the world's mangrove forests. *Environ. Conserv.* **2002**, *29*, 331–349. [CrossRef]
17. Allen, J.A.; Ewel, K.C.; Jack, J. Patterns of natural and anthropogenic disturbance of the mangroves on the Pacific Island of Kosrae. *Wetl. Ecol. Manag.* **2001**, *9*, 291–301. [CrossRef]
18. Giri, C.; Zhu, Z.; Tieszen, L.; Singh, A.; Gillette, S.; Kelmelis, J. Mangrove forest distributions and dynamics (1975–2005) of the tsunami-affected region of Asia. *J. Biogeogr.* **2008**, *35*, 519–528. [CrossRef]
19. Baillie, J.E.; Hilton-Taylor, C.; Stuart, S.N. *A Global Species Assessment*; International Union for Conservation of Nature (IUCN): Gland, Switzerland, 2004.
20. Kathiresan, K.; Rajendran, N. Mangrove ecosystems of the Indian Ocean region. *Indian J. Mar. Sci.* **2005**, *34*, 104–113.
21. Sandilyan, S.; Kathiresan, K. Mangrove conservation: A global perspective. *Biodivers. Conserv.* **2012**, *21*, 3523–3542. [CrossRef]

22. Shanker, K. *Biodiversity of Mangrove Ecosystems*; Medknow Publications: Mumbai, India, 2005.

23. Kathiresan, K.; Qasim, S.Z. *Biodiversity of Mangrove Ecosystems*; Hindustan Publishing: New Delhi, India, 2005.

24. Kathiresan, K. Importance of mangrove forest of India. *J. Coast. Environ.* **2010**, *1*, 11–26.

25. Kathiresan, K. Why are mangroves degrading? *Curr. Sci.* **2002**, *83*, 1246–1249.

26. Pandey, P.C.; Anand, A.; Srivastava, P.K. Spatial Distribution of Mangrove Forest species and Biomass Assessment Using Field Inventory and Earth Observation Hyperspectral data. *Biodivers. Conserv.* **2019**, *28*, 2143–2162. [CrossRef]

27. Yang, C.; Liu, J.; Zhang, Z.; Zhang, Z. Estimation of the carbon stock of tropical forest vegetation by using remote sensing and GIS. In Proceedings of the IGARSS 2001. Scanning the Present and Resolving the Future. In Proceedings of the IEEE 2001 International Geoscience and Remote Sensing Symposium (Cat. No. 01CH37217), Sydney, Australia, 9–13 July 2001; pp. 1672–1674.

28. Ramankutty, N.; Gibbs, H.K.; Achard, F.; Defries, R.; Foley, J.A.; Houghton, R. Challenges to estimating carbon emissions from tropical deforestation. *Glob. Chang. Biol.* **2007**, *13*, 51–66. [CrossRef]

29. Atmadja, S.; Verchot, L. A review of the state of research, policies and strategies in addressing leakage from reducing emissions from deforestation and forest degradation (REDD+). *Mitig. Adapt. Strateg. Glob. Chang.* **2012**, *17*, 311–336. [CrossRef]

30. Minang, P.A.; Van Noordwijk, M. Design challenges for achieving reduced emissions from deforestation and forest degradation through conservation: Leveraging multiple paradigms at the tropical forest margins. *Land Use Policy* **2013**, *31*, 61–70. [CrossRef]

31. CIFOR. Global Comparative Study on REDD+ Subnational REDD+ Initiatives. Available online: https://www.cifor.org/gcs/modules/redd-subnationalinitiatives/ (accessed on 25 May 2018).

32. Atwood, T.B.; Connolly, R.M.; Almahasheer, H.; Carnell, P.E.; Duarte, C.M.; Lewis, C.J.E.; Irigoien, X.; Kelleway, J.J.; Lavery, P.S.; Macreadie, P.I. Global patterns in mangrove soil carbon stocks and losses. *Nat. Clim. Chang.* **2017**, *7*, 523. [CrossRef]

33. Heumann, B.W. An object-based classification of mangroves using a hybrid decision tree—Support vector machine approach. *Remote Sens.* **2011**, *3*, 2440–2460. [CrossRef]

34. Chaube, N.R.; Lele, N.; Misra, A.; Murthy, T.; Manna, S.; Hazra, S.; Panda, M.; Samal, R. Mangrove species discrimination and health assessment using AVIRIS-NG hyperspectral data. *Curr. Sci.* **2019**, *116*, 1136. [CrossRef]

35. Kumar, T.; Panigrahy, S.; Kumar, P.; Parihar, J.S. Classification of floristic composition of mangrove forests using hyperspectral data: Case study of Bhitarkanika National Park, India. *J. Coast. Conserv.* **2013**, *17*, 121–132. [CrossRef]

36. Ashokkumar, L.; Shanmugam, S. Hyperspectral band selection and classification of Hyperion image of Bhitarkanika mangrove ecosystem, eastern India. *Proc. SPIE* **2014**, *9239*, 923914.

37. Padma, S.; Sanjeevi, S. Jeffries Matusita-Spectral Angle Mapper (JM-SAM) spectral matching for species level mapping at Bhitarkanika, Muthupet and Pichavaram mangroves. *Int. Arch. Photogramm. Remote Sens. Spat. Inf. Sci.* **2014**, *40*, 1403. [CrossRef]

38. Everitt, J.; Yang, C.; Judd, F.; Summy, K. Use of archive aerial photography for monitoring black mangrove populations. *J. Coast. Res.* **2010**, *26*, 649–653. [CrossRef]

39. Lam-Dao, N.; Pham-Bach, V.; Nguyen-Thanh, M.; Pham-Thi, M.-T.; Hoang-Phi, P. Change detection of land use and riverbank in Mekong Delta, Vietnam using time series remotely sensed data. *J. Resour. Ecol.* **2011**, *2*, 370–375.

40. Satyanarayana, B.; Mohamad, K.A.; Idris, I.F.; Husain, M.-L.; Dahdouh-Guebas, F. Assessment of mangrove vegetation based on remote sensing and ground-truth measurements at Tumpat, Kelantan Delta, East Coast of Peninsular Malaysia. *Int. J. Remote Sens.* **2011**, *32*, 1635–1650. [CrossRef]

41. Pattanaik, C.; Prasad, S.N. Assessment of aquaculture impact on mangroves of Mahanadi delta (Orissa), East coast of India using remote sensing and GIS. *Ocean Coast. Manag.* **2011**, *54*, 789–795. [CrossRef]

42. Rahman, A.F.; Dragoni, D.; Didan, K.; Barreto-Munoz, A.; Hutabarat, J.A. Detecting large scale conversion of mangroves to aquaculture with change point and mixed-pixel analyses of high-fidelity MODIS data. *Remote Sens. Environ.* **2013**, *130*, 96–107. [CrossRef]

43. Pu, R.; Bell, S. A protocol for improving mapping and assessing of seagrass abundance along the West Central Coast of Florida using Landsat TM and EO-1 ALI/Hyperion images. *ISPRS J. Photogramm. Remote Sens.* **2013**, *83*, 116–129. [CrossRef]

44. Lucas, R.; Rebelo, L.-M.; Fatoyinbo, L.; Rosenqvist, A.; Itoh, T.; Shimada, M.; Simard, M.; Souza-Filho, P.W.; Thomas, N.; Trettin, C. Contribution of L-band SAR to systematic global mangrove monitoring. *Mar. Freshw. Res.* **2014**, *65*, 589–603. [CrossRef]

45. Vu, T.D.; Takeuchi, W.; Van, N.A. Carbon stock calculating and forest change assessment toward REDD+ activities for the mangrove forest in Vietnam. *Trans. Jpn. Soc. Aeronaut. Space Sci. Aerosp. Technol. Jpn.* **2014**, *12*. [CrossRef]

46. Thomas, N.; Lucas, R.; Itoh, T.; Simard, M.; Fatoyinbo, L.; Bunting, P.; Rosenqvist, A. An approach to monitoring mangrove extents through time-series comparison of JERS-1 SAR and ALOS PALSAR data. *Wetl. Ecol. Manag.* **2015**, *23*, 3–17. [CrossRef]

47. Garcia, R.; Hedley, J.; Tin, H.; Fearns, P. A method to analyze the potential of optical remote sensing for benthic habitat mapping. *Remote Sens.* **2015**, *7*, 13157–13189. [CrossRef]

48. Son, N.T.; Thanh, B.X.; Da, C.T. Monitoring mangrove forest changes from multi-temporal Landsat data in Can Gio Biosphere Reserve, Vietnam. *Wetlands* **2016**, *36*, 565–576. [CrossRef]

49. Nardin, W.; Locatelli, S.; Pasquarella, V.; Rulli, M.C.; Woodcock, C.E.; Fagherazzi, S. Dynamics of a fringe mangrove forest detected by Landsat images in the Mekong River Delta, Vietnam. *Earth Surf. Process. Landf.* **2016**, *41*, 2024–2037. [CrossRef]

50. Viennois, G.; Proisy, C.; Feret, J.-B.; Prosperi, J.; Sidik, F.; Rahmania, R.; Longépé, N.; Germain, O.; Gaspar, P. Multitemporal analysis of high-spatial-resolution optical satellite imagery for mangrove species mapping in Bali, Indonesia. *IEEE J. Sel. Top. Appl. Earth Obs. Remote Sens.* **2016**, *9*, 3680–3686. [CrossRef]

51. Pham, L.T.; Brabyn, L. Monitoring mangrove biomass change in Vietnam using SPOT images and an object-based approach combined with machine learning algorithms. *ISPRS J. Photogramm. Remote Sens.* **2017**, *128*, 86–97. [CrossRef]

52. Benson, L.; Glass, L.; Jones, T.; Ravaoarinorotsihoarana, L.; Rakotomahazo, C. Mangrove carbon stocks and ecosystem cover dynamics in southwest Madagascar and the implications for local management. *Forests* **2017**, *8*, 190. [CrossRef]

53. Bullock, E.L.; Fagherazzi, S.; Nardin, W.; Vo-Luong, P.; Nguyen, P.; Woodcock, C.E. Temporal patterns in species zonation in a mangrove forest in the Mekong Delta, Vietnam, using a time series of Landsat imagery. *Cont. Shelf Res.* **2017**, *147*, 144–154. [CrossRef]

54. Mondal, P.; Trzaska, S.; de Sherbinin, A. Landsat-derived estimates of mangrove extents in the sierra leone coastal landscape complex during 1990–2016. *Sensors* **2018**, *18*, 12. [CrossRef]

55. Wang, M.; Cao, W.; Guan, Q.; Wu, G.; Wang, F. Assessing changes of mangrove forest in a coastal region of southeast China using multi-temporal satellite images. *Estuar. Coast. Shelf Sci.* **2018**, *207*, 283–292. [CrossRef]

56. Abdel-Hamid, A.; Dubovyk, O.; Abou El-Magd, I.; Menz, G. Mapping Mangroves Extents on the Red Sea Coastline in Egypt using Polarimetric SAR and High Resolution Optical Remote Sensing Data. *Sustainability* **2018**, *10*, 646. [CrossRef]

57. Pan, Z.; Glennie, C.; Fernandez-Diaz, J.C.; Starek, M. Comparison of bathymetry and seagrass mapping with hyperspectral imagery and airborne bathymetric lidar in a shallow estuarine environment. *Int. J. Remote Sens.* **2016**, *37*, 516–536. [CrossRef]

58. Warfield, A.D.; Leon, J.X. Estimating Mangrove Forest Volume Using Terrestrial Laser Scanning and UAV-Derived Structure-from-Motion. *Drones* **2019**, *3*, 32. [CrossRef]

59. Green, E.; Clark, C.; Mumby, P.; Edwards, A.; Ellis, A. Remote sensing techniques for mangrove mapping. *Int. J. Remote Sens.* **1998**, *19*, 935–956. [CrossRef]

60. Wang, L.; Sousa, W.P. Distinguishing mangrove species with laboratory measurements of hyperspectral leaf reflectance. *Int. J. Remote Sens.* **2009**, *30*, 1267–1281. [CrossRef]

61. Yang, C.; Everitt, J.H.; Fletcher, R.S.; Jensen, R.R.; Mausel, P.W. Evaluating AISA+ hyperspectral imagery for mapping black mangrove along the South Texas Gulf Coast. *Photogramm. Eng. Remote Sens.* **2009**, *75*, 425–435. [CrossRef]

62. Held, A.; Ticehurst, C.; Lymburner, L.; Williams, N. High resolution mapping of tropical mangrove ecosystems using hyperspectral and radar remote sensing. *Int. J. Remote Sens.* **2003**, *24*, 2739–2759. [CrossRef]

63. Cao, J.; Leng, W.; Liu, K.; Liu, L.; He, Z.; Zhu, Y. Object-based mangrove species classification using unmanned aerial vehicle hyperspectral images and digital surface models. *Remote Sens.* **2018**, *10*, 89. [CrossRef]

64. Hirano, A.; Madden, M.; Welch, R. Hyperspectral image data for mapping wetland vegetation. *Wetlands* **2003**, *23*, 436–448. [CrossRef]

65. Koedsin, W.; Vaiphasa, C. Discrimination of tropical mangroves at the species level with EO-1 Hyperion data. *Remote Sens.* **2013**, *5*, 3562–3582. [CrossRef]
66. Kamal, M.; Phinn, S. Hyperspectral data for mangrove species mapping: A comparison of pixel-based and object-based approach. *Remote Sens.* **2011**, *3*, 2222–2242. [CrossRef]
67. Odisha, W.O. Bhitarkanika Wildlife Sanctuary. Available online: https://www.wildlife.odisha.gov.in/WebPortal/PA_Bhitarkanika.aspx (accessed on 28 May 2018).
68. Pandey, P.C.; Tate, N.J.; Balzter, H. Mapping tree species in coastal portugal using statistically segmented principal component analysis and other methods. *IEEE Sens. J.* **2014**, *14*, 4434–4441. [CrossRef]
69. Pattanaik, C.; Reddy, C.; Dhal, N.; Das, R. Utilisation of Mangrove Forests in Bhitarkanika Wildlife Sanctuary, Orissa. *Indian J. Tradit. Know.* **2008**, *7*, 598–603.
70. Boardman, J.W. *Automating Spectral Unmixing of AVIRIS Data Using Convex Geometry Concepts*; NASA: Wahington, DC, USA, 1993.
71. Research Systems. *ENVI Tutorials*; Research Systems: 2000; Harris Geospatial Solutions: Broomfield, CO, USA; Available online: https://www.harrisgeospatial.com/docs/tutorials.html (accessed on 4 December 2019).
72. Kruse, F.A.; Lefkoff, A.; Boardman, J.; Heidebrecht, K.; Shapiro, A.; Barloon, P.; Goetz, A. The spectral image processing system (SIPS)—Interactive visualization and analysis of imaging spectrometer data. *Remote Sens. Environ.* **1993**, *44*, 145–163. [CrossRef]
73. Elatawneh, A.C.; Kalaitzidis, G.P.; Schneider, T. Evaluation of Diverse Classification Approaches for Land Use/Cover Mapping in a Mediterranean Region Utilizing Hyperion Data. *Int. J. Digit. Earth* **2012**, 1–23. [CrossRef]
74. Petropoulos, G.K.P.; Vadrevu, G.; Xanthopoulos, G.K.; Scholze, M. A Comparison of Spectral Angle Mapper and Artificial Neural Network Classifiers Combined with Landsat TM Imagery Analysis for Obtaining Burnt Area Mapping. *Sensors* **2010**, *10*, 1967–1985. [CrossRef] [PubMed]
75. Brown, S.; Gillespie, A.J.; Lugo, A.E. Biomass estimation methods for tropical forests with applications to forest inventory data. *For. Sci.* **1989**, *35*, 881–902.
76. Negi, J.; Sharma, S.; Sharma, D. Comparative assessment of methods for estimating biomass in forest ecosystem. *Indian For.* **1988**, *114*, 136–144.
77. Luckman, A.; Baker, J.; Kuplich, T.M.; Yanasse, C.D.C.F.; Frery, A.C. A study of the relationship between radar backscatter and regenerating tropical forest biomass for spaceborne SAR instruments. *Remote Sens. Environ.* **1997**, *60*, 1–13. [CrossRef]
78. Schroeder, P.; Brown, S.; Mo, J.; Birdsey, R.; Cieszewski, C. Biomass estimation for temperate broadleaf forests of the United States using inventory data. *For. Sci.* **1997**, *43*, 424–434.
79. Vargas-Larreta, B.; López-Sánchez, C.A.; Corral-Rivas, J.J.; López-Martínez, J.O.; Aguirre-Calderón, C.G.; Álvarez-González, J.G. Allometric equations for estimating biomass and carbon stocks in the temperate forests of North-Western Mexico. *Forests* **2017**, *8*, 269. [CrossRef]
80. Komiyama, A.; Jintana, V.; Sangtiean, T.; Kato, S. A common allometric equation for predicting stem weight of mangroves growing in secondary forests. *Ecol. Res.* **2002**, *17*, 415–418. [CrossRef]
81. Komiyama, A.; Poungparn, S.; Kato, S. Common allometric equations for estimating the tree weight of mangroves. *J. Trop. Ecol.* **2005**, *21*, 471–477. [CrossRef]
82. Alves, D.; Soares, J.V.; Amaral, S.; Mello, E.; Almeida, S.; da Silva, O.F.; Silveira, A. Biomass of primary and secondary vegetation in Rondônia, Western Brazilian Amazon. *Glob. Chang. Biol.* **1997**, *3*, 451–461. [CrossRef]
83. Brown, S. *Estimating Biomass and Biomass Change of Tropical Forests: A Primer*; Food & Agriculture Organization: Rome, Italy, 1997; Volume 134.
84. Negi, J.; Manhas, R.; Chauhan, P. Carbon allocation in different components of some tree species of India: A new approach for carbon estimation. *Curr. Sci.* **2003**, *85*, 1528–1531.
85. Vicharnakorn, P.; Shrestha, R.; Nagai, M.; Salam, A.; Kiratiprayoon, S. Carbon stock assessment using remote sensing and forest inventory data in Savannakhet, Lao PDR. *Remote Sens.* **2014**, *6*, 5452–5479. [CrossRef]
86. Mattsson, E.; Ostwald, M.; Nissanka, S.; Pushpakumara, D. Quantification of carbon stock and tree diversity of homegardens in a dry zone area of Moneragala district, Sri Lanka. *Agrofor. Syst.* **2015**, *89*, 435–445. [CrossRef]
87. Sheffield, C. Selecting Band Combinations from Multi Spectral Data. *Photogramm. Eng. Remote Sens.* **1985**, *58*, 681–687.

88. Tucker, C.J. Red and photographic infrared linear combinations for monitoring vegetation. *Remote Sens. Environ.* **1979**, *8*, 127–150. [CrossRef]

89. Tomar, V.; Kumar, P.; Rani, M.; Gupta, G.; Singh, J. A satellite-based biodiversity dynamics capability in tropical forest. *Electron. J. Geotech. Eng.* **2013**, *18*, 1171–1180.

90. Huete, A.; Didan, K.; Miura, T.; Rodriguez, E.P.; Gao, X.; Ferreira, L.G. Overview of the radiometric and biophysical performance of the MODIS vegetation indices. *Remote Sens. Environ.* **2002**, *83*, 195–213. [CrossRef]

91. Heute, A.; Liu, H.; Batchily, K.; Van Leeuwen, W. A comparison of vegetation indices over a global set of TM images for EOS-MODIS. *Remote Sens. Environ.* **1997**, *59*, 440–451. [CrossRef]

92. Matsushita, B.; Yang, W.; Chen, J.; Onda, Y.; Qiu, G. Sensitivity of the enhanced vegetation index (EVI) and normalized difference vegetation index (NDVI) to topographic effects: A case study in high-density cypress forest. *Sensors* **2007**, *7*, 2636–2651. [CrossRef]

93. Gedan, K.B.; Silliman, B.R.; Bertness, M.D. Centuries of human-driven change in salt marsh ecosystems. *Annu. Rev. Mar. Sci.* **2009**, *1*, 117–141. [CrossRef] [PubMed]

94. Morris, J.T.; Sundareshwar, P.; Nietch, C.T.; Kjerfve, B.; Cahoon, D.R. Responses of coastal wetlands to rising sea level. *Ecology* **2002**, *83*, 2869–2877. [CrossRef]

95. Adam, E.; Mutanga, O.; Abdel-Rahman, E.M.; Ismail, R. Estimating standing biomass in papyrus (Cyperus papyrus L.) swamp: Exploratory of in situ hyperspectral indices and random forest regression. *Int. J. Remote Sens.* **2014**, *35*, 693–714. [CrossRef]

96. Santin-Janin, H.; Garel, M.; Chapuis, J.-L.; Pontier, D. Assessing the performance of NDVI as a proxy for plant biomass using non-linear models: A case study on the Kerguelen archipelago. *Polar Biol.* **2009**, *32*, 861–871. [CrossRef]

97. Wicaksono, P.; Danoedoro, P.; Hartono; Nehren, U. Mangrove biomass carbon stock mapping of the Karimunjawa Islands using multispectral remote sensing. *Int. J. Remote Sens.* **2016**, *37*, 26–52. [CrossRef]

remote sensing

MDPI

Communication

Applicability of Smoothing Techniques in Generation of Phenological Metrics of *Tectona grandis* L. Using NDVI Time Series Data

Ramandeep Kaur M. Malhi [1], G. Sandhya Kiran [1,*], Mangala N. Shah [2], Nirav V. Mistry [2], Viral H. Bhavsar [2], Chandra Prakash Singh [3], Bimal Kumar Bhattacharya [3], Philip A. Townsend [4] and Shiv Mohan [5]

[1] Ecophysiology and RS-GIS Laboratory, Department of Botany, Faculty of Science, The Maharaja Sayajirao University of Baroda, Vadodara 390002, Gujarat, India; deep_malhi56@yahoo.co.in
[2] Department of Statistics, Faculty of Science, The Maharaja Sayajirao University of Baroda, Vadodara 390002, Gujarat, India; sudhasha_00@yahoo.com (M.N.S.); wayuwantnirav@gmail.com (N.V.M.); viralbhavsar10@gmail.com (V.H.B.)
[3] Space Applications Centre, Indian Space Research Organisation, Ahmedabad 380015, Gujarat, India; cpsingh@sac.isro.gov.in (C.P.S.); bkbhattacharya@sac.isro.gov.in (B.K.B.)
[4] Department of Forest and Wildlife Ecology, University of Wisconsin, Madison, WI 53706, USA; ptownsend@wisc.edu
[5] PLANEX, Physical Research Laboratory, Ahmedabad 380059, Gujarat, India; shivmohan.isro@gmail.com
* Correspondence: sandhyakiran60@yahoo.com

Citation: Malhi, R.K.M.; Kiran, G.S.; Shah, M.N.; Mistry, N.V.; Bhavsar, V.H.; Singh, C.P.; Bhattarcharya, B.K.; Townsend, P.A.; Mohan, S. Applicability of Smoothing Techniques in Generation of Phenological Metrics of *Tectona grandis* L. Using NDVI Time Series Data. *Remote Sens.* **2021**, *13*, 3343. https://doi.org/10.3390/rs13173343

Academic Editor: Arturo Sanchez-Azofeifa

Received: 6 June 2021
Accepted: 9 August 2021
Published: 24 August 2021

Publisher's Note: MDPI stays neutral with regard to jurisdictional claims in published maps and institutional affiliations.

Abstract: Information on phenological metrics of individual plant species is meager. Phenological metrics generation for a specific plant species can prove beneficial if the species is ecologically or economically important. Teak, a dominating tree in most regions of the world has been focused on in the present study due to its multiple benefits. Forecasts on such species can attain a substantial improvement in their productivity. MODIS NDVI time series when subjected to statistical smoothing techniques exhibited good output with Tukey's smoothing (TS) with a low RMSE of 0.042 compared to single exponential (SE) and double exponential (DE). Phenological metrics, namely, the start of the season (SOS), end of the season (EOS), maximum of the season (MAX), and length of the season (LOS) were generated using Tukey-smoothed MODIS NDVI data for the years 2003–2004 and 2013–2014. Post shifts in SOS and EOS by 14 and 37 days respectively with a preshift of 28 days in MAX were observed in the year 2013–2014. Preshift in MAX was accompanied by an increase in greenness exhibiting increased NDVI value. LOS increased by 24 days in the year 2013–2014, showing an increase in the duration of the season of teak. Dates of these satellite-retrieved phenological occurrences were validated with ground phenological data calculated using crown cover assessment. The present study demonstrated the potential of a spatial approach in the generation of phenometrics for an individual plant species, which is significant in determining productivity or a crucial trophic link for a given region.

Keywords: phenology; NDVI; smoothing; MODIS

1. Introduction

The occurrence dates of phenophases such as blooming, full leaf expansion, leaf coloration, or senescence are keys for the determination of phenological metrics, viz., start of the season (SOS), end of the season (EOS), and maximum of the season (MAX) of any tree [1]. These metrics can prove to be of crucial importance in the tree's productivity assessment. They can also prove significant in modeling and monitoring climate change [2], explaining the seasonal changes, determining net terrestrial carbon dioxide flux [3–6], and assessing leaf cycle functioning.

Information is meager on phenological data of individual plant species in many parts of the world [7]. More focus is placed on generating phenological metrics of forest types [8–11] rather than on a particular natural species [12,13]. Hence, evaluating the capability of satellite data for effective and reliable phenological metrics generation for one individual forest species, namely, *Tectona grandis* L. (teak) will be of great importance. Teak, a deciduous tree of the Verbenaceae family, is considered a valuable timber and has been widely used in India for more than 200 years due to its elegance and durability. In addition, it is an economically important tree as it has a tight grain and high oil content and tensile strength and can also be grown under various climatic regimes. Moreover, this tree occupies major worldwide areas and is also a dominant tree of tropical dry deciduous forests (DDF) of the Narmada district. Due to a paucity of ground data on this system, there is a great need to develop remote sensing techniques to characterize its biology. Phenological metrics generation of teak growing in the DDF can provide useful information about both spatial and temporal patterns of its productivity. Such forecasts can attain a substantial improvement in its productivity.

Appraisal of phenophases of any plant species mainly involves two approaches; ground-based phenology measurements and satellite-based phenology monitoring [14]. Ground-based measurements entail direct traditional insitu visual recordings of phenological events [15–19], manual periodic photography, or fixed-position camera-based digital repeated photography [20–22]. These methods, though they provide good outputs, have their limitations concerning lack of synoptic coverage and time [23]. An alternative space approach overcomes such limitations both with respect to time and coverage. Recent studies, therefore, are based on using the potential of spatial data in understanding plant phenology [24–31]. Specifically, time series normalized difference vegetation index (NDVI) data derived from moderate resolution imaging spectroradiometer (MODIS) [32–38] and advanced very high resolution radiometer (AVHRR) [39–44] are largely used for the generation of phenological metrics. This satellite-based vegetation index is recognized as an effective tool in the quantification of vegetation greenness [45–47], canopy carbon [48–52], and water fluxes [53–55]. It has proved its utility in monitoring the phenological changes in vegetation across large spatial scales and over long time periods. Many researchers have used NDVI and enhanced vegetation index (EVI) time series in the identification of different phenological parameters at both single pixel scale and large spatial scales by developing various methodologies [25,56–61]. Studies are also available wherein Landsat time series data have been used for retrieving phenological metrics [62–64]. Some have built statistical relationships between NDVI or EVI and climatic parameters for monitoring the entire growing season [65–68]. The phenological parameters derived from vegetation indices may not correspond directly to conventional ground-based phenological events but do provide indications for understanding these phenophases which now have become pertinent in climate change analysis [69,70]. In addition, differences among biomes and their drivers of phenology (e.g., dry deciduous vs. cold deciduous) may necessitate different interpretations of image-derived phenophases. As such, detailed comparisons of these satellite measures and ground-based phenological events are needed across a range of biomes with different ecological drivers.

The time series NDVI or EVI data when subjected to smoothing techniques remove the outliers [71] occurring due to cloud contamination, atmospheric perturbations, variable illumination, and viewing geometry. These include methods like moving average time series (MATS) [72], a weighted least-squares linear regression approach [73]. Fourier harmonics-based methods (classical and discrete) [74–78], threshold-based methods [79–81], curve fitting methods or fitting of polynomial functions [82,83], point of inflection methods [81,84], simple sliding windows [85], piecewise linear regression [86,87], single exponential [88], double exponential [89], Tukey smoothing [88], etc.

The objective of the present work is to: (1) investigate three statistical smoothing techniques, viz., single exponential (SE), double exponential (DE), and Tukey's Smoothing (TS) for diminishing noise effects and removing outliers in MODIS NDVI time series; (2) use the best reconstructed time series NDVI MODIS data to generate different phenological metrics, namely, SOS, EOS, MAX, and LOS of teak growing in Narmada forests, Gujarat, India.

2. Materials and Methods

2.1. In Situ Phenological Data Collection

The study area selected for the present study is DDF of Narmada district, Gujarat, India (21.86 N, 73.56 E) (Figure 1). This area exhibits the dominance of teak trees. Occurrence dates of different phenophase events for the year 2003–2004 were obtained from Sardar Sarovar Narmada Limited (SSNL) report [90]. The data for the year 2013–2014 were generated by monitoring ten 30 m × 30 m (900 m^2) homogenous teak patches at regular intervals of 20–25 days from May 2013 till April 2014. Analysis of within-population phenophase frequency at every sampling date was performed based on the modified qualitative method of Ohshan [91]. The population was randomly selected and labeled before the beginning of the sampling for ten healthy adult teak trees. During each field visit, the degree of incidence was determined for the different phenophases of teak, viz., leaf initiation, maximum greenness, and leaf fall, for which each tree was examined precisely. A frequency index was assigned for particular phenophases at the time of their presence on a tree, depending on the percentage of the crown where it occurred: 1 = presence less than 5% (leaf fall), 2 = presence between 5–25% (leaf initiation), and 3 = presence over 25% of the crown (maximum greenness).The average of the ten sampled plants' frequency indicesrepresents the calculation of the incidence of a phenophase in the whole population for each date. These ground-based phenological observations were then compared to phenological metrics derived using time series satellite data for the validation purpose.

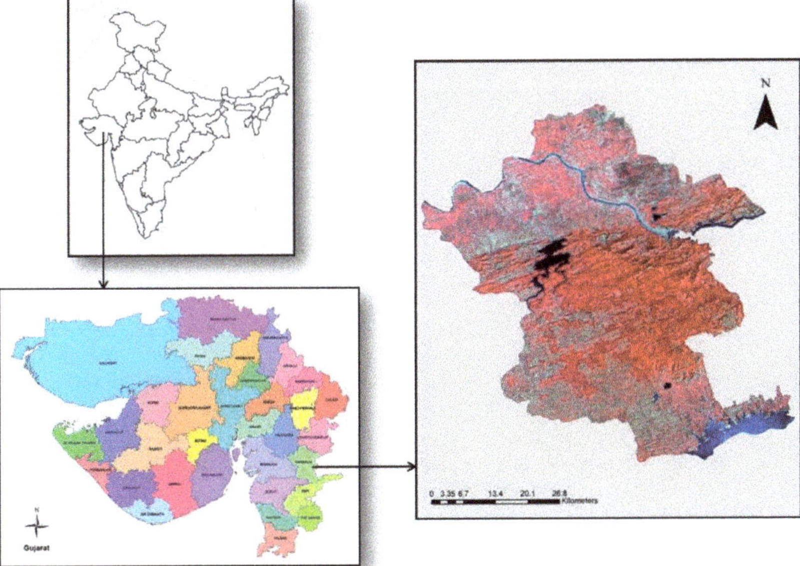

Figure 1. A location map showing the study area, Narmada district, Gujarat, India; image used in the study area map is a Landsat 11 RGB composite.

2.2. Satellite Data

Annual spatiotemporal MODIS vegetation indices-terra (MOD13A1)'s NDVI product of the year 2003–2004 and 2013–2014 with repetivity of 16 days, and 500 m spatial resolution was used to extract the phenological metrics of teak. Before the extraction of these metrics, NDVI time series data were subjected to different smoothing techniques.

2.2.1. Application of Smoothing Techniques

Different smoothing algorithms were applied to the time series NDVI MODIS data to diminish noise effects: single exponential (SE), double exponential (DE), and Tukey's smoothing (TS). All three techniques are in ascending order in terms of sophistication.

Single Exponential Smoothing (SE)

SE is a simple and accessible tool for smoothing time series data. A simple average calculation is used to assign exponentially decreasing weights, starting with the most recent observations. New observations are weighted more in the average calculation than earlier observations. This method is for univariate data and does not include trend or seasonality. It uses only one parameter named alpha (α), even known as smoothing factor or smoothing coefficient. This smoothing technique is widely employed due to its simplicity and success. The notation for the same is as given below:

$$S_{t+1} = \alpha y_t + (1 - \alpha)S_t \tag{1}$$

where S_i is the smoothed value of time series at time i, y_i is the actual value of time series at time i, and α is the smoothing constant and $0.0 < \alpha < 1.0$.

Double Exponential Smoothing (DE)

DE is used on data sets involving seasonality and for handling trend analysis. It is an extension to exponential smoothing, adding explicit support for trends in the univariate time series. It is used when there is a linear trend in the data. This involves an additional smoothing factor along with alpha (α) parameter. This is to control the decay of the influence of the change in a trend called beta (β).

For data exhibiting linear trend as:

$$y_t = b_0 + b_1 t + e_t \tag{2}$$

where b_0 and b_1 can change with time at a slow pace. The basic equations named Holt's method are as below:

$$\mu_t = \alpha y_t + (1 - \alpha)(\mu_{t-1} + T_{t-1}) \tag{3}$$

$$T_t = \beta(\mu_t - \mu_{t-1}) + (1 - \beta)T_{t-1} \tag{4}$$

where μ_t is the exponentially smoothed value of time series at time t, y_t is the actual observation of time series at time t, T_t is the trend estimate, α is the exponential smoothing constant for the data, and β is the smoothing constant for the trend.

Tukey's Smoothing (TS)

TS uses running medians to provide flexible but straightforward curves and is a robust smoother. Median smoothing methods were introduced by Tukey in 1977 for extracting smooth patterns which tend to hide due to non-linear spikes in time series data [92]. Such filtering smooths any existing volatile behavior that occurs in trends or seasonal behavior. This method smooths out the data acquired from equally spaced, linearly ordered intervals such as every year, every month, every quarter, etc.

NDVI data reconstructed using all three types of smoothing techniques were compared based on their potential in removing outliers.

Performance assessment of the smoothing techniques was also carried out by calculating root mean squared error (RMSE). RMSE between observed raw NDVI time series data and predicted smoothed NDVI time series data was used for evaluating the performance of each smoothing technique. The best optimal technique that would reconstruct the best denoised data sets was considered to be the one generating the least RMSE. The best-reconstructed data was used further for extracting phenological metrics of teak by detecting the inflection point (i.e., date) when the NDVI time series begins to ascend or descend for the specific year. Occurrence dates obtained using these smoothed NDVI time series were compared to the ground-based phenological observation and these along with RMSE were considered for selecting the optimum smoothing technique.

2.2.2. Determination of Phenological Metrics from NDVI Time Series Data

The intra-annual variations in the NDVI time series were used as the base for determining SOS, MAX, EOS, and LOS, using the *NDVI ratio*. This is the derivative method where the maximum value of *NDVI ratio* corresponds to the greatest change of NDVI time series. Equation (5) is given as:

$$NDVI\,ratio(t) = \frac{NDVI(t+1) - NDVI(t)}{NDVI(t)} \tag{5}$$

where *NDVI(t)* is the NDVI value at time *t*, and *NDVI ratio (t)* is the calculated relative change at time *t*. SOS was determined as the time *t* or the day with the maximum *NDVI ratio* [93]. Likewise, EOS date was determined as the time *t* or the day having the minimum *NDVI ratio*. The time or duration between SOS and EOS with *NDVI ratio* closest to zero was identified as MAX date. Furthermore, the LOS was obtained by defining the period between SOS and EOS in each grid point.

3. Results and Discussion

RMSEs generated to check the potential of SE, DE, and TS are 0.072, 0.097, and 0.048, respectively. Comparison of reconstructed NDVI time series using SE, DE, and TS techniques showed that both SE and DE smoothed higher values but could not encompass all the outliers (Figures 2 and 3). These methods are comparatively less effective in accounting for missing values or correcting outliers. TS smoothing approach is identified as a more effective and robust smoothing method, bringing out high-quality NDVI time series as it fairly handled the outliers (Figure 4). The scatterplot diagram distinctly highlights the presence of outliers that needed to be removed to generate precise output. Statistically, the smoothing techniques are supposed to remove these outliers. At this point, the selection of Tukey's technique is performed in comparison to the other two smoothing techniques as it is considered to be a resistant smoothing technique which uses running medians. Despite the fact that it lacks mathematical generalization, the main purpose of the smoothing technique is to provide a general idea of relatively slow changes in values with little attention to the close matching of data values. Phenological attributes exhibit such characteristics. Further, running medians are thought to be fast exploratory tools to allow a quick view of data components. On comparing with ground data, TS again exhibited the greatest effectiveness over SE and DE (Figure 5, Table 1). The Tukey-smoothed NDVI data provided greater vegetation phenology information, i.e., even minor changes in phenological dates which cannot be correctly identified from raw NDVI data can be obtained from Tukey-smoothed NDVI data. TS retrieved phenological data and ground phenological data showed very close values, thereby validating the TS results. Thus, TS potentially proved to be the optimal technique to best reconstruct the NDVI time series data and hence TS reconstructed NDVI time series is used for delineating shifts in SOS, EOS, MAX, and LOS of teak.

Table 1. Phenological metrics derived using different data at different time periods.

Phenophases		SOS	EOS	MAX
Ground data	2003–2004	22 July	10 February	15 September
	2013–2014	30 July	6 March	31 August
	2015–2016	1 August	23 March	27 August
Tukey-smoothed NDVI	2003–2004	28 July	13 February	26 September
	2013–2014	11 August	22 March	29 August

MODIS data analyzed for the monitoring of the phenological metrics of teak for the years 2003–2004 and 2013–2014 showed significant patterns despite their coarse resolution.

Annual time series of NDVI data generated for teak enabled the differentiation of various phenological metrics such as the SOS (Figures 6 and 7), EOS (Figures 8 and 9), MAX (Figures 10 and 11), and LOS (Figures 12 and 13). Such data applications have proven to be useful in several studies on global environmental change [94–98].

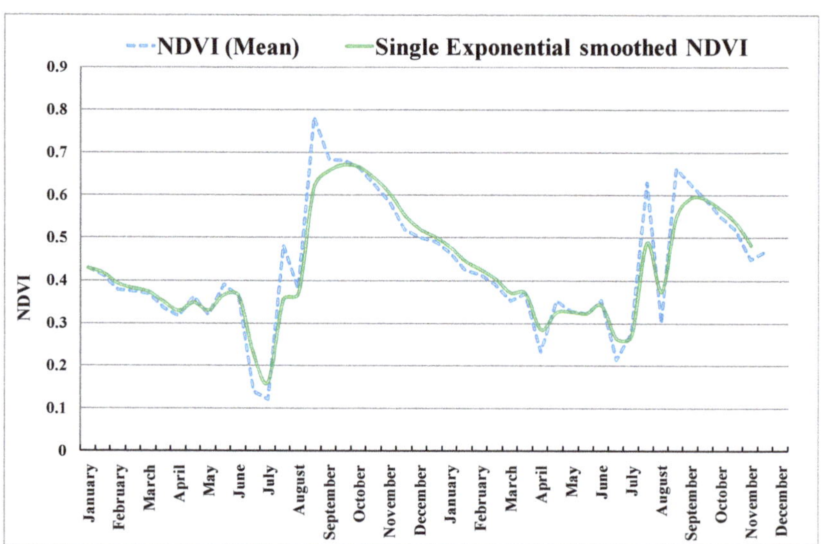

Figure 2. NDVI time series for year 2013–2014 reconstructed using single exponential smoothing.

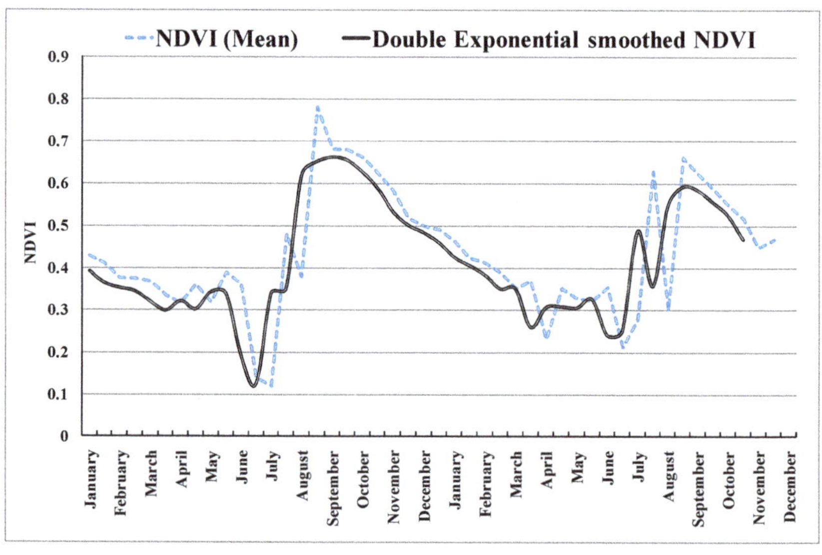

Figure 3. NDVI time series for year 2013–2014 reconstructed using double exponential smoothing.

Differences in phenological patterns are notable. SOS and EOS in DDF teak were delayed by 14 and 37 days, respectively, in the year 2013–2014 (SOS date—11 August 2013; EOS date—22 March 2014) when compared to 2003–2004 (SOS date—28 July 2003; EOS date—13 February 2004). The SOS in the year 2003–2004 was observed in late July which shifted to mid-August in 2013–2014. The EOS was observed in mid-February in 2003–2004 while in 2013–2014, the season ended in the end of March. Advancement of 28 days in MAX of teak was observed in the year 2013–2014 compared to 2003–2004. Greenness in teak in the district during the year 2003–2004 reaches its peak at the end of September, while in 2013–2014, teak reached its maximum greenness at the end of August. Results highlighted the fact that teak reached its MAX early in 2013–2014, indicating the shift in the phenology of the tree. LOS increased by 24 days in the year 2013–2014 (LOS—224 days) (Figure 14) compared to the year 2003–2004 (LOS—200 days) (Figure 13). Except for MAX, the year 2003–2004 showed earlier dates than the year 2013–2014 for the occurrence of all phenological metrics.

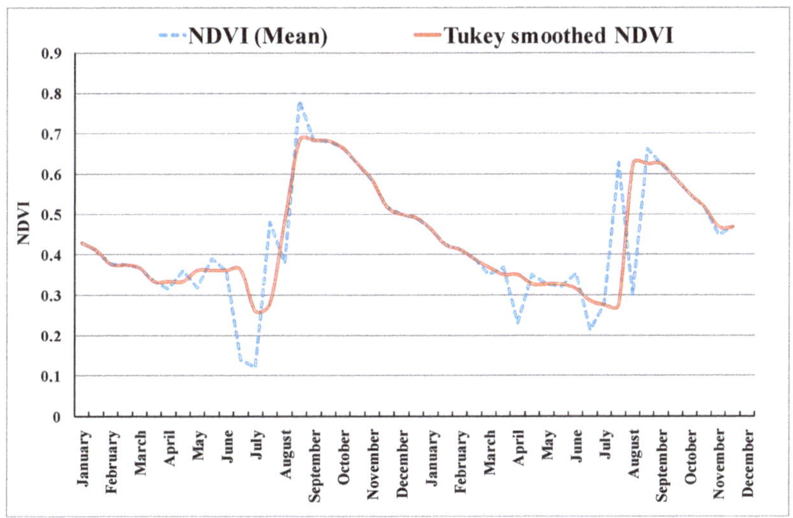

Figure 4. NDVI time series for the year 2013–2014 reconstructed using Tukey's smoothing.

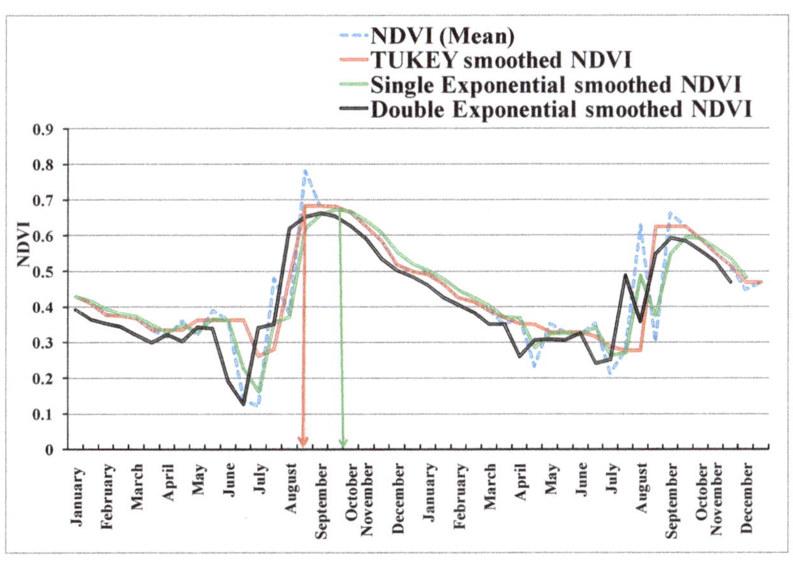

Figure 5. Different smoothing techniques' utility in derivation of maximum of the season.

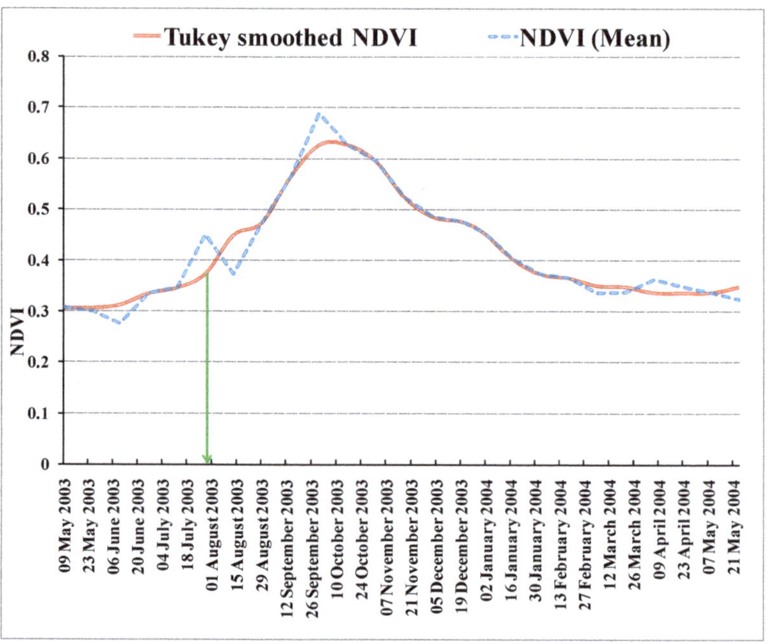

Figure 6. Raw MODIS NDVI and TS-smoothed reconstructed NDVI time series of year 2003–2004 for derivation of start of the season in *Tectona grandis* indicated by green arrow.

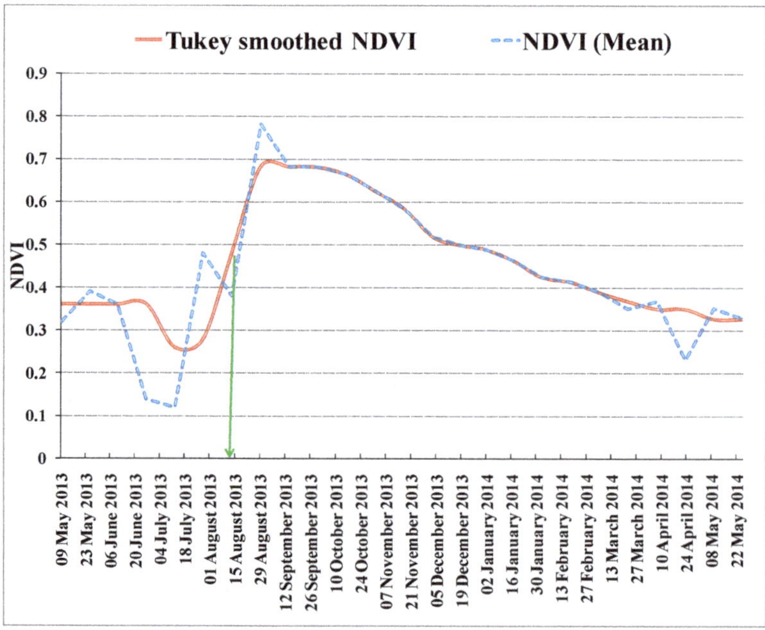

Figure 7. Raw MODIS NDVI and TS-smoothed reconstructed NDVI time series of year 2013–2014 for derivation of start of the season in *Tectona grandis* indicated by green arrow.

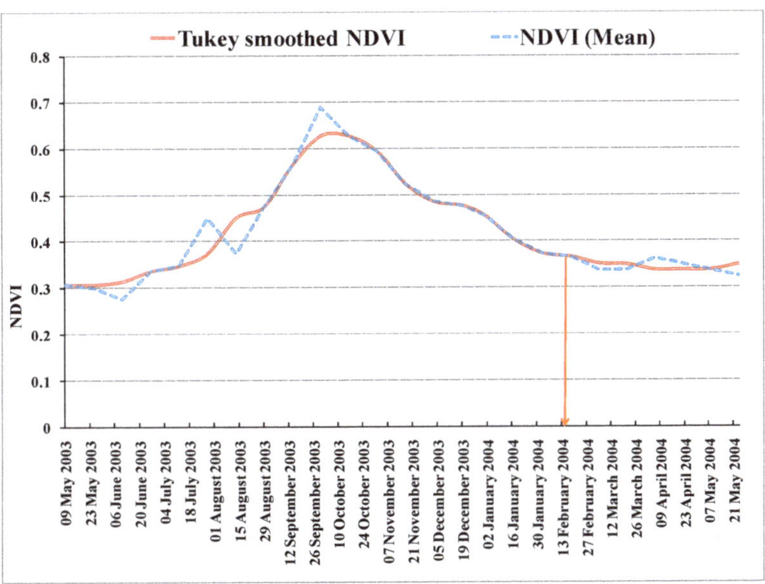

Figure 8. Raw MODIS NDVI and TS-smoothed reconstructed NDVI time series of year 2003–2004 for derivation of end of the season in *Tectona grandis* indicated by orange arrow.

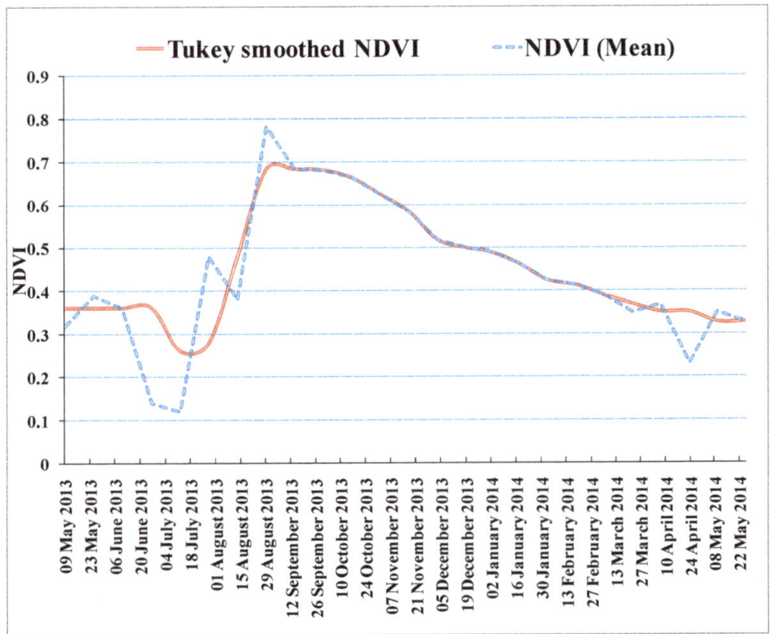

Figure 9. Raw MODIS NDVI and TS-smoothed reconstructed NDVI time series of year 2013–2014 for derivation ofend of the season in *Tectona grandis* indicated by orange arrow.

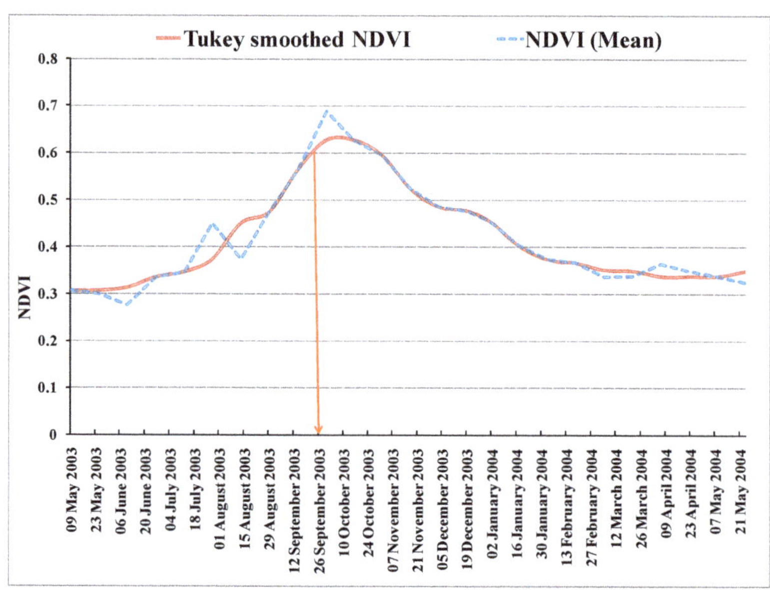

Figure 10. Raw MODIS NDVI and TS-smoothed reconstructed NDVI time series of year 2003–2004 for derivation of maximum of the season in *Tectona grandis* indicated by orange arrow.

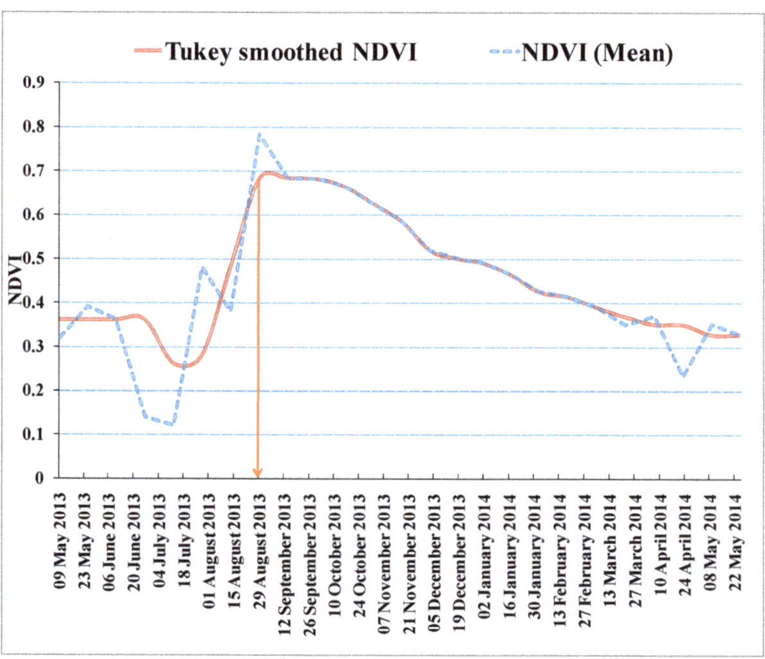

Figure 11. Raw MODIS NDVI and TS-smoothed reconstructed NDVI time series of year 2013–2014 for derivation of maximum of the season in *Tectona grandis* indicated by orange arrow.

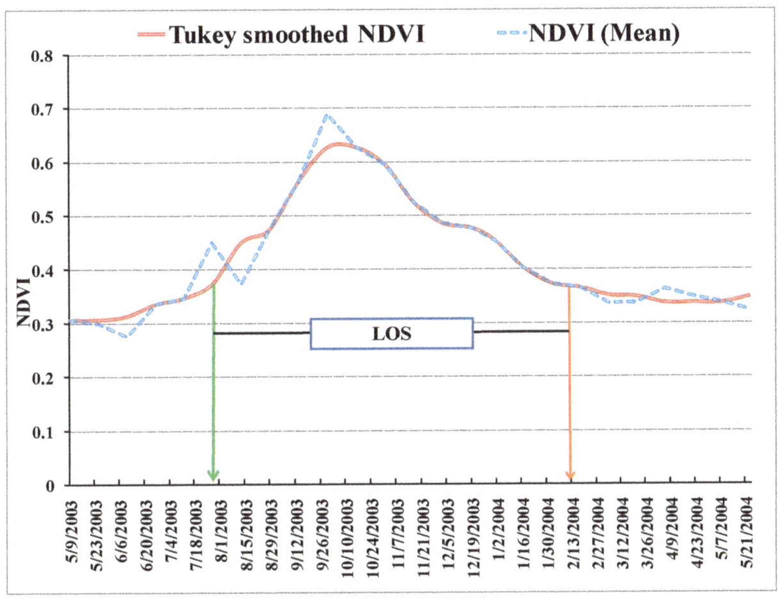

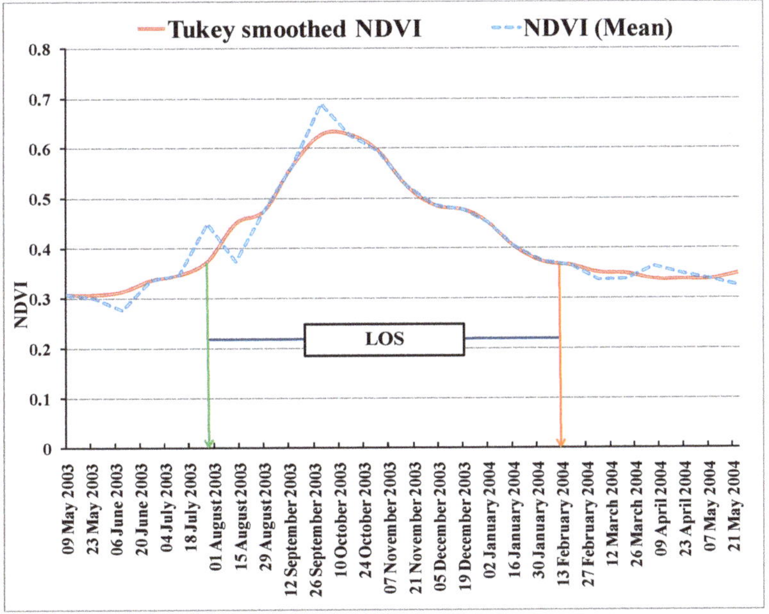

Figure 12. Raw MODIS NDVI and TS-smoothed reconstructed NDVI time series of year 2003–2004 for derivation of length of the season in *Tectona grandis* indicated between green and orange arrows.

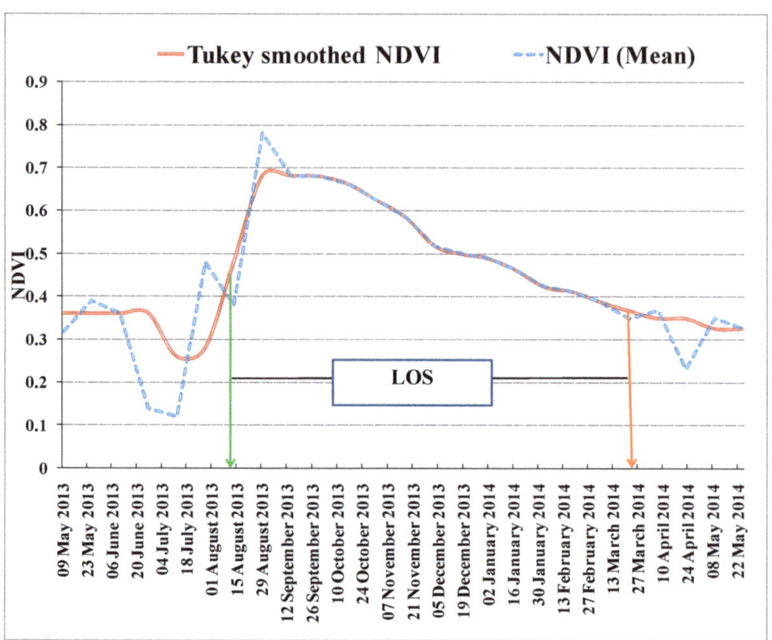

Figure 13. Raw MODIS NDVI and TS-smoothed reconstructed NDVI time series of year 2013–2014 for derivation oflength of the season in *Tectona grandis* indicated between green and orange arrows.

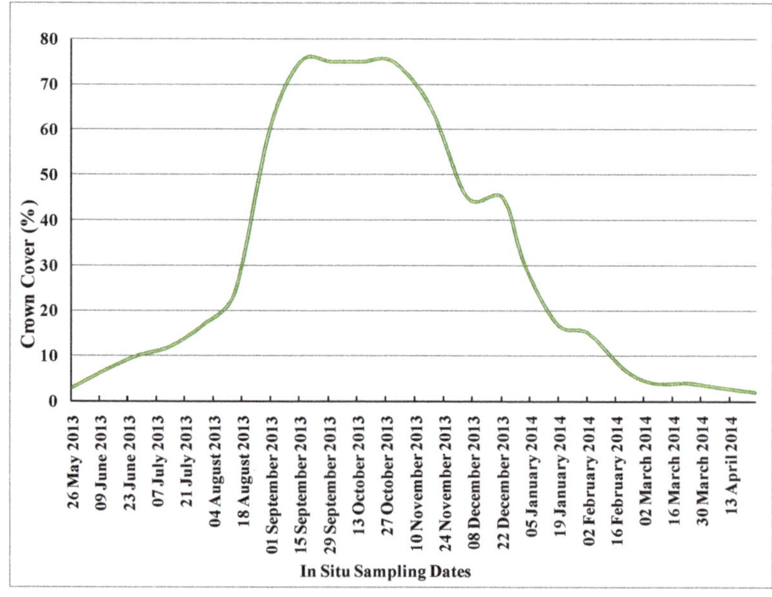

Figure 14. Crown cover in percentage during in situ sampling period.

For the validation of the phenological metrics generated from MODIS TS NDVI time series, the dates of different phenological occurrences are compared to ground measured crown cover. In situ crown cover assessment of 2013–2014 provides the information on different phenophases (Table 1). A plot of crown cover versus sampling dates matches the NDVI curve of the period (Figure 14). Results of the crown cover assessment can be used for the validation of the results generated through MODIS data. Average crown cover between 5 and 25% that corresponds to leaf initiation on ground and SOS on satellite data was observed on 30 July 2013. Maximum greenness or MAX that corresponds to crown cover over 25% was noted on 27 August 2013 and leaf fall on the ground, i.e., less than 5% crown cover that corresponds to EOS was recorded on 6 March 2014. These ground-based phenological observations are comparable to phenological metrics derived using time series data, thereby validating our results. Results of in situ observations also show that teak trees considered for the present study shifted their phenophases between the years 2003–2004 and 2013–2014. Leaf initiation on the ground showed post shift of 8 days. In situ leaf fall recorded showed a delay of 24 days. Maximum greenness noted on the ground also compliments satellite-retrieved MAX results which showed a preshift of 15 days. The results highlight that phenological metrics from satellite data such as SOS, EOS, and MAX with temporal dynamics similar to those of ground phenology measurements such as leaf initiation, leaf fall, and maximum greenness through crown cover assessment can be generated.

Substantial interannual variability in the start and end of the growing season of the teak of DDF forests can be explained by studying many factors such as year-to-year variability in weather, especially climatic factors including rainfall, temperature, elevated CO_2, altered precipitation regimes, etc. Such types of variability may be responsible for the observed shifts in phenology which ultimately can be robust indicators of the impacts of climate change. This makes derivation of phenological metrics imperative as they can serve as important inputs for better understanding of the climatic drivers controlling phenology. These alterations or shifts in phenophases can influence the climate system at both global and regional levels through feedback mechanisms of surface albedo, CO_2 fluxes, and evaporation [99]. Improved inputs of phenological measurements into global biosphere models conducted via satellite data will also enhance the understanding of temporal and spatial global carbon dynamics. This will be more useful for the areas where there exists a lack of ground phenological observations due to inaccessibility [38].

4. Conclusions

The present study tested the potential of three statistical techniques, viz., single exponential, double exponential, and Tukey smoothing. Comparison of these three techniques demonstrates the effectiveness of the Tukey smoothing technique in denoising the NDVI time series and removing the outliers. Statistical Tukey smoothing enhances the visibility of unique patterns present in the data and thereby aids in identifying distinct decadal shifts in the phenological metrics. Significant shifts in the growing seasons of DDF teak of Narmada district from the years 2003–2004 to 2013–2014 are identified on the basis of changes in SOS, EOS, LOS, and MAX delineated using Tukey-smoothed MODIS NDVI time series data. These shifts are validated by ground phenology measurements that are calculated using crown cover assessment. Earlier studies using remote sensing more commonly focused on phenological studies of forest types but the present study highlights the fact that satellite data has great capability and can be used for generating phenological metrics of individual plant species as well.

Author Contributions: Conceptualization, G.S.K. and R.K.M.M.; methodology, R.K.M.M., M.N.S. and P.A.T.; software, N.V.M., V.H.B. and R.K.M.M.; validation, N.V.M., V.H.B., R.K.M.M., S.M.; N.V.M., V.H.B. and R.K.M.M.; investigation, R.K.M.M., N.V.M., V.H.B. and G.S.K.; data curation, G.S.K.; writing—original draft preparation, R.K.M.M. and G.S.K.; writing—review and editing, G.S.K., P.A.T. and S.M.; visualization, S.M.; supervision, C.P.S. and B.K.B.; project administration, G.S.K.; funding acquisition, G.S.K. and M.N.S. All authors have read and agreed to the published version of the manuscript.

Funding: This research was funded by ISRO RESPOND, project id OGP134″ and we give thanks to MoEFCC for their added support.

Institutional Review Board Statement: Ethical review and approval were waived for this study.

Informed Consent Statement: Not applicable.

Data Availability Statement: Not applicable.

Acknowledgments: In this section, I acknowledge the funds received from ISRO RESPOND for carrying out the present work. I also extend my thanks to MoEF&CC for data and field work support.

Conflicts of Interest: Authors declare no conflict of interest.

References

1.   Prabakaran, C.; Singh, C.; Panigrahy, S.; Parihar, J. Retrieval of forest phenological parameters from remote sensing-based NDVI time-series data. *Curr. Sci.* **2013**, *105*, 795–802.
2.   Kaduk, J.; Heimann, M. A prognostic phenology scheme for global terrestrial carbon cycle models. *Clim. Res.* **1996**, *6*, 1–19. [CrossRef]
3.   Hall, C.A.; Ekdahl, C.A.; Wartenberg, D.E. A fifteen-year record of biotic metabolism in the Northern Hemisphere. *Nature* **1975**, *255*, 136. [CrossRef]
4.   Keeling, C.D.; Chin, J.; Whorf, T. Increased activity of northern vegetation inferred from atmospheric CO_2 measurements. *Nature* **1996**, *382*, 146. [CrossRef]
5.   D'Arrigo, R.; Jacoby, G.C.; Fung, I.Y. Boreal forests and atmosphere—Biosphere exchange of carbon dioxide. *Nature* **1987**, *329*, 321. [CrossRef]
6.   White, M.; Running, S.W.; Thornton, P.E. The impact of growing-season length variability on carbon assimilation and evapotranspiration over 88 years in the eastern US deciduous forest. *Int. J. Biometeorol.* **1999**, *42*, 139–145. [CrossRef] [PubMed]
7.   Chmielewski, F.-M.; Müller, A.; Bruns, E. Climate changes and trends in phenology of fruit trees and field crops in Germany, 1961–2000. *Agric. For. Meteorol.* **2004**, *121*, 69–78. [CrossRef]
8.   Lee, B.; Kim, E.; Lim, J.-H.; Seo, B.; Chung, J.-M. Detecting Vegetation Phenology in Various Forest Types Using Long-Term MODIS Vegetation Indices. In Proceedings of the IGARSS 2018 IEEE International Geoscience and Remote Sensing Symposium, Valencia, Spain, 22–27 July 2018; pp. 5243–5246.
9.   Mohanta, M.R.; Suresh, H.; Sahu, S.C. A Review on Plant Phenology Study in Different Forest Types of India. *Indian For.* **2020**, *146*, 1137–1148.
10. Klosterman, S.; Hufkens, K.; Gray, J.; Melaas, E.; Sonnentag, O.; Lavine, I.; Mitchell, L.; Norman, R.; Friedl, M.; Richardson, A. Evaluating remote sensing of deciduous forest phenology at multiple spatial scales using PhenoCam imagery. *Biogeosciences* **2014**, *11*, 4305–4320. [CrossRef]
11. Prasad, V.K.; Badarinath, K.; Eaturu, A. Spatial patterns of vegetation phenology metrics and related climatic controls of eight contrasting forest types in India—Analysis from remote sensing datasets. *Theor. Appl. Climatol.* **2007**, *89*, 95–107. [CrossRef]
12. Shukla, R.; Ramakrishnan, P. Phenology of trees in a sub-tropical humid forest in north-eastern India. *Vegetatio* **1982**, *49*, 103–109. [CrossRef]
13. Nanda, A.; Suresh, H.S.; Krishnamurthy, Y.L. Phenology of a tropical dry deciduous forest of Bhadra wildlife sanctuary, southern India. *Ecol. Process.* **2014**, *3*, 1–12. [CrossRef]
14. Morisette, J.T.; Richardson, A.D.; Knapp, A.K.; Fisher, J.I.; Graham, E.A.; Abatzoglou, J.; Wilson, B.E.; Breshears, D.D.; Henebry, G.M.; Hanes, J.M. Tracking the rhythm of the seasons in the face of global change: Phenological research in the 21st century. *Front. Ecol. Environ.* **2009**, *7*, 253–260. [CrossRef]
15. Schnelle, F.; Volkert, E. Internationale phänologische gärten Stationen eines grundnetzes für internationale phänologische beobachtungen. *Agric. Meteorol.* **1964**, *1*, 22–29. [CrossRef]
16. Newman, G.; Wiggins, A.; Crall, A.; Graham, E.; Newman, S.; Crowston, K. The future of citizen science: Emerging technologies and shifting paradigms. *Front. Ecol. Environ.* **2012**, *10*, 298–304. [CrossRef]
17. Schwartz, M.D.; Betancourt, J.L.; Weltzin, J.F. From Caprio's lilacs to the USA National Phenology Network. *Front. Ecol. Environ.* **2012**, *10*, 324–327. [CrossRef]
18. van Vliet, A.J.; Bron, W.A.; Mulder, S.; van der Slikke, W.; Odé, B. Observed climate-induced changes in plant phenology in the Netherlands. *Reg. Environ. Chang.* **2014**, *14*, 997–1008. [CrossRef]
19. Vrieling, A.; Meroni, M.; Darvishzadeh, R.; Skidmore, A.K.; Wang, T.; Zurita-Milla, R.; Oosterbeek, K.; O'Connor, B.; Paganini, M. Vegetation phenology from Sentinel-2 and field cameras for a Dutch barrier island. *Remote Sens. Environ.* **2018**, *215*, 517–529. [CrossRef]
20. Richardson, A.D.; Braswell, B.H.; Hollinger, D.Y.; Jenkins, J.P.; Ollinger, S.V. Near-surface remote sensing of spatial and temporal variation in canopy phenology. *Ecol. Appl.* **2009**, *19*, 1417–1428. [CrossRef]
21. Ide, R.; Oguma, H. Use of digital cameras for phenological observations. *Ecol. Inform.* **2010**, *5*, 339–347. [CrossRef]
22. Nagai, S.; Maeda, T.; Gamo, M.; Muraoka, H.; Suzuki, R.; Nasahara, K.N. Using digital camera images to detect canopy condition of deciduous broad-leaved trees. *Plant Ecol. Divers.* **2011**, *4*, 79–89. [CrossRef]
23. Jeganathan, C.; Dash, J.; Atkinson, P.M. Characterising the spatial pattern of phenology for the tropical vegetation of India using multi-temporal MERIS chlorophyll data. *Landsc. Ecol.* **2010**, *25*, 1125–1141. [CrossRef]

24. Justice, C.O.; Townshend, J.; Holben, B.; Tucker, E.C. Analysis of the phenology of global vegetation using meteorological satellite data. *Int. J. Remote Sens.* **1985**, *6*, 1271–1318. [CrossRef]

25. Reed, B.C.; Brown, J.F.; VanderZee, D.; Loveland, T.R.; Merchant, J.W.; Ohlen, D.O. Measuring phenological variability from satellite imagery. *J. Veg. Sci.* **1994**, *5*, 703–714. [CrossRef]

26. White, M.A.; Hoffman, F.; Hargrove, W.W.; Nemani, R.R. A global framework for monitoring phenological responses to climate change. *Geophys. Res. Lett.* **2005**, *32*, 1–4. [CrossRef]

27. Atkinson, P.M.; Jeganathan, C.; Dash, J.; Atzberger, C. Inter-comparison of four models for smoothing satellite sensor time-series data to estimate vegetation phenology. *Remote Sens. Environ.* **2012**, *123*, 400–417. [CrossRef]

28. Jiang, N.; Zhu, W.; Zheng, Z.; Chen, G.; Fan, D. A comparative analysis between GIMSS NDVIg and NDVI3g for monitoring vegetation activity change in the northern hemisphere during 1982–2008. *Remote Sens.* **2013**, *5*, 4031–4044. [CrossRef]

29. Shen, M.; Tang, Y.; Desai, A.R.; Gough, C.; Chen, J. Can EVI-derived land-surface phenology be used as a surrogate for phenology of canopy photosynthesis? *Int. J. Remote Sens.* **2014**, *35*, 1162–1174. [CrossRef]

30. Löw, M.; Koukal, T. Phenology Modelling and Forest Disturbance Mapping with Sentinel-2 Time Series in Austria. *Remote Sens.* **2020**, *12*, 4191. [CrossRef]

31. Dixon, D.J.; Callow, J.N.; Duncan, J.M.; Setterfield, S.A.; Pauli, N. Satellite prediction of forest flowering phenology. *Remote Sens. Environ.* **2021**, *255*, 112197. [CrossRef]

32. Clerici, N.; Weissteiner, C.J.; Gerard, F. Exploring the use of MODIS NDVI-based phenology indicators for classifying forest general habitat categories. *Remote Sens.* **2012**, *4*, 1781–1803. [CrossRef]

33. Rankine, C.; Sánchez-Azofeifa, G.; Guzmán, J.A.; Espirito-Santo, M.; Sharp, I. Comparing MODIS and near-surface vegetation indexes for monitoring tropical dry forest phenology along a successional gradient using optical phenology towers. *Environ. Res. Lett.* **2017**, *12*, 105007. [CrossRef]

34. Lu, L.; Kuenzer, C.; Wang, C.; Guo, H.; Li, Q. Evaluation of three MODIS-derived vegetation index time series for dryland vegetation dynamics monitoring. *Remote Sens.* **2015**, *7*, 7597–7614. [CrossRef]

35. Böttcher, K.; Aurela, M.; Kervinen, M.; Markkanen, T.; Mattila, O.-P.; Kolari, P.; Metsämäki, S.; Aalto, T.; Arslan, A.N.; Pulliainen, J. MODIS time-series-derived indicators for the beginning of the growing season in boreal coniferous forest—A comparison with CO_2 flux measurements and phenological observations in Finland. *Remote Sens. Environ.* **2014**, *140*, 625–638. [CrossRef]

36. St Peter, J.; Hogland, J.; Hebblewhite, M.; Hurley, M.; Hupp, N.; Proffitt, K. Linking phenological indices from digital cameras in Idaho and Montana to MODIS NDVI. *Remote Sens.* **2018**, *10*, 1612. [CrossRef]

37. Wu, C.; Gonsamo, A.; Gough, C.M.; Chen, J.M.; Xu, S. Modeling growing season phenology in North American forests using seasonal mean vegetation indices from MODIS. *Remote Sens. Environ.* **2014**, *147*, 79–88. [CrossRef]

38. Hamunyela, E.; Verbesselt, J.; Roerink, G.; Herold, M. Trends in spring phenology of western European deciduous forests. *Remote Sens.* **2013**, *5*, 6159–6179. [CrossRef]

39. Goward, S.N.; Tucker, C.J.; Dye, D.G. North American vegetation patterns observed with the NOAA-7 advanced very high resolution radiometer. *Vegetatio* **1985**, *64*, 3–14. [CrossRef]

40. Cleland, E.E.; Chuine, I.; Menzel, A.; Mooney, H.A.; Schwartz, M.D. Shifting plant phenology in response to global change. *Trends Ecol. Evol.* **2007**, *22*, 357–365. [CrossRef]

41. Richardson, A.D.; Jenkins, J.P.; Braswell, B.H.; Hollinger, D.Y.; Ollinger, S.V.; Smith, M.-L. Use of digital webcam images to track spring green-up in a deciduous broadleaf forest. *Oecologia* **2007**, *152*, 323–334. [CrossRef]

42. Piao, S.; Ciais, P.; Friedlingstein, P.; Peylin, P.; Reichstein, M.; Luyssaert, S.; Margolis, H.; Fang, J.; Barr, A.; Chen, A. Net carbon dioxide losses of northern ecosystems in response to autumn warming. *Nature* **2008**, *451*, 49. [CrossRef] [PubMed]

43. Atzberger, C.; Klisch, A.; Mattiuzzi, M.; Vuolo, F. Phenological metrics derived over the European continent from NDVI3g data and MODIS time series. *Remote Sens.* **2014**, *6*, 257–284. [CrossRef]

44. Piao, S.; Tan, J.; Chen, A.; Fu, Y.H.; Ciais, P.; Liu, Q.; Janssens, I.A.; Vicca, S.; Zeng, Z.; Jeong, S.-J. Leaf onset in the northern hemisphere triggered by daytime temperature. *Nat. Commun.* **2015**, *6*, 6911. [CrossRef]

45. Huete, A.; Didan, K.; van Leeuwen, W.; Miura, T.; Glenn, E. MODIS vegetation indices. In *Land Remote Sensing and Global Environmental Change*; Springer: Berlin/Heidelberg, Germany, 2010; pp. 579–602.

46. Miura, T.; Smith, C.Z.; Yoshioka, H. Validation and analysis of Terra and Aqua MODIS, and SNPP VIIRS vegetation indices under zero vegetation conditions: A case study using Railroad Valley Playa. *Remote Sens. Environ.* **2021**, *257*, 112344. [CrossRef]

47. Albarakat, R.; Lakshmi, V. Comparison of normalized difference vegetation index derived from Landsat, MODIS, and AVHRR for the Mesopotamian marshes between 2002 and 2018. *Remote Sens.* **2019**, *11*, 1245. [CrossRef]

48. Situmorang, J.P.; Sugianto, S.; Darusman, D. Estimation of Carbon Stock Stands using EVI and NDVI vegetation index in production forest of lembah Seulawah sub-district, Aceh Indonesia. *Aceh Int. J. Sci. Technol.* **2016**, *5*, 126–139.

49. Le Maire, G.; Marsden, C.; Nouvellon, Y.; Grinand, C.; Hakamada, R.; Stape, J.-L.; Laclau, J.-P. MODIS NDVI time-series allow the monitoring of Eucalyptus plantation biomass. *Remote Sens. Environ.* **2011**, *115*, 2613–2625. [CrossRef]

50. Habib, S.; Al-Ghamdi, S.G. Estimation of Above-Ground Carbon-Stocks for Urban Greeneries in Arid Areas: Case Study for Doha and FIFA World Cup Qatar 2022. *Front. Environ. Sci.* **2021**, *9*, 186. [CrossRef]

51. Gang, B.; Bao, Y. Remotely sensed estimate of biomass carbon stocks in Xilingol grassland using MODIS NDVI data. In Proceedings of the 2013 International Conference on Mechatronic Sciences, Electric Engineering and Computer (MEC), Shengyang, China, 20–22 December 2013; pp. 676–679.

52. Anand, A.; Pandey, P.C.; Petropoulos, G.P.; Pavlides, A.; Srivastava, P.K.; Sharma, J.K.; Malhi, R.K.M. Use of hyperion for mangrove forest carbon stock assessment in Bhitarkanika forest reserve: A contribution towards blue carbon initiative. *Remote Sens.* **2020**, *12*, 597. [CrossRef]

53. Che, X.; Feng, M.; Jiang, H.; Song, J.; Jia, B. Downscaling MODIS surface reflectance to improve water body extraction. *Adv. Meteorol.* **2015**, *2015*, 424291. [CrossRef]

54. Xie, F.; Fan, H. Deriving drought indices from MODIS vegetation indices (NDVI/EVI) and Land Surface Temperature (LST): Is data reconstruction necessary? *Int. J. Appl. Earth Obs. Geoinf.* **2021**, *101*, 102352. [CrossRef]

55. Mo, X.; Liu, S.; Lin, Z.; Wang, S.; Hu, S. Trends in land surface evapotranspiration across China with remotely sensed NDVI and climatological data for 1981–2010. *Hydrol. Sci. J.* **2015**, *60*, 2163–2177. [CrossRef]

56. Testa, S.; Soudani, K.; Boschetti, L.; Mondino, E.B. MODIS-derived EVI, NDVI and WDRVI time series to estimate phenological metrics in French deciduous forests. *Int. J. Appl. Earth Obs. Geoinf.* **2018**, *64*, 132–144. [CrossRef]

57. Peng, D.; Wu, C.; Li, C.; Zhang, X.; Liu, Z.; Ye, H.; Luo, S.; Liu, X.; Hu, Y.; Fang, B. Spring green-up phenology products derived from MODIS NDVI and EVI: Intercomparison, interpretation and validation using National Phenology Network and AmeriFlux observations. *Ecol. Indic.* **2017**, *77*, 323–336. [CrossRef]

58. Karkauskaite, P.; Tagesson, T.; Fensholt, R. Evaluation of the plant phenology index (PPI), NDVI and EVI for start-of-season trend analysis of the Northern Hemisphere boreal zone. *Remote Sens.* **2017**, *9*, 485. [CrossRef]

59. Wang, C.; Li, J.; Liu, Q.; Zhong, B.; Wu, S.; Xia, C. Analysis of differences in phenology extracted from the enhanced vegetation index and the leaf area index. *Sensors* **2017**, *17*, 1982. [CrossRef]

60. Verhegghen, A.; Bontemps, S.; Defourny, P. A global NDVI and EVI reference data set for land-surface phenology using 13 years of daily SPOT-VEGETATION observations. *Int. J. Remote Sens.* **2014**, *35*, 2440–2471. [CrossRef]

61. Osunmadewa, B.A.; Gebrehiwot, W.Z.; Csaplovics, E.; Adeofun, O.C. Spatio-temporal monitoring of vegetation phenology in the dry sub-humid region of Nigeria using time series of AVHRR NDVI and TAMSAT datasets. *Open Geosci.* **2018**, *10*, 1–11. [CrossRef]

62. Melaas, E.K.; Sulla-Menashe, D.; Gray, J.M.; Black, T.A.; Morin, T.H.; Richardson, A.D.; Friedl, M.A. Multisite analysis of land surface phenology in North American temperate and boreal deciduous forests from Landsat. *Remote Sens. Environ.* **2016**, *186*, 452–464. [CrossRef]

63. Snyder, K.A.; Huntington, J.L.; Wehan, B.L.; Morton, C.G.; Stringham, T.K. Comparison of Landsat and Land-Based Phenology Camera Normalized Difference Vegetation Index (NDVI) for Dominant Plant Communities in the Great Basin. *Sensors* **2019**, *19*, 1139. [CrossRef] [PubMed]

64. White, K.; Pontius, J.; Schaberg, P. Remote sensing of spring phenology in northeastern forests: A comparison of methods, field metrics and sources of uncertainty. *Remote Sens. Environ.* **2014**, *148*, 97–107. [CrossRef]

65. Zhou, L.; Tucker, C.J.; Kaufmann, R.K.; Slayback, D.; Shabanov, N.V.; Myneni, R.B. Variations in northern vegetation activity inferred from satellite data of vegetation index during 1981 to 1999. *J. Geophys. Res. Atmos.* **2001**, *106*, 20069–20083. [CrossRef]

66. Yang, L.; Wylie, B.K.; Tieszen, L.L.; Reed, B.C. An analysis of relationships among climate forcing and time-integrated NDVI of grasslands over the US northern and central Great Plains. *Remote Sens. Environ.* **1998**, *65*, 25–37. [CrossRef]

67. Yang, W.; Yang, L.; Merchant, J. An assessment of AVHRR/NDVI-ecoclimatological relations in Nebraska, USA. *Int. J. Remote Sens.* **1997**, *18*, 2161–2180. [CrossRef]

68. Suepa, T.; Qi, J.; Lawawirojwong, S.; Messina, J.P. Understanding spatio-temporal variation of vegetation phenology and rainfall seasonality in the monsoon Southeast Asia. *Environ. Res.* **2016**, *147*, 621–629. [CrossRef] [PubMed]

69. Fu, C.; Wen, G. Variation of ecosystems over East Asia in association with seasonal, interannual and decadal monsoon climate variability. *Clim. Chang.* **1999**, *43*, 477–494. [CrossRef]

70. Justice, C.; Holben, B.; Gwynne, M. Monitoring East African vegetation using AVHRR data. *Int. J. Remote Sens.* **1986**, *7*, 1453–1474. [CrossRef]

71. Lloyd, D. A phenological classification of terrestrial vegetation cover using shortwave vegetation index imagery. *Int. J. Remote. Sens.* **1990**, *11*, 2269–2279. [CrossRef]

72. Chen, X.; Xu, C.; Tan, Z. An analysis of relationships among plant community phenology and seasonal metrics of Normalized Difference Vegetation Index in the northern part of the monsoon region of China. *Int. J. Biometeorol.* **2001**, *45*, 170–177. [CrossRef]

73. Cai, Z.; Jönsson, P.; Jin, H.; Eklundh, L. Performance of smoothing methods for reconstructing NDVI time-series and estimating vegetation phenology from MODIS data. *Remote Sens.* **2017**, *9*, 1271. [CrossRef]

74. Hermance, J.F.; Jacob, R.W.; Bradley, B.A.; Mustard, J.F. Extracting phenological signals from multiyear AVHRR NDVI time series: Framework for applying high-order annual splines with roughness damping. *IEEE Trans. Geosci. Remote Sens.* **2007**, *45*, 3264–3276. [CrossRef]

75. Moody, A.; Johnson, D.M. Land-surface phenologies from AVHRR using the discrete Fourier transform. *Remote Sens. Environ.* **2001**, *75*, 305–323. [CrossRef]

76. Roerink, G.; Menenti, M.; Verhoef, W. Reconstructing cloudfree NDVI composites using Fourier analysis of time series. *Int. J. Remote Sens.* **2000**, *21*, 1911–1917. [CrossRef]

77. Jakubauskas, M.E.; Legates, D.R.; Kastens, J.H. Harmonic analysis of time-series AVHRR NDVI data. *Photogramm. Eng. Remote Sens.* **2001**, *67*, 461–470.

78. Wagenseil, H.; Samimi, C. Assessing spatio-temporal variations in plant phenology using Fourier analysis on NDVI time series: Results from a dry savannah environment in Namibia. *Int. J. Remote Sens.* **2006**, *27*, 3455–3471. [CrossRef]
79. Atzberger, C.; Eilers, P.H. Evaluating the effectiveness of smoothing algorithms in the absence of ground reference measurements. *Int. J. Remote Sens.* **2011**, *32*, 3689–3709. [CrossRef]
80. Verger, A.; Filella, I.; Baret, F.; Peñuelas, J. Vegetation baseline phenology from kilometric global LAI satellite products. *Remote Sens. Environ.* **2016**, *178*, 1–14. [CrossRef]
81. Ma, Y.; Niu, X.; Liu, J. A comparison of different methods for studying vegetation phenology in Central Asia. In *Geo-Informatics in Resource Management and Sustainable Ecosystem*; Springer: Berlin/Heidelberg, Germany, 2015; pp. 301–307.
82. De Beurs, K.M.; Henebry, G.M. Land surface phenology and temperature variation in the International Geosphere–Biosphere Program high-latitude transects. *Glob. Chang. Biol.* **2005**, *11*, 779–790. [CrossRef]
83. Jonsson, P.; Eklundh, L. Seasonality extraction by function fitting to time-series of satellite sensor data. *IEEE Trans. Geosci. Remote Sens.* **2002**, *40*, 1824–1832. [CrossRef]
84. Yu, B.; Shang, S. Multi-year mapping of maize and sunflower in Hetao irrigation district of China with high spatial and temporal resolution vegetation index series. *Remote Sens.* **2017**, *9*, 855. [CrossRef]
85. Xu, X.; Conrad, C.; Doktor, D. Optimising phenological metrics extraction for different crop types in Germany using the moderate resolution imaging spectrometer (MODIS). *Remote Sens.* **2017**, *9*, 254. [CrossRef]
86. Klisch, A.; Royer, A.; Lazar, C.; Baruth, B.; Genovese, G. Extraction of phenological parameters from temporally smoothed vegetation indices. *Methods* **2006**, *3*, 5.
87. Zhang, X.; Friedl, M.A.; Schaaf, C.B.; Strahler, A.H.; Hodges, J.C.; Gao, F.; Reed, B.C.; Huete, A. Monitoring vegetation phenology using MODIS. *Remote Sens. Environ.* **2003**, *84*, 471–475. [CrossRef]
88. Goodman, M.L. A new look at higher-order exponential smoothing for forecasting. *Oper. Res.* **1974**, *22*, 880–888. [CrossRef]
89. Carreño-Conde, F.; Sipols, A.E.; de Blas, C.S.; Mostaza-Colado, D. A forecast model applied to monitor crops dynamics using vegetation indices (Ndvi). *Appl. Sci.* **2021**, *11*, 1859. [CrossRef]
90. Sabnis, S.; Amin, J. *Eco-Environmental Studies of Sardar Sarovar Environs*; Report of Eco-Environment and Wildlife Management Studies Project; M.S. University of Baroda Press: Baroda, India, 1992.
91. Pilar, C.-D.; Gabriel, M.-M. Phenological pattern of fifteen Mediterranean phanaerophytes from shape Quercus ilex communities of NE-Spain. *Plant Ecol.* **1998**, *139*, 103–112. [CrossRef]
92. Tukey, J.W. *Exploratory Data Analysis*; Addison-Wesley Publishing Company: Reading, MA, USA; Menlo Park, CA, USA, 1977; Volume 2.
93. Jeong, S.J.; HO, C.H.; GIM, H.J.; Brown, M.E. Phenology shifts at start vs. end of growing season in temperate vegetation over the Northern Hemisphere for the period 1982–2008. *Glob. Chang. Biol.* **2011**, *17*, 2385–2399. [CrossRef]
94. Schucknecht, A.; Erasmi, S.; Niemeyer, I.; Matschullat, J. Assessing vegetation variability and trends in north-eastern Brazil using AVHRR and MODIS NDVI time series. *Eur. J. Remote Sens.* **2013**, *46*, 40–59. [CrossRef]
95. Hentze, K.; Thonfeld, F.; Menz, G. Evaluating crop area mapping from MODIS time-series as an assessment tool for Zimbabwe's "fast track land reform programme". *PLoS ONE* **2016**, *11*, e0156630. [CrossRef]
96. Reddy, G.P.O.; Kumar, N.; Sahu, N.; Srivastava, R.; Singh, S.K.; Naidu, L.G.K.; Chary, G.R.; Biradar, C.M.; Gumma, M.K.; Reddy, B.S. Assessment of spatio-temporal vegetation dynamics in tropical arid ecosystem of India using MODIS time-series vegetation indices. *Arab. J. Geosci.* **2020**, *13*, 1–13. [CrossRef]
97. Pervez, S.; McNally, A.; Arsenault, K.; Budde, M.; Rowland, J. Vegetation Monitoring Optimization With Normalized Difference Vegetation Index and Evapotranspiration Using Remote Sensing Measurements and Land Surface Models Over East Africa. *Front. Clim.* **2021**, *3*, 589981. [CrossRef]
98. Dagnachew, M.; Kebede, A.; Moges, A.; Abebe, A. Effects of climate variability on normalized difference vegetation index (NDVI) in the Gojeb river catchment, omo-gibe basin, Ethiopia. *Adv. Meteorol.* **2020**, *2020*, 8263246. [CrossRef]
99. Menzel, A. Phenology: Its importance to the global change community. *Clim. Chang.* **2002**, *54*, 379–385. [CrossRef]

remote sensing

Article

Satellite Based Fraction of Absorbed Photosynthetically Active Radiation Is Congruent with Plant Diversity in India

Swapna Mahanand [1], Mukunda Dev Behera [1,2,*], Partha Sarathi Roy [3], Priyankar Kumar [2], Saroj Kanta Barik [4] and Prashant Kumar Srivastava [5]

[1] School of Water Resources, IIT Kharagpur, Kharagpur 721302, India; swapna.mahanand@atree.org
[2] SAM Lab, Centre for Oceans, Rivers, Atmosphere and Land Sciences, IIT Kharagpur, Kharagpur 721302, India; priyankar7200@iitkgp.ac.in
[3] World Resources Institute, New Delhi 110016, India; Parth.roy@wri.org
[4] CSIR-National Botanical Research Institute, Lucknow 226001, India; skbarik@nbri.res.in
[5] Institute of Environment and Sustainable Development, Banaras Hindu University, Varanasi 221005, India; prashant.iesd@bhu.ac.in
* Correspondence: mdbehera@coral.iitkgp.ac.in

Abstract: A dynamic habitat index (DHI) based on satellite derived biophysical proxy (fraction of absorbed photosynthetically active radiation, FAPAR) was used to evaluate the vegetation greenness pattern across deserts to alpine ecosystems in India that account to different biodiversity. The cumulative (DHI-cum), minimum (DHI-min), and seasonal (DHI-sea) DHI were generated using Moderate Resolution Imaging Spectroradiometer (MODIS)-based FAPAR. The higher DHI-cum and DHI-min represented the biodiversity hotspots of India, whereas the DHI-sea was higher in the semi-arid, the Gangetic plain, and the Deccan peninsula. The arid and the trans-Himalaya are dominated with grassland or barren land exhibit very high DHI-sea. The inter-year correlation demonstrated an increase in vegetation greenness in the semi-arid region, and continuous reduction in greenness in the Northeastern region. The DHI components validated using field-measured plant richness data from four biogeographic regions (semi-arid, eastern Ghats, the Western Ghats, and Northeast) demonstrated good congruence. DHI-cum that represents the annual greenness strongly correlated with the plant richness ($R^2 = 0.90$, p-value < 0.001), thereby emerging as a suitable indicator for assessing plant richness in large-scale biogeographic studies. Overall, the FAPAR-based DHI components across Indian biogeographic regions provided understanding of natural variability of the greenness pattern and its congruence with plant diversity.

Keywords: dynamic habitat index; moderate resolution imaging spectroradiometer; plant richness; Indian biogeographic region

Citation: Mahanand, S.; Behera, M.D.; Roy, P.S.; Kumar, P.; Barik, S.K.; Srivastava, P.K. Satellite Based Fraction of Absorbed Photosynthetically Active Radiation is Congruent with Plant Diversity in India. *Remote Sens.* **2021**, *13*, 159. https://doi.org/10.3390/rs13020159

Received: 16 November 2020
Accepted: 30 December 2020
Published: 6 January 2021

Publisher's Note: MDPI stays neutral with regard to jurisdictional claims in published maps and institutional affiliations.

1. Introduction

Biodiversity has a powerful influence on ecosystem dynamics and functions at various geographical scales [1,2]. Global biodiversity observations are needed to provide a better understanding of the distribution of biodiversity, to better identify high priority areas for conservation, and to maintain essential ecosystem goods and services [3]. The gradual decline in biodiversity endangers essential ecosystem services and risks unacceptable environmental consequences [4]. The traditional in situ biodiversity monitoring practice is insufficient to solve the problems associated with biodiversity conservation [5].

Remote sensing technology has provided an effective and evident way to address biodiversity patterns at different geographical scales [6]. Space-borne platforms operate different earth observation satellites, which presents the potential to prepare conservation responses that are commensurate with the scale of conservation [7]. Satellite sensors measure the reflected solar energy emitted from the ground that determines the radiation-interception characteristics of plant canopies linked to photosynthesis [8]. The difference between the carbon assimilated by plant leaves during photosynthesis is a quantitative measure of plant growth and carbon uptake, which represents the vegetation productivity [9,10]. Due to dynamic environmental conditions, vegetation productivity varies with time and space [11,12].

Satellite-derived biophysical proxies provide clues about diversity patterns as they are used for productivity estimation and quantification of spatial heterogeneity of vegetation [13]. Plant richness is a straight forward indicator of plant diversity, which is directly associated with habitat heterogeneity [14,15]. Forest type and species composition plays a vital role to correlate between satellite-derived biophysical proxy and plant richness [16]. Comparison of temporal scales exhibit that the annual pattern of correlation faired more than seasonal (i.e., monsoon) between satellite-derived biophysical proxy and plant richness in the Western Ghats, India [17]. Chitale et al. [18] obtained a high correlation between satellite-derived biophysical proxies and plant richness for open canopy vegetation classes with low species richness (grasslands, scrubs, and dry deciduous forests) followed by vegetation classes with moderately dense canopy in the Western Ghats, Indo-Burma and Himalayan regions in India.

Satellite observations have captured an increase in vegetation growth and productivity mostly in Asia, Africa, and Europe due to agricultural intensification and other human activities [19–21]. In India, the vegetation greenness has increased much more due to agriculture (82%) compared with forests (4.4%) [22]. Also, it has been reported that the crop production has improved up to six times in the past five decades [23]. Agricultural intensification has increased the vegetation greenness in the Western Himalaya, while reduced pre-monsoon moisture levels has resulted a decrease in the greenness pattern in the eastern Himalaya [24]. A study analyzed the seasonal NDVI trend from 2000 to 2014, which reported greening-up due to the rain-fed cultivated area in the lower elevation, while browning-up trend has been consistent along the elevational range that holds closed needle-leaved forests and alpine scrublands in the Uttarakhand Himalaya [25]. A reduction in vegetation greenness with warmer temperatures was reported for trans-Himalayan and western Indian regions [26,27]. Chakraborty et al. [28] analyzed seasonal greenness trends in different forest types of India and found changes in protected areas across India (Simlipal Wildlife Sanctuary, Rajaji National Park, Achanakmar Wildlife Sanctuary, Sundarbans Biosphere Reserve).

The satellite proxies (fraction of absorbed photosynthetically active radiation (FAPAR), leaf area index (LAI), etc.) derived using multiple bands have been found to be better estimators compared to other satellite-derived biophysical proxies (normalized difference vegetation index—NDVI, enhanced vegetation index—EVI, etc.) of habitat conditions and therefore used to explain greenness [29]. The dynamic habitat index (DHI) based on satellite-derived biophysical proxies has been found to provide a good approximation of the habitat conditions [30,31]. The DHI has three components, and each of these is relevant to an hypothesis: (1) DHI-cumulative (DHI-cum) is used to evaluate the available energy hypothesis, according to which greater energy availability is associated with higher productivity and hence with greater biodiversity [32–34]. (2) DHI-minimum (DHI-min) is a proxy for the environmental stress hypothesis, according to which the biodiversity is greater where there is a higher minimum productivity throughout the year [35–37]. (3) DHI-seasonal (DHI-sea) is used to evaluate the environmental stability hypothesis, according to which the biodiversity is greater where the intra-annual variability in productivity is lower [38]. The Moderate Resolution Imaging Spectroradiometer (MODIS) sensor has been used widely to derive the DHI components that are of significance in explaining biodiversity. MODIS-FAPAR data have been used to calculate the DHI to evaluate the relationships between habitat heterogeneity and faunal diversity in the Canadian province of Ontario, United States and Australia [30,39,40]. FAPAR DHI-min characterized 70% of the arid region of the Australian continent as having low greenness cover [41]. Areas under agricultural crops had moderate annual FAPAR levels, large variations in greenness and low annual minimum cover. With increasing annual FAPAR, the greenness cover and the plant diversity increased [42]. In contrast, high annual FAPAR was found in the Western Ghats with low annual variation [17]. Further, the DHI can be integrated with the basic ecological drivers, i.e., environmental heterogeneity (climatic, geophysical variables), habitat productivity, and land cover change, to map and monitor changes in biodiversity [43,44]. However, DHI-min and DHI-sea represented the climatic and anthropogenic induced changes, hence can be useful in addressing land degradation and development [4].

Though the remote sensing proxy-derived DHI explains vegetation productivity ranges and greenness patterns, the validation and pattern analysis need to be performed using sampled plant data. A dynamic habitat index (DHI) based on a satellite derived biophysical proxy (fraction of absorbed photosynthetically active radiation, FAPAR) was used to evaluate the vegetation greenness pattern across deserts to alpine ecosystems in India. Further, the cumulative (DHI-cum), seasonal (DHI-sea), and minimum (DHI-min) DHI were used to establish correlation with biodiversity in varied biogeographic regions of India during 2001–2015. This study outcome would demonstrate the variability in greenness cover across the Indian biogeographic regions, providing valuable insights towards the conservation and management plan.

2. Materials and Methods

2.1. Study Area

The entire Indian nation that consists of the mainland with islands on either side, i.e., the Andaman & Nicobar Islands (in the east) and the Lakshadweep Islands (in the west) was selected as the study area. The Indian mainland has the largest peninsula, extending 3219 km from north to south and 2977 km from east to west, with a geographic extent of 3,287,263 km^2 [45]. Across the 10 biogeographic regions (Figure 1a), India accommodates a wide range of vegetation with grassland and pastureland in arid and semi-arid regions to the broadleaved and alpine forest in Himalaya and Northeast regions. The climatic profile varies from temperate in the north to monsoonal in the south with large variation in precipitation pattern across its length and breadth. The country accommodates diverse topography, soil, and climate.

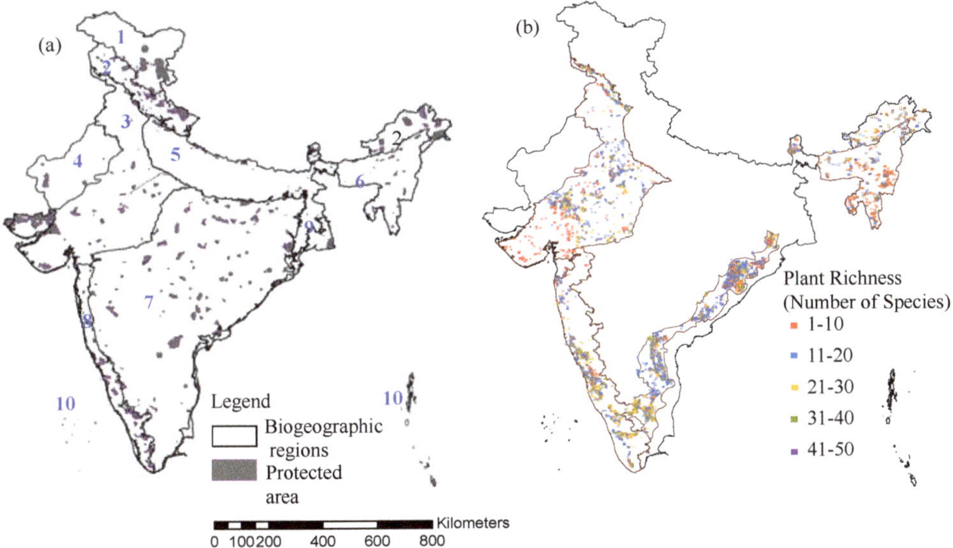

Figure 1. (**a**) Biogeographic regions of India where, 1-trans-Himalaya, 2-Himalaya, 3-semi-arid, 4-arid, 5-Gangetic plain, 6-Northeast, 7-Deccan peninsula, 8-Western Ghats, 9-Coasts, 10-Islands; and Protected Areas; (**b**) Spatial distribution of plant richness from 0.04 ha nested quadrats that ranged from 1 to 50 in four selected biogeographic regions of India (semi-arid, eastern Ghats, Western Ghats, and Northeast) for this study.

2.2. Satellite Data

National Aeronautics and Space Administration (NASA) launched the TERRA (1999) and AQUA (2001) satellites, which aboard the MODIS sensors for global carbon cycle monitoring [46]. The MODIS instruments have 36 spectral bands (0.4 μm to 14.4 μm) at varying spatial resolutions (2 bands at 250 m, 5 bands at 500 m, and 29 bands at 1 km) [47]. NASA provides a suite of atmospherically, geo-registered, data products of MODIS on a routine basis, including FAPAR [48]. The geo-rectified and atmospherically corrected MODIS-FAPAR (MOD15A2H) data were downloaded for the period of 2001–2015 (https://earthdata.nasa.gov/), with spatial resolution of 500 m and temporal resolution of 8 days. All the image-processing tasks such as mosaicking, projections, masking, and normalization were performed using the MRT and ArcGIS (10.5 version) tools, and the temporal resolution was brought to a monthly time frame.

2.3. Generation of DHI Components and Their Composite

Monthly maximum of MODIS FAPAR values was the basic input data set to compute the three relevant annual indices for the DHI analysis. DHI-cum that represents the cumulative annual productivity of a year, was derived as the arithmetic summation of the monthly FAPAR data and the index vary from 0 to 12 [49]. DHI-min that represents the lowest monthly productivity in a year was derived as the arithmetic minima of the monthly FAPAR data and the index varies from 0 to 1. Similarly, DHI-sea that represents the seasonal variability in greenness, was computed by dividing the standard deviation and mean of the monthly FAPAR data for a year and the index vary from $-\infty$ to $+\infty$ [49]. A DHI composite image was derived by assigning DHI-cum, DHI-min and DHI-sea to green, blue, red color plains respectively for visualization and indication on the greenness cover and productivity. Further, the inter-annual variability (2001–2015) of the three DHI components were adjudged by estimating the correlation, regression and standard deviation [4].

Correlation coefficient:

$$\frac{\sum(x - \bar{x})\,(y - \bar{y})}{\sqrt{\sum((x - \bar{x})^2}\,\sqrt{\sum(y - \bar{y})^2}} \tag{1}$$

where x represents number of year and y represents the pixel values corresponding to DHI components (DHI-cum or DHI-min or DHI-sea).

Regression coefficient:

$$\frac{N \times \sum_{n=1}^{N} n \times DHI_i - \left(\sum_{n=1}^{N} n\right)\left(\sum_{n=1}^{N} DHI_i\right)}{\left(N \times \sum_{n=1}^{N} n^2 - \left(\sum_{n=1}^{N} n\right)^2\right)} \tag{2}$$

where N and n represents the total number of years and individual year from 2001 to 2015, respectively and DHI_i represents the pixel values correspond to DHI-cum or DHI-min or DHI-sea.

Standard deviation (SD):

$$\frac{\sqrt{\sum(DHI_i - \mu)^2}}{N} \tag{3}$$

where DHI_i represents the yearly pixel values correspond to DHI-cum or DHI-min or DHI-sea, μ represents the mean of those pixel values from 2001 to 2015, and N represents the total number of years evaluated.

2.4. DHI Components and Plant Richness

The plant data for the study sites was procured from a national project entitled 'Biodiversity Characterization at Landscape Level (BCLL)' that was carried out during 1997–2012 [50]. The stratified random sampling was considered to lay nested quadrats of size 20×20 m^2 for trees (>15 cm, circumference at breast height-cbh) and lianas, two 5×5 m^2 plots for shrubs and saplings (>5 cm and <10 cm cbh), and four 1×1 m^2

plots for herbs and seedlings. The database has a record of 15,656 geo-tagged field plots and 6222 unique species from 10 biogeographic regions of India.

Four biogeographic regions (semi-arid, eastern Ghats, Western Ghats, and Northeast) having varied moisture level, plant richness and environmental heterogeneity were chosen for analysis of pattern between the plant richness and the DHI components. These regions encompass 7365 geo-tagged nested quadrats from the BCLL database. The unique species count from each nested quadrat represented the plant richness for the respective quadrat. The pixel values of averaged DHI components were linked using corresponding plant richness data from 7365 nested quadrat locations for the four biogeographic regions using ArcGIS platform [51]. Firstly, the yearly variations in DHIs from 2001 to 2015 in the four biogeographic regions were visualized using 3-D scatter plots. It was generated by utilizing the mean of each DHI components for each year that calculated the Euclidean distance for the selected biogeographic regions of India. Secondly, the distribution of plant richness from 7365 quadrats were plotted in box diagram along the average of DHIs. In the box plot, mean values denote the level of accuracy, whereas R^2 and *p*-values indicate the significance of the plant richness distribution for each DHI component. The standard deviation of three DHI components were randomly chosen to demonstrate the variation in greenness cover at a test site (93.84°E, 27.49°N) in the Northeastern region.

3. Results

3.1. Visualizing Greenness Pattern Using DHI Components along Indian Biogeographic Regions

The DHI-cum that represents the annual cumulative greenness, varied from 4 to 8, 2 to 8, and 2 to 6 along the western Himalaya, eastern Ghats, and the Gangetic plain regions respectively, during 2001–2015 (Figure 2). However, the DHI-cum range was less for the semi-arid and Deccan peninsula (0 to 4), and more for the Himalayan (0 to 11), Northeast (0 to 11), and Western Ghats (2 to 11) regions. Interestingly, the eastern part of the Deccan peninsula demonstrated DHI-cum variation between 0 to 6 and 0 to 12 in alternate years during 2001–2015 (Figure 2). The grasslands or scrublands on very fertile lands across the riverbeds demonstrated annual greenness of 2 to 6, which are mostly seasonal contributions. The DHI-cum reflected between 0 and 2 for non-vegetated classes as snow cover, rivers, and deserts. Thus, Indian heterogeneous biogeographic regions clearly demonstrated DHI-cum variation between 0 and 12 with varied greenness.

The DHI-min exhibits the status for minimum greenness sustained throughout a particular year. The DHI-min ranged from 0.4 to 1.0 found across the highly diverse regions, i.e., western Himalaya, Western Ghats and eastern Ghats and Northeast during 2001–2015 (Figure 3). In contrast, the lower range (0 to 0.2) of DHI-min recorded along the Deccan peninsula, arid, semi-arid and trans-Himalaya. The variation in DHI-min for the two largest river basins were also observed less, i.e., Brahmaputra basin (0.2–0.3), Gangetic plain (0.1–0.2) (Figure 3). The Terai region showed an increase in greenness cover between 0.1 and 0.3 during 2001 to 2015 (Figure 3). The semi-arid, arid and trans-Himalaya region showed a minimum greenness cover of 0 to 0.2. The DHI-min indicates the natural forest that sustains throughout a year including the evergreen, alpine and tundra forest observed in the Himalaya, Western Ghats, eastern Ghats, and Northeast. On the other hand, the grassland, agricultural land or barren land showed lower green cover, which is significantly low for snow, water and desert (Figure 3). This indicates that DHI-min is useful in demonstrating the proportion of various forest types ranging from grassland/agricultural to evergreen/alpine forests in different biogeographic zones.

The DHI-sea varied between 1.0 and 3.5 in the trans-Himalaya, while 0.3 and 1.0 for the Deccan peninsula and Gangetic plain (Figure 4). On the contrary, lower DHI-sea (between 0.2 and 0.3) is observed for riverbeds of Ganges, Brahmaputra, and Mahanadi Rivers. Interestingly, the DHI-sea is observed very less (0.01–0.2) for the eastern Ghats, Western Ghats, Northeast, and western Himalaya (Figure 4). The higher range of DHI-sea is observed for biogeographic regions with a higher proportion of grasslands, scrublands, pasture, and cultivated lands, while lower values were observed for the regions encompassing more natural forests.

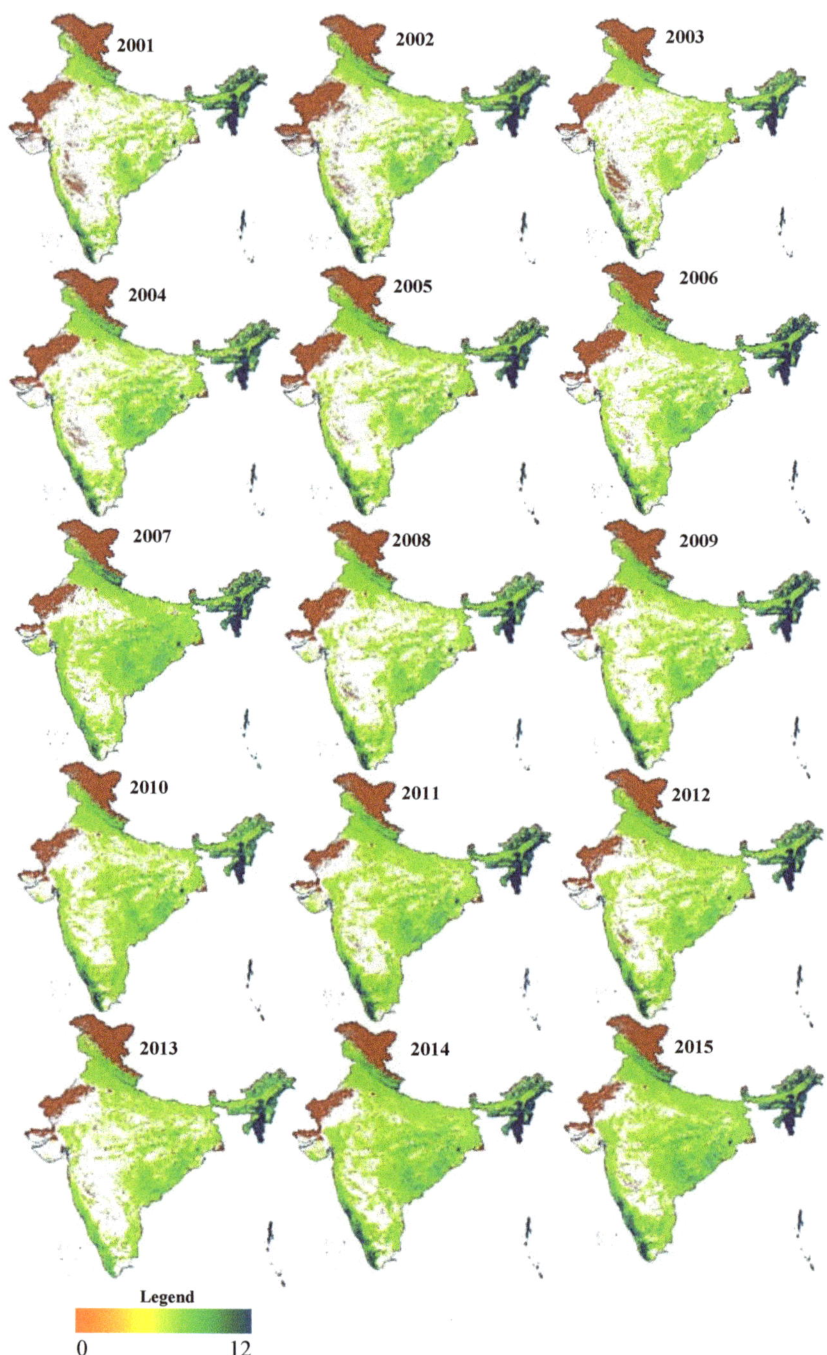

Figure 2. Cumulative dynamic habitat index (DHI-cum) derived from the monthly sum of FAPAR for a particular year is utilized to map the variation in annual greenness from 2001 to 2015 in India.

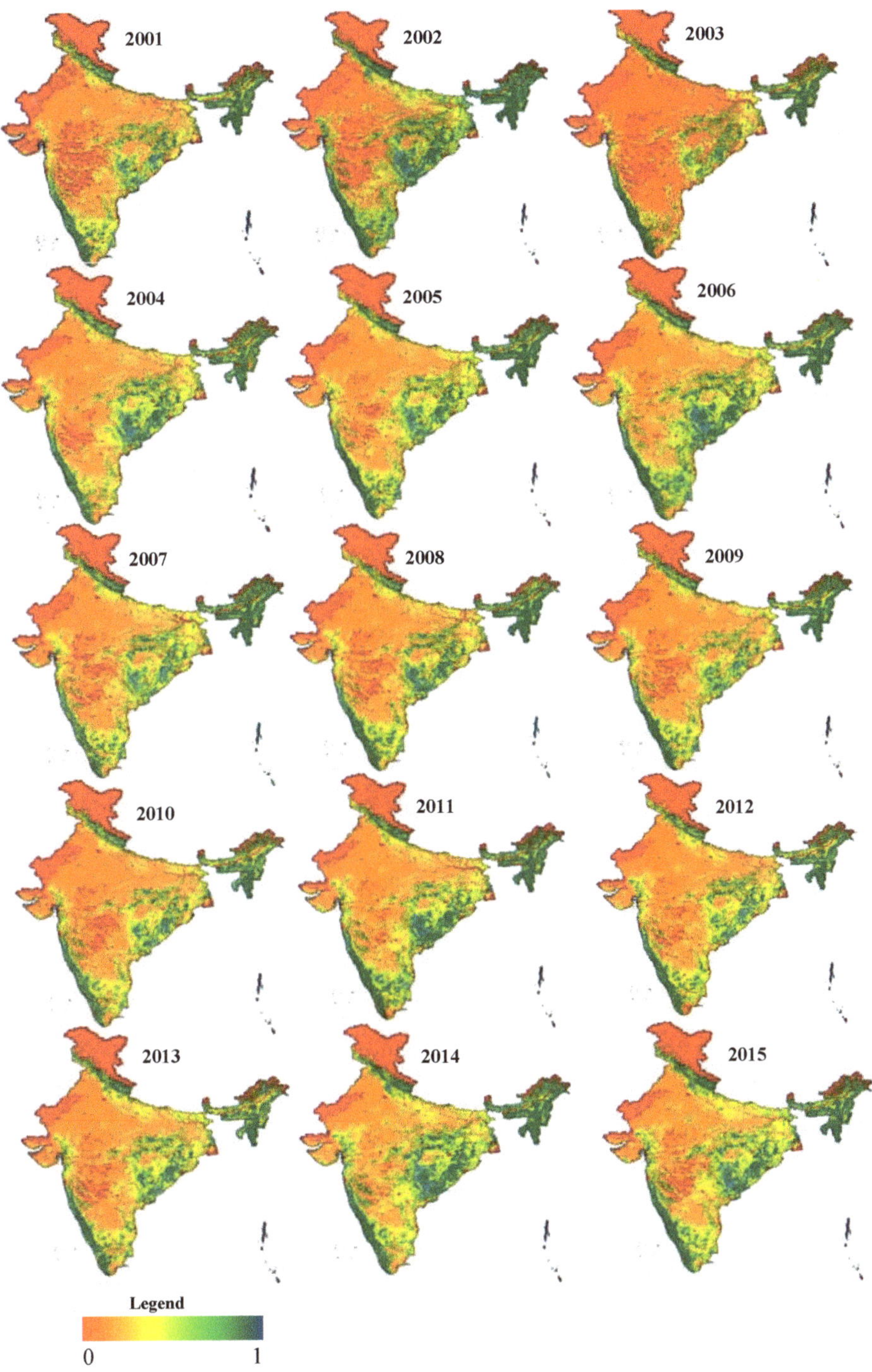

Figure 3. Minimum dynamic habitat index (DHI-min) represents the minimum monthly FAPAR value in a particular year and is utilized to map the variation of minimum greenness cover from 2001 to 2015 in India.

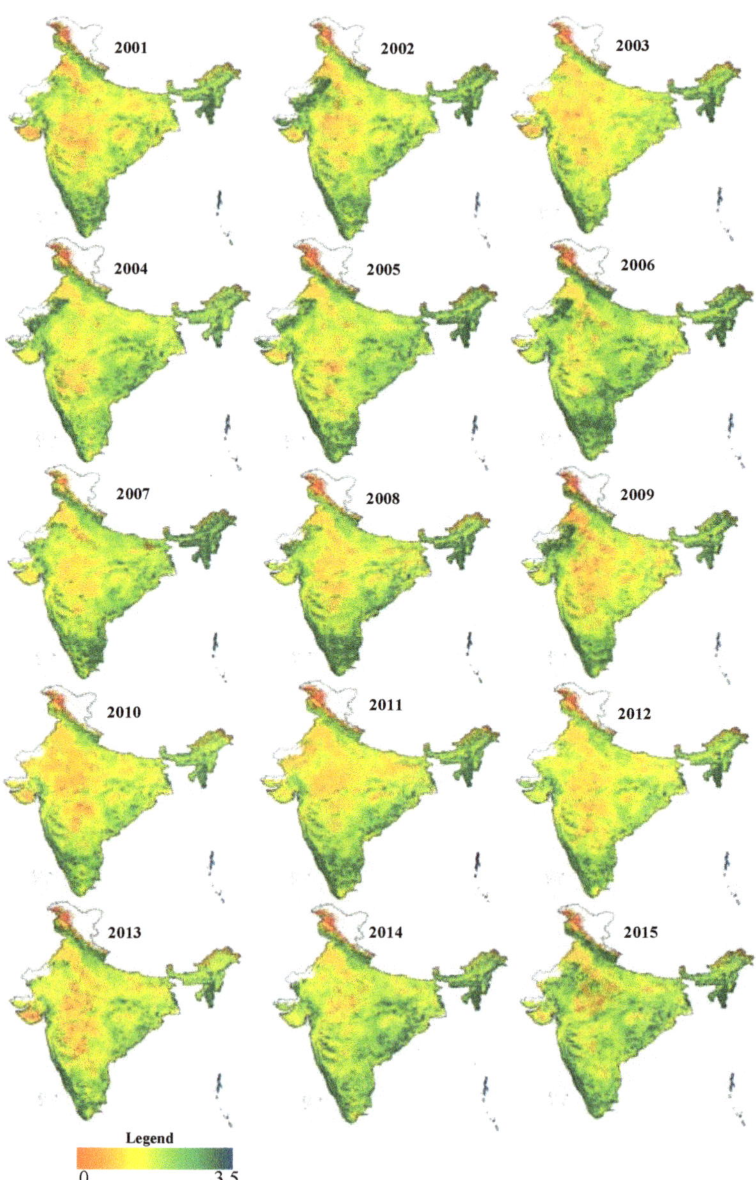

Figure 4. Seasonal dynamic habitat index (DHI-sea) is the ratio between mean and standard deviation of monthly FAPAR for a particular year was utilized to map the seasonal greenness variability from 2001 to 2015.

The DHIs composite illustrate the relative importance of each component at pixel level during 2001–2015. The pallet assigned to cyan color observed across Northeast, Western Ghats, and western Himalaya represents higher DHI-cum and DHI-min values and lower DHI-sea (Figure 5). The DHIs composite with reddish-brown color in the arid and semi-arid exhibits high DHI-Cum and DHI-sea, while low DHI-min. The greenish brown pallet in the Deccan peninsula and semi-arid region indicates high DHI-cum, low DHI-min, and moderate DHI-sea. The Gangetic plain along the river channel exhibits yellowish orange to greenish brown pallet, while the composite seems to be varied from light cyan to light green along the Terai region (Figure 5). The reddish-brown (i.e., high DHI-sea) gradually changed to greenish brown pallet (i.e., high DHI-cum) during the period 2001–2015 for the semi-arid, Deccan peninsula and Gangetic plain (Figure 5). On the other hand, he cyan pallet extant was found replaced with the greenish pallet is indicating gradual decline in natural forest cover along the Himalaya, eastern Ghats, and Northeast during 2001–2015 (Figure 5).

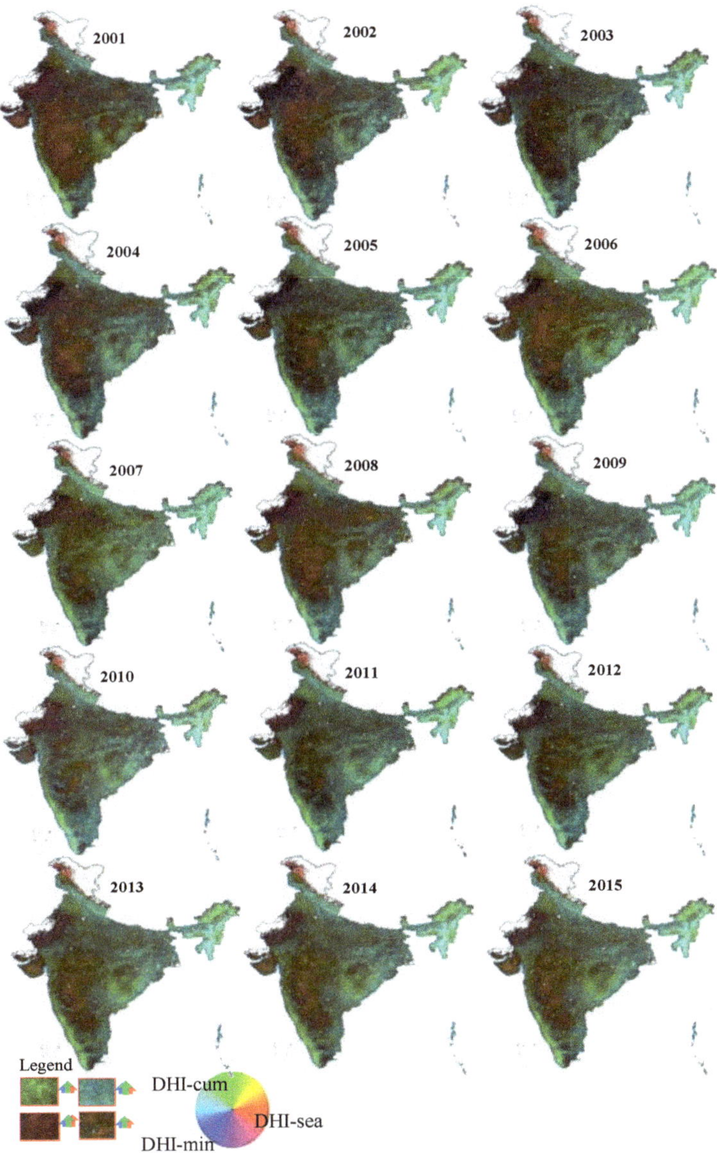

Figure 5. The composite image of three DHI components were mapped from 2001 to 2015, where, DHI-cum was assigned with the green band, DHI-min was assigned with the blue band, and DHI-sea was assigned with the red band.

3.2. Significance of Greenness Variability from 2001 to 2015 Using DHI Components

The inter-year correlation (2001–2015) of DHI-cum found to be positive that varied between 0.8 and 1.0 for the semi-arid and Deccan peninsula (Figure 6a, (i)). On the other hand, a negative correlation of DHI-cum ranged between −0.6 and −1.0 observed for the trans-Himalayan region. The areas with zero/negative correlation of DHI-cum from 2001 to 2015 mostly belong to the Northeast, Gangetic plain and Deccan peninsula regions (Figure 6a, (i)). The DHI-min exhibited a negative correlation (−0.01 to −1.0) in the majority of Indian landscapes, indicating a reduction in minimum greenness cover from 2001 to 2015. The arid and trans-Himalaya were characterized by highly negative correlation values (−0.7 to −1.0) for DHI-min (Figure 6a, (ii)). Importantly, the negative correlation of DHI-min (0 to −0.4) for Northeast has raised an alarming concern indicating the loss of natural forests (Figure 6a (ii)). Similarly, a negative correlation (0 to −0.4) of DHI-min was observed in a few areas of the Gangetic plain and Deccan peninsula regions. The DHI-sea exhibited a positive correlation range between 0.4 and 0.8 for the semi-arid, Northeast, Gangetic plain, and Deccan peninsula during 2001 to 2015 (Figure 6a, (ii)). However, the DHI-sea demonstrated a negative correlation (−0.1 to −0.6) for the eastern Ghats, Western Ghats, western Himalaya, and Deccan peninsula, which could be attributed to insignificant alteration in the vegetation greenness in these regions (Figure 6a, (iii)).

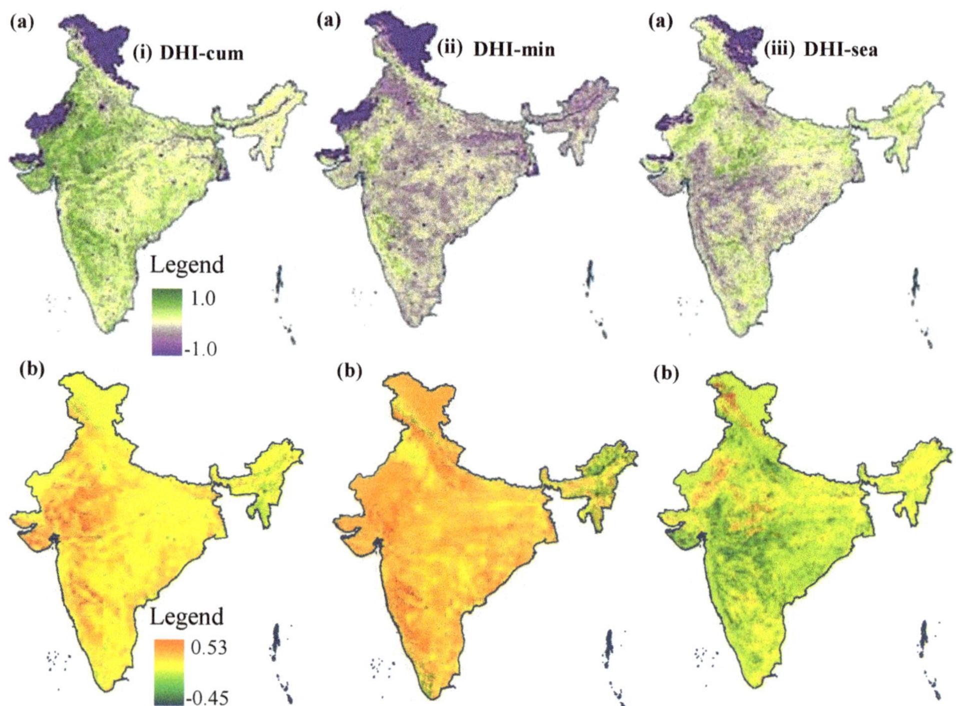

Figure 6. The inter-year (2001–2015) correlation coefficient and regression coefficient calculated for each DHI components are shown as: (i) DHI-cum (**a,b**); (ii) DHI-min (**a,b**); (iii) DHI-sea (**a,b**), respectively.

The regression coefficient of DHI-cum that indicated the slope of annual greenness during 2001 to 2015 found positive and ranged between 0.02 and 0.53 for the semi-arid and Deccan peninsula regions (Figure 6b, (i)). Similarly, the eastern Ghats and western Himalayan region exhibited a positive slope (0.02–0.05). However, a negative slope (−0.45 to −0.01) for DHI-cum was observed for the Gangetic plain and Northeastern regions.

The regression coefficient of DHI-min was found negative for the majority of the Indian landscapes (Figure 6b, (ii)). Importantly, the negative slope across the Northeast represents a gradual decline in the minimum greenness. The slopes of the arid and the trans-Himalaya were also found negative (−0.002). On the other hand, a steady slope towards increasing minimum greenness (0 to 0.01) recorded for the semi-arid, Western Ghats, and Deccan peninsula regions. Moreover, the higher range of positive regression coefficient of DHI-min (0.01 to 0.06) recorded among the protected areas mostly occurs within highly diverse zones. For DHI-sea, the positive regression coefficient ranging from 0.004 to 0.02 indicates a consistent increase in the vegetation greenness of the semi-arid, Deccan peninsula, Gangetic plain and Northeast (Figure 6b, (iii)). In contrast, the regression coefficient of DHI-sea varies from −0.12 to −0.002 for the Western Ghats, Deccan peninsula, semi-arid, and Gangetic plain regions indicating lower variability in the vegetation greenness (Figure 6b, (iii)).

The standard deviation of the three DHI components (DHI-cum SD, DHI-min SD, DHI-sea SD) summarizes their variability from the mean during 2001–2015 (Figure 7). For DHI-Cum SD, the majority of the coverage in Indian Biogeographic regions showed a range of 0.25–0.5. The DHI-cum SD ranged between 0.5 and 0.75 recorded from the Northeast, Western Ghats, semi-arid, and Deccan peninsula. The DHI-cum SD varies between 0.75 and 1.0 found in small patches across the semi-arid and Deccan peninsula (Figure 7a). The DHI-cum SD with higher range subjected to occurrence of more variability in the annual vegetation greenness during the analyzed time period. The DHI-min SD, which indicates the variation in minimum greenness cover during 2001 to 2015 found to be ranging from 0.03 to 0.06 for most of the biogeographic regions (Figure 7b). The grasslands and scrublands dominated the semi-arid region, representing low range up to 0.03 for DHI-min SD. The natural forested regions such as the Western Ghats and Northeast regions comparably shared a higher variability that ranged from 0.12 to 0.15 for the DHI-min SD (Figure 7b). For DHI-sea SD, the majority of the Indian landscape exhibited variability in vegetation greenness between the range of 0 and 0.1. Comparatively, the higher range of DHI-sea SD (0.05 to 0.15) for the semi-arid and Northeast express more variation in vegetation greenness of these regions (Figure 7c).

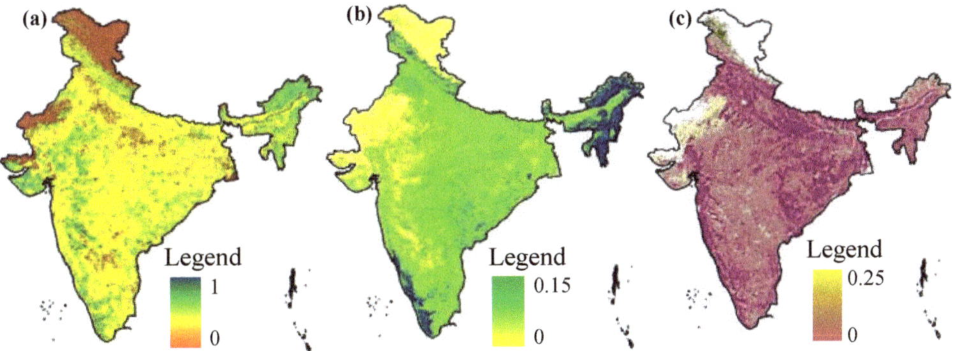

Figure 7. Representing the standard deviation map of the three DHI components from 2001 to 2015 in India, where: (**a**) DHI-cum SD; (**b**) DHI-min SD and (**c**) DHI-sea SD.

3.3. Validation of DHI Components Using Plant Richness Data

There were substantial differences between changes in the DHI components of the four biogeographic regions, i.e., the semi-arid, the eastern Ghats, the Western Ghats, and the Northeastern regions over the period 2001–2015 (Figure 8, Table 1). The lowest and the highest value for DHI-min and DHI-cum were demonstrated by semi-arid and the Northeastern regions respectively. The 3-D schematic of the semi-arid showed that it had a higher range of seasonality (DHI-sea > 0.4) and a lower range of greenness (DHI-cum < 4) compared with the other three biogeographic regions (Table 1). The higher range of DHI-cum and DHI-min indicate the dense canopy and vegetation greenness in the Western Ghats and Northeast, whereas high seasonality in the semi-arid and eastern Ghats indicates the dominance of scrubland, grassland, or deciduous forests. As per the forest type and canopy cover of the Northeastern region, DHI-sea was slightly high, and thereby it refers towards human activities intervening natural forest ecosystem.

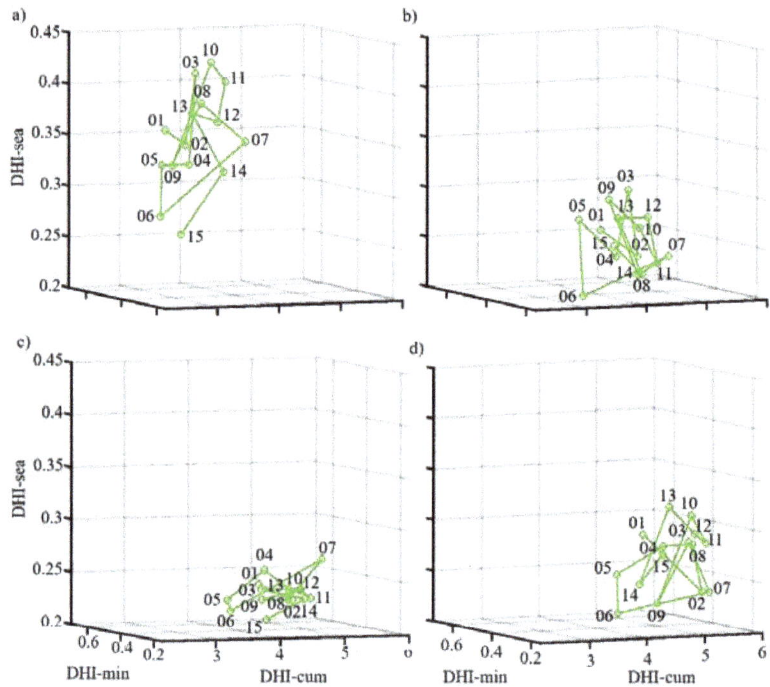

Figure 8. Schematic representation of three DHI components using the 3-D scatter plots (DHI-cum in x-axis, DHI-min in y-axis, DHI-sea in z-axis) during 2001–2015 for the four biogeographic regions of India, where: (**a**) semi-arid; (**b**) eastern Ghats; (**c**) Western Ghats; (**d**) Northeastern region.

Table 1. Distribution of DHI values of all components in four biogeographic regions.

BG Regions	DHI-Cum	DHI-Min	DHI-Sea
Semi-Arid	2.15–3.54	0.15–0.27	0.27–0.44
Eastern Ghats	3.43–4.9	0.28–0.48	0.2–0.31
Western Ghats	3.83–5.1	0.31–0.52	0.21–0.27
Northeast	4.5–6.0	0.33–0.61	0.21–0.32

The DHI components validated using the plant richness data showed DHI-cum, 0–10.11; DHI-min, 0–0.7; and DHI-sea, 0.07–3.32 (Figure 9a, (i–iii)). The pattern of plant richness along each DHI components showed a positive trend with skewed distribution, such as DHI-cum, R^2 = 0.90; DHI-min, R^2 = 0.71; and for DHI-sea, R^2 = 0.74 at *p*-value < 0.001 (Figure 9b, (i–iii)).

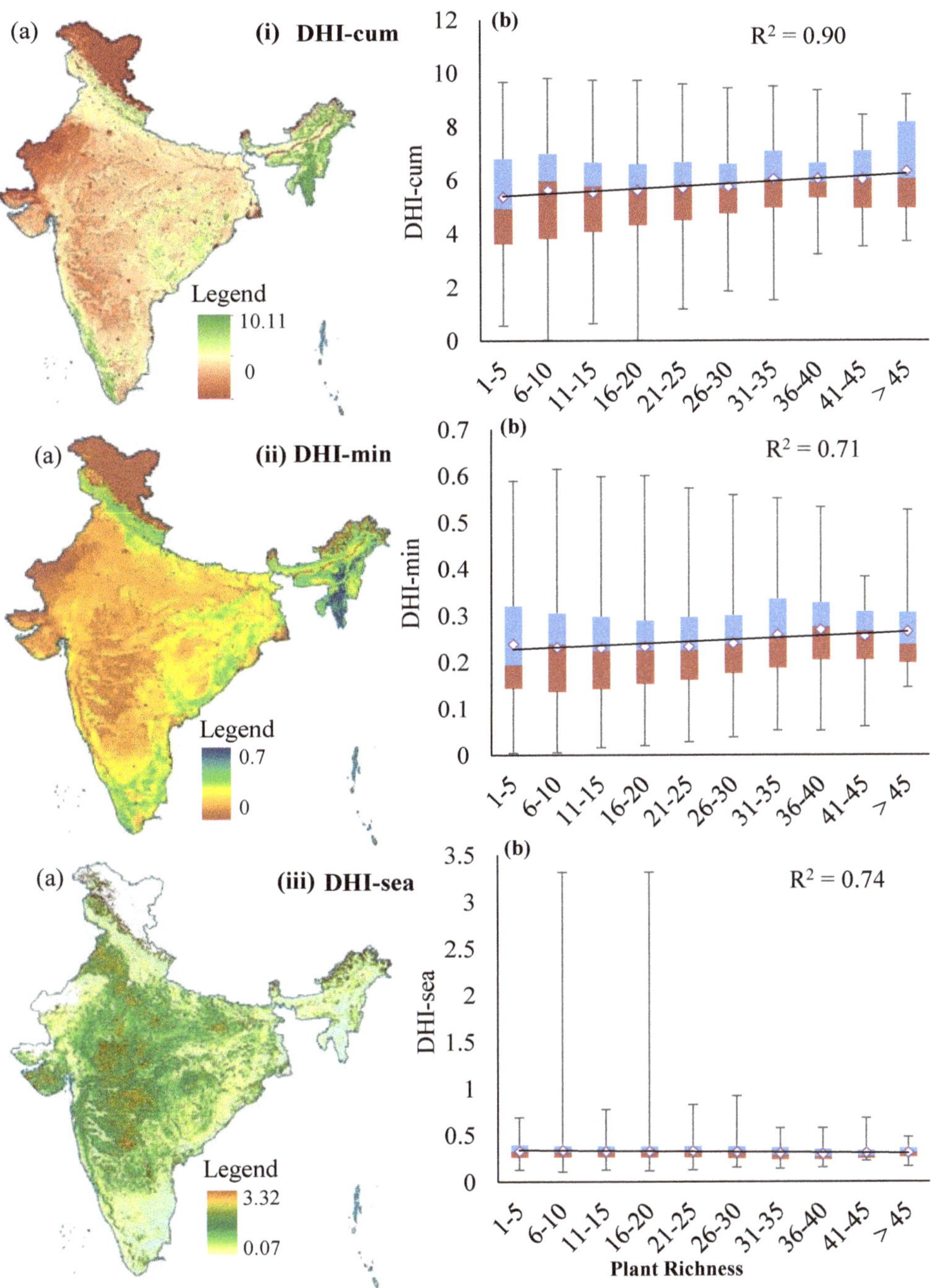

Figure 9. The three DHI components (i) DHI-cum; (ii) DHI-min, and (iii) DHI-sea were used: (**a**) to map the greenness variation for India, and (**b**) the box plots against the field-measured plant richness data from the four Biogeographic regions of India.

4. Discussion

Among the existing satellite-derived biophysical proxies, FAPAR has the advantage of multiple bands and the physically processed algorithm, and has the capability to address the greenness variability of habitats [29]. Initial approaches utilizing FAPAR-based DHI have evaluated the congruence of habitat heterogeneity and faunal diversity in Canada, United Sites and Australia and China [30,39,40]. DHI-min highlights the vegetation productivity and proportions of evergreen and deciduous vegetation cover [52], while DHI-sea is sensitive to the seasonal variations in greenness and is useful in distinguishing among major forest types. Overall, DHI components provide insights upon species composition, change, and diversity within a given area may be quantitatively produced over large areas and over-time [31].

The varied forest types and biogeographical regions offered a suitable test site to evaluate and validate the significance of the satellite-derived DHI. The DHI is different from other remote sensing indices because it is well grounded in the ecological theory of biodiversity patterns [53]. The satellite proxy-based DHI has three major components, and each of them share a key feature about the habitat conditions. For example, status of annual greenness by DHI-cum, minimum greens cover over a year by DHI-min and variation in greenness cover by DHI-Sea. The three DHI components calculated from 2001 to 2015 show distinct variation in greenness across all the Indian biogeographic regions. This infers that from grassland to the alpine forest has been subjected to hold the variability in the greenness pattern, which may be due to the direct or indirect influence of climatic alteration and human activities [4,54].

Indian biogeographic regions include a number of global biodiversity hotspots (the Himalaya, Sundaland, the Northeast, and the Western Ghats) lie partly or entirely within India and have forest cover with evergreen, semi-evergreen, and deciduous species, the DHI-cum and DHI-min values of these regions are greater. Importantly, because of the permanent foliage of evergreen forests, the greenness and productivity are high throughout the year. Thus, the annual greenness of evergreen forests is high, and there is little seasonality [31,52]. The trans-Himalaya has highly variable climate and forest types, ranging from sub-tropical evergreen to dense conifer forests, as well as grasslands. The DHI-cum and DHI-min ranges of the region are intermediate and that the DHI-sea value is high. Human activities, i.e., urbanization and agriculture intensification from lower to middle elevation could be the major cause for the high seasonality and annual variability in greenness [55]. In addition, the warming climate and reduced monsoon affect variability in greenness in the Himalaya region [24].

Although, diversity of the Northeastern region is rich, there were patches representing high DHI-sea values due to the human activities that include deforestation, shifting cultivation, agriculture, and settlement. Also, the availability of moisture before the monsoon and average rainfall in the Northeastern region are reported to be low [24]. In contrast, semi-arid, Deccan peninsula and the Gangetic plain regions with more settlements, agricultural land and less forest cover, demonstrated moderate DHI-cum, DHI-min, and DHI-sea values during 2001–2015. The intensity of agriculture in most of the Indian biogeographic regions is the reason for the abrupt greening and enhanced plant productivity followed by the consistent browning noted [56]. However, the arid and trans-Himalayan biogeographic regions are mostly covered by scrubland, snow and barren stretches and have low to moderate DHI-cum, DHI-min, and high DHI-sea values (Figures 2–4). In biogeographic regions at higher latitudes and altitudes, which are covered by snow in winter, the FAPAR value approaches 0, while locations that have no significant snow cover showed FAPAR > 0.

Composite map of the three DHI components rightly explained the behavior of each component at pixel level (Figure 5). Very diverse and forested regions, such as the western Himalaya, Northeastern, Western Ghats, and eastern Ghats regions, are very distinct (high DHI-cum and DHI-min values and low DHI-sea values). The low seasonality is the indicative for more diversity and dense canopy dominated by the evergreen and semi-evergreen species. The areas with high seasonal variability are mostly irrigated pastures, barren land or grasslands. So, biogeographic regions, such as semi-arid, arid, Deccan peninsula, and trans-Himalaya, with thin forest cover demonstrated moderate annual and minimum greenness cover, while high seasonality (Figure 5). The seasonal variation of each pixel indicates the integrated influence of climate, topography, and land use on the status of the vegetation greenness [49].

The inter-year positive correlation of DHI-cum is very distinct for the Deccan peninsula and semi-arid region. This may be explained by increased productivity resulting from increased agricultural activities and climatic variability [27]. The western Himalaya and Western Ghats exhibited mixed representation of high to low positive correlation. The increase in the DHI-min values in the Western Ghats and western Himalaya may be due to afforestation and conservation measures. The weak yet positive correlation of annual cumulative greenness, whereas weak negative correlation of minimum greenness indicate that the agricultural activities may be the source for increasing vegetation in the eastern Himalaya and Northeast region. On the other hand, the gradual decline in natural forests is clearly visible through these variability in vegetation greenness, and this information will be crucial for the conservation and management team to avoid severe environmental consequences [57]. The increasing minimum cover in the semi-arid region may be due to agriculture practices that brought abrupt greenness. The negative correlation of DHI-sea in Northeast is because the region is experiencing a continuous decrease in greenness subjected to various human activities [58]. Additionally, the reduction in pre-monsoon moisture availability and precipitation results in reduced greenness [24]. Increasing temperatures have reduced the greenness of the arid region and the trans-Himalaya.

The regression coefficient confirms that the intensity of change in greenness is different for each DHI component. The higher regression coefficient of DHI-cum in the semi-arid and the Deccan peninsular region can be explained by expansion of agricultural activity with increased soil water availability [59]. However, in the Northeastern region with random deforestation and shifting cultivation, the forested areas were continuously exploited as per the negative regression coefficient of DHI-cum. A test site of Northeastern India had also recorded with reduced forest cover and natural vegetation [60], while DHI-sea showed variability in greenness (Supplementary Figure S1). The semi-arid and the Northeastern region demonstrated higher DHI-sea values because of the patchy landscape, environmental conditions and anthropogenic activities [61].

The standard deviation of the three DHI components (2001–2015) was greater for less forested regions in the semi-arid, arid, Deccan peninsula and the trans-Himalayan regions. Moreover, the standard deviation of three DHI components was the maximum in the Northeast, where deforestation and shifting cultivation have affected the greenness and subsequent FAPAR as observed in the past 15 years. Overall, plant richness best correlates with annual greenness represented as DHI-cum. Further, it infers that the habitat productivity is well correlated with plant richness, as observed by Connell and Orias [35]. Moreover, the MODIS-FAPAR-based DHI-cum also exhibit a positive correlation with the richness of birds and other animals [4,49]. Therefore, DHI-cum is observed to be the most important univariate predictor out of the three DHIs in explaining the richness of plant species as exemplified in this study for India. The biogeographic regions with dominant forest cover generally have high productivity, assuring the cumulative DHI as a good indicator of plant diversity.

5. Conclusions

The present study demonstrated the capability of the DHI, based on a satellite-derived biophysical proxy (i.e., FAPAR), to identify changes in the vegetation greenness. The semi-arid, the Deccan peninsula, and the Gangetic plain may be experiencing much variability in greenness cover. The study also showed that biodiversity hotspots could be distinguished using DHI components. Hence, the results of the present study could be useful in prioritizing and planning conservation measures for the biogeographic regions of India.

This study was a maiden attempt to correlate the DHI components with field sampled plant richness that highlighted the importance of DHI-cum in explaining plant richness. Thus, DHI-cum can be used as a rapid indicator to evaluate the plant richness pattern in large-scale biogeographic studies. As various natural and anthropogenic disturbances threaten biodiversity due to a loss of habitats, which has led to growing interest in the search for rapid proxies for large-scale use in conservation, management, and monitoring. This study provides baseline information for stakeholders seeking to monitor biodiversity in large areas. Future studies should focus on causal analysis of the decrease in vegetation greenness at a local scale, particularly partitioning climatic and anthropogenic influences.

6. Highlights

- Characterized spatiotemporal variability of FAPAR-based DHI components (2001–2015) for India.
- Individual as well as composites of DHI components very well differentiated the biogeographic regions of India with high/low biodiversity levels.
- The inter-year correlation and regression of DHIs exhibited gradual decrease in vegetation greenness for Northeastern region, while the semi-arid and Deccan peninsular regions showed abrupt increase in vegetation greenness and seasonality.
- DHI-cum representing the annual greenness was strongly correlated with the plant richness thereby emerging as a suitable indicator to monitor the plant diversity.

Supplementary Materials: The following are available online at https://www.mdpi.com/2072-4292/13/2/159/s1, Figure S1: A test site (93.84°E, 27.49°N) in Northeastern region demonstrated the variation in greenness cover through the standard deviation of three DHI components during 2001–2015, i.e., (**a**) DHI-cum SD; (**b**) DHI-min SD; (**c**) DHI-sea SD.

Author Contributions: Conceptualization, M.D.B.; data curation, M.D.B. and P.S.R.; formal analysis, S.M. and P.K.; methodology, S.M. and P.K.; software, S.M.; supervision, M.D.B.; validation, S.M.; visualization, S.M. and P.K.; writing—original draft, S.M.; writing—review and editing, M.D.B., P.S.R., S.K.B. and P.K.S. All authors have read and agreed to the published version of the manuscript.

Funding: S.M. received a Senior Research Fellowship (National Eligibility Test) from the University Grant Commission to pursue PhD. P.K. received fellowship from the Ministry of Education (formerly, Ministry of Human Resources Development, MHRD) to pursue M. Tech. degree in Earth System Science and Technology at CORAL, IIT Kharagpur.

Institutional Review Board Statement: Not applicable.

Informed Consent Statement: Not applicable.

Data Availability Statement: The data are available at https://earthdata.nasa.gov/, https://bis.iirs.gov.in/.

Acknowledgments: We acknowledge utilization of a comprehensive plant database that was generated through a national project on 'Biodiversity characterization at landscape-BCLL level'. S. M. acknowledges the University Grant Commission for providing financial assistance in form of a Senior Research Fellowship (National Eligibility Test). P.K. acknowledges the Ministry of Education (formerly, Ministry of Human Resources Development, MHRD) for providing financial assistance to pursue M Tech. degree in Earth System Science and Technology at CORAL, IIT Kharagpur. We acknowledge the facilities provided by the authorities of Indian Institute of Technology Kharagpur, to undertake this study.

Conflicts of Interest: The authors declare no conflict of interest.

References

1. Foley, J.A.; Levis, S.; Costa, M.H.; Cramer, W.; Pollard, D. Incorporating dynamic vegetation cover within global climate models. *Ecol. Appl.* **2000**, *10*, 1620–1632. [CrossRef]
2. Cui, L.; Shi, J. Temporal and spatial response of vegetation NDVI to temperature and precipitation in eastern China. *J. Geogr. Sci.* **2010**. [CrossRef]
3. Wang, R.; Gamon, J.A. Remote sensing of terrestrial plant biodiversity. *Remote Sens. Environ.* **2019**, *231*, 111218. [CrossRef]

4. Zhang, C.; Cai, D.; Guo, S.; Guan, Y.; Fraedrich, K.; Nie, Y.; Liu, X.; Bian, X. Spatial-temporal dynamics of China's terrestrial biodiversity: A dynamic habitat index diagnostic. *Remote Sens.* **2016**, *8*, 227. [CrossRef]
5. Erdelen, W.R. Shaping the Fate of Life on Earth: The Post-2020 Global Biodiversity Framework. *Glob. Policy* **2020**. [CrossRef]
6. Šímová, I.; Li, Y.M.; Storch, D. Relationship between species richness and productivity in plants: The role of sampling effect, heterogeneity and species pool. *J. Ecol.* **2013**. [CrossRef]
7. Jetz, W.; Cavender-Bares, J.; Pavlick, R.; Schimel, D.; Davis, F.W.; Asner, G.P.; Guralnick, R.; Kattge, J.; Latimer, A.M.; Moorcroft, P.; et al. Monitoring plant functional diversity from space. *Nat. Plants* **2016**, *2*, 16024. [CrossRef]
8. Tucker, C.J.; Sellers, P.J. Satellite remote sensing of primary production. *Int. J. Remote Sens.* **1986**, *7*, 1395–1416. [CrossRef]
9. Waring, R.H.; Running, S.W. Forest ecosystems: Analysis at multiple scales. *Choice Rev. Online* **1998**. [CrossRef]
10. Fensholt, R.; Sandholt, I.; Rasmussen, M.S.; Stisen, S.; Diouf, A. Evaluation of satellite based primary production modelling in the semi-arid Sahel. *Remote Sens. Environ.* **2006**, *105*, 173–188. [CrossRef]
11. Anderson, R.G.; Canadell, J.G.; Randerson, J.T.; Jackson, R.B.; Hungate, B.A.; Baldocchi, D.D.; Ban-Weiss, G.A.; Bonan, G.B.; Caldeira, K.; Cao, L.; et al. Biophysical considerations in forestry for climate protection. *Front. Ecol. Environ.* **2011**, *9*, 174–182. [CrossRef]
12. Jackson, R.B.; Randerson, J.T.; Canadell, J.G.; Anderson, R.G.; Avissar, R.; Baldocchi, D.D.; Bonan, G.B.; Caldeira, K.; Diffenbaugh, N.S.; Field, C.B.; et al. Protecting climate with forests. *Environ. Res. Lett.* **2008**. [CrossRef]
13. Oindo, B.O.; Skidmore, A.K. Interannual variability of NDVI and species richness in Kenya. *Int. J. Remote Sens.* **2002**. [CrossRef]
14. Huston, M.A. Biological diversity: The coexistence of species on changing landscapes. *Biol. Divers. Coexistence Species Chang. Landsc.* **1994**. [CrossRef]
15. Muchoney, D.M. Earth observations for terrestrial biodiversity and ecosystems. *Remote Sens. Environ.* **2008**. [CrossRef]
16. Thakur, T.K.; Padwar, G.K.; Patel, D.K.; Bijalwan, A. Monitoring land use, species composition and diversity of moist tropical environ in Achanakmaar Amarkantak Biosphere reserve, India using satellite data. *Biodivers. Int. J.* **2019**. [CrossRef]
17. Mahanand, S.; Behera, M.D. Relationship between Field-Based Plant Species Richness and Satellite-Derived Biophysical Proxies in the Western Ghats, India. *Proc. Natl. Acad. Sci. India Sect. A Phys. Sci.* **2017**, *87*. [CrossRef]
18. Chitale, V.S.; Behera, M.D.; Roy, P.S. Deciphering plant richness using satellite remote sensing: A study from three biodiversity hotspots. *Biodivers. Conserv.* **2019**. [CrossRef]
19. Zhou, L.; Tucker, C.J.; Kaufmann, R.K.; Slayback, D.; Shabanov, N.V.; Myneni, R.B. Variations in northern vegetation activity inferred from satellite data of vegetation index during 1981 to 1999. *J. Geophys. Res. Atmos.* **2001**, *106*, 20069–20083. [CrossRef]
20. Nemani, R.R.; Keeling, C.D.; Hashimoto, H.; Jolly, W.M.; Piper, S.C.; Tucker, C.J.; Myneni, R.B.; Running, S.W. Climate-driven increases in global terrestrial net primary production from 1982 to 1999. *Science* **2003**. [CrossRef]
21. Fensholt, R.; Langanke, T.; Rasmussen, K.; Reenberg, A.; Prince, S.D.; Tucker, C.; Scholes, R.J.; Le, Q.B.; Bondeau, A.; Eastman, R.; et al. Greenness in semi-arid areas across the globe 1981-2007 - an Earth Observing Satellite based analysis of trends and drivers. *Remote Sens. Environ.* **2012**, *121*, 144–158. [CrossRef]
22. Chen, C.; Park, T.; Wang, X.; Piao, S.; Xu, B.; Chaturvedi, R.K.; Fuchs, R.; Brovkin, V.; Ciais, P.; Fensholt, R.; et al. China and India lead in greening of the world through land-use management. *Nat. Sustain.* **2019**, *2*, 122–129. [CrossRef]
23. Mondal, P.; Jain, M.; Robertson, A.W.; Galford, G.L.; Small, C.; DeFries, R.S. Winter crop sensitivity to inter-annual climate variability in central India. *Clim. Chang.* **2014**. [CrossRef]
24. Mishra, N.B.; Mainali, K.P. Greening and browning of the Himalaya: Spatial patterns and the role of climatic change and human drivers. *Sci. Total Environ.* **2017**, *587–588*, 326–339. [CrossRef] [PubMed]
25. Mishra, N.B.; Chaudhuri, G. Spatio-temporal analysis of trends in seasonal vegetation productivity across Uttarakhand, Indian Himalayas, 2000-2014. *Appl. Geogr.* **2015**, *56*, 29–41. [CrossRef]
26. de Jong, R.; Verbesselt, J.; Schaepman, M.E.; de Bruin, S. Trend changes in global greening and browning: Contribution of short-term trends to longer-term change. *Glob. Chang. Biol.* **2012**, *18*, 642–655. [CrossRef]
27. Murthy, K.; Bagchi, S. Spatial patterns of long-term vegetation greening and browning are consistent across multiple scales: Implications for monitoring land degradation. *Land Degrad. Dev.* **2018**, *29*, 2485–2495. [CrossRef]
28. Chakraborty, A.; Seshasai, M.V.R.; Reddy, C.S.; Dadhwal, V.K. Persistent negative changes in seasonal greenness over different forest types of India using MODIS time series NDVI data (2001–2014). *Ecol. Indic.* **2018**, *85*, 887–903. [CrossRef]
29. Rocchini, D. Effects of spatial and spectral resolution in estimating ecosystem α-diversity by satellite imagery. *Remote Sens. Environ.* **2007**, *111*, 423–434. [CrossRef]
30. Mackey, B.; Bryan, J.; Randall, L. *Australia's Dynamic Habitat Template for 2003*; ANU Research Publications: Canberra, Australia, 2003.
31. Coops, N.C.; Wulder, M.A.; Iwanicka, D. Demonstration of a satellite-based index to monitor habitat at continental-scales. *Ecol. Indic.* **2009**, *9*, 948–958. [CrossRef]
32. Wright, D.H. Species-Energy Theory: An Extension of Species-Area Theory. *Oikos* **1983**. [CrossRef]
33. Currie, D.J.; Mittelbach, G.G.; Cornell, H.V.; Field, R.; Guégan, J.F.; Hawkins, B.A.; Kaufman, D.M.; Kerr, J.T.; Oberdorff, T.; O'Brien, E.; et al. Predictions and tests of climate-based hypotheses of broad-scale variation in taxonomic richness. *Ecol. Lett.* **2004**, *7*, 1121–1134. [CrossRef]
34. Hurlbert, A.H. Linking species-area and species-energy relationships in Drosophila microcosms. *Ecol. Lett.* **2006**. [CrossRef] [PubMed]
35. Connell, J.H.; Orias, E. The Ecological Regulation of Species Diversity. *Am. Nat.* **1964**. [CrossRef]

36. Hu, G.; Jin, Y.; Liu, J.; Yu, M. Functional diversity versus species diversity: Relationships with habitat heterogeneity at multiple scales in a subtropical evergreen broad-leaved forest. *Ecol. Res.* **2014**. [CrossRef]
37. Mason, N.W.H.; Mouillot, D.; Lee, W.G.; Wilson, J.B. Functional richness, functional evenness and functional divergence: The primary components of functional diversity. *Oikos* **2005**. [CrossRef]
38. Williams, S.E.; Middleton, J. Climatic seasonality, resource bottlenecks, and abundance of rainforest birds: Implications for global climate change. *Divers. Distrib.* **2008**. [CrossRef]
39. Coops, N.C.; Fontana, F.M.A.; Harvey, G.K.A.; Nelson, T.A.; Wulder, M.A. Monitoring of a national-scale indirect indicator of biodiversity using a long time-series of remotely sensed imagery. *Can. J. Remote Sens.* **2014**. [CrossRef]
40. Nelson, T.A.; Coops, N.C.; Wulder, M.A.; Perez, L.; Fitterer, J.; Powers, R.; Fontana, F. Predicting climate change impacts to the canadian boreal forest. *Diversity* **2014**, *6*, 133–157. [CrossRef]
41. Berry, S.; Mackey, B.; Brown, T. Potential applications of remotely sensed vegetation greenness to habitat analysis and the conservation of dispersive fauna. In Proceedings of the Pacific Conservation Biology. 2007. Available online: https://www.publish.csiro.au/PC/PC070120 (accessed on 4 January 2021).
42. Berry, S.L.; Roderick, M.L. Estimating mixtures of leaf functional types using continental-scale satellite and climatic data. *Glob. Ecol. Biogeogr.* **2002**. [CrossRef]
43. Cramer, M.J.; Willig, M.R. Habitat heterogeneity, species diversity and null models. *Oikos* **2005**. [CrossRef]
44. Turner, M.G.; Donato, D.C.; Romme, W.H. Consequences of spatial heterogeneity for ecosystem services in changing forest landscapes: Priorities for future research. *Landsc. Ecol.* **2013**. [CrossRef]
45. Barik, S.K.; Behera, M.D. Studies on ecosystem function and dynamics in Indian sub-continent and emerging applications of satellite remote sensing technique. *Trop. Ecol.* **2020**, *61*, 1–4. [CrossRef]
46. Zhao, M.; Heinsch, F.A.; Nemani, R.R.; Running, S.W. Improvements of the MODIS terrestrial gross and net primary production global data set. *Remote Sens. Environ.* **2005**. [CrossRef]
47. Heinsch, F.A.; Zhao, M.; Running, S.W.; Kimball, J.S.; Nemani, R.R.; Davis, K.J.; Bolstad, P.V.; Cook, B.D.; Desai, A.R.; Ricciuto, D.M.; et al. Evaluation of remote sensing based terrestrial productivity from MODIS using regional tower eddy flux network observations. *IEEE Trans. Geosci. Remote Sens.* **2006**. [CrossRef]
48. Tian, Y.; Zhang, Y.; Knyazikhin, Y.; Myneni, R.B.; Glassy, J.M.; Dedieu, G.; Running, S.W. Prototyping of MODIS LAI and FPAR algorithm with LASUR and LANDSAT data. *IEEE Trans. Geosci. Remote Sens.* **2000**. [CrossRef]
49. Coops, N.C.; Wulder, M.A.; Duro, D.C.; Han, T.; Berry, S. The development of a Canadian dynamic habitat index using multi-temporal satellite estimates of canopy light absorbance. *Ecol. Indic.* **2008**. [CrossRef]
50. Roy, P.S.; Kushwaha, S.P.S.; Murthy, M.S.R.; Roy, A.; Kushwaha, D.; Reddy, C.S.; Behera, M.D.; Mathur, V.B.; Padalia, H.; Saran, S.; et al. *Biodiversity Characterisation at Landscape Level: National Assessment*; Indian Institute of Remote Sensing: Dehradun, India, 2012; ISBN 8190141880.
51. Tripathi, P.; Dev Behera, M.; Roy, P.S. Optimized grid representation of plant species richness in India-Utility of an existing national database in integrated ecological analysis. *PLoS ONE* **2017**, *12*, e0173774. [CrossRef]
52. Schwartz, M.D.; Ahas, R.; Aasa, A. Onset of spring starting earlier across the Northern Hemisphere. *Glob. Chang. Biol.* **2006**. [CrossRef]
53. Gaston, K.J.; Blackburn, T.M. Pattern and Process in Macroecology. 2007. Available online: https://www.wiley.com/en-ag/Pattern+and+Process+in+Macroecology-p-9780470999592 (accessed on 4 January 2021).
54. Kleijn, D.; Rundlöf, M.; Scheper, J.; Smith, H.G.; Tscharntke, T. Does conservation on farmland contribute to halting the biodiversity decline? *Trends Ecol. Evol.* **2011**, *26*, 474–481. [CrossRef]
55. Negi, V.S.; Joshi, B.C.; Pathak, R.; Rawal, R.S.; Sekar, K.C. Assessment of fuelwood diversity and consumption patterns in cold desert part of Indian Himalaya: Implication for conservation and quality of life. *J. Clean. Prod.* **2018**, *196*, 23–31. [CrossRef]
56. Vicente-Serrano, S.M.; Cabello, D.; Tomás-Burguera, M.; Martín-Hernández, N.; Beguería, S.; Azorin-Molina, C.; Kenawy, A. El Drought variability and land degradation in semiarid regions: Assessment using remote sensing data and drought indices (1982-2011). *Remote Sens.* **2015**, *7*, 4391–4423. [CrossRef]
57. MEA. Ecosystems and Human Well-Being. Synthesis. 2005. Available online: http://www.bioquest.org/wp-content/blogs.dir/files/2009/06/ecosystems-and-health.pdf (accessed on 4 January 2021).
58. Saikia, A. NDVI variability in North East India. *Scottish Geogr. J.* **2009**. [CrossRef]
59. Roy, P.S.; Roy, A.; Joshi, P.K.; Kale, M.P.; Srivastava, V.K.; Srivastava, S.K.; Dwevidi, R.S.; Joshi, C.; Behera, M.D.; Meiyappan, P.; et al. Development of decadal (1985-1995-2005) land use and land cover database for India. *Remote Sens.* **2015**, *7*, 2401–2430. [CrossRef]
60. Pasha, S.V.; Behera, M.D.; Mahawar, S.K.; Barik, S.K.; Joshi, S.R. Assessment of shifting cultivation fallows in Northeastern India using Landsat imageries. *Trop. Ecol.* **2020**. [CrossRef]
61. Roy, P.S.; Behera, M.D. Assessment of biological richness in different altitudinal zones in the Eastern Himalayas, Arunachal Pradesh, India. *Curr. Sci.* **2005**, *88*, 250–257.

 *remote sensing*

Article

Spaceborne Multifrequency PolInSAR-Based Inversion Modelling for Forest Height Retrieval

Shashi Kumar [1,*], Himanshu Govil [2], Prashant K. Srivastava [3], Praveen K. Thakur [1] and Satya P. S. Kushwaha [1]

[1] Indian Institute of Remote Sensing, Dehradun 248001, India; praveen@iirs.gov.in (P.K.T.) spskushwaha@iirs.gov.in (S.P.S.K.)

[2] Department of Applied Geology, National Institute of Technology, Raipur 492010, India; hgovil.geo@nitrr.ac.in

[3] Institute of Environment and Sustainable Development, Banaras Hindu University, Uttar Pradesh 221005, India; prashant.iesd@bhu.ac.in

* Correspondence: shashi@iirs.gov.in; Tel.: +91-135-2524119

Received: 22 October 2020; Accepted: 7 December 2020; Published: 10 December 2020

Abstract: Spaceborne and airborne polarimetric synthetic-aperture radar interferometry (PolInSAR) data have been extensively used for forest parameter retrieval. The PolInSAR models have proven their potential in the accurate measurement of forest vegetation height. Spaceborne monostatic multifrequency data of different SAR missions and the Global Ecosystem Dynamics Investigation (GEDI)-derived forest canopy height map were used in this study for vegetation height retrieval. This study tested the performance of PolInSAR complex coherence-based inversion models for estimating the vegetation height of the forest ranges of Doon Valley, Uttarakhand, India. The inversion-based forest height obtained from the three-stage inversion (TSI) model had higher accuracy than the coherence amplitude inversion (CAI) model-based estimates. The vegetation height values of GEDI-derived canopy height map did not show good relation with field-measured forest height values. It was found that, at several locations, GEDI-derived forest height values underestimated the vegetation height. The statistical analysis of the GEDI-derived estimates with field-measured height showed a high root mean square error (RMSE; 5.82 m) and standard error (SE; 5.33 m) with a very low coefficient of determination (R^2; 0.0022). An analysis of the spaceborne-mission-based forest height values suggested that the L-band SAR has great potential in forest height retrieval. TSI-model-based forest height values showed lower p-values, which indicates the significant relation between modelled and field-measured forest height values. A comparison of the results obtained from different SAR systems is discussed, and it is observed that the L-band-based PolInSAR inversion gives the most reliable result with low RMSE (2.87 m) and relatively higher R^2 (0.53) for the linear regression analysis between the modelled tree height and the field data. These results indicate that higher wavelength PolInSAR datasets are more suitable for tree canopy height estimation using the PolInSAR inversion technique.

Keywords: spaceborne SAR; multifrequency; GEDI; PolInSAR inversion; forest height

1. Introduction

Spaceborne remote sensing technique is an important tool to measure different parameters of forest vegetation to understand the forest carbon fluxes for the modelling of gross primary production and net ecosystem production [1]. The carbon cycle of the Earth is the most important parameter, which regulates the optimum climatic conditions suitable for all life forms on the Earth [2–5]. However, the recent loss of equilibrium of the Earth's carbon cycle, mainly caused due to human activity, has likely triggered recent climate changes which could be harmful to the biosphere of the Earth [6–8].

Because of these reasons, accurate tracking and measures to achieve a better understanding of the Earth's carbon cycle are very important for sustaining life. The spaceborne synthetic-aperture radar (SAR) systems can provide high-resolution information of the Earth objects with large coverage in one scene. Earth-observation-based active and passive sensors have also been used as a tool to retrieve forest parameters for inaccessible areas [9–13]. Nowadays, remote sensing is used as a primary source to do forest mapping and monitoring at regular intervals [14]. Among these technologies, SAR remote sensing has an advantage of forest canopy and cloud penetration capability due to the long-wavelength range of electromagnetic (EM) waves compared to optical remote sensing and the active nature of the SAR system, which allows for night operation. The SAR systems can be operated in different imaging modes to acquire the data according to the requirement of the application to retrieve structural and biophysical parameters of a forest area [15]. Several studies have shown the potential of SAR remote sensing for forest mapping and monitoring [16–19]. Polarimetric SAR modelling for retrieval of forest parameters has become a successful tool to retrieve scattering contributed by different scatterers within small forest patches for structural and biophysical characterization of the forest [20–23]. Forest tree height is directly related to biomass (AGB) and carbon stock [24–27]. The height of the forest trees needs to be measured at regular intervals to understand the spatiotemporal changes in the aboveground biomass and carbon stock.

The interferometric technique of SAR-based sensing has shown great potential to retrieve forest height with reliable accuracy [28–30]. The potential of SAR tomography has also been investigated successfully for forest height retrieval and scattering power retrieval at different height levels [31–34]. The requirement of a large number of interferometric pairs is a major limitation of the SAR tomography and sometimes it becomes difficult to implement this technology for forest height retrieval in the absence of several repeat passes of the data. Polarimetric SAR interferometry (PolInSAR) involves all of the possible polarimetric combinations and is hence better than SAR interferometry in resolving the coherence optimization problem to obtain the optimum scattering mechanisms related to different scattering mechanisms from the canopy [35].

Polarimetric acquisitions of an interferometric pair provide the capability to improve the interferogram quality for optimum coherence estimates with highly accurate phase information [36]. Several studies have been carried in the last few years to establish the capability of PolInSAR for forest height measurement [37–40]. The development of inversion models with PolInSAR data can provide a useful approach to retrieve forest height from SAR-based remote sensing [41–44]. The polarimetric SAR interferometry-based modelling approach for vegetation height of a forest area has been successfully implemented with airborne SAR data [42,43,45–47]. Fewer studies have been carried out to retrieve forest tree height from PolInSAR inversion using spaceborne SAR data [48]. The normal baseline component of the interferometric acquisition plays a significant role in vertical wavenumber generation for inversion-based models of PolInSAR data [49,50]. For a spaceborne bistatic or airborne SAR system, the system parameters could be easily planned with higher accuracy to derive vertical baseline components However, in monostatic SAR systems, the large spatial baseline of interferometric acquisition over a forest area results in very low coherence. The InSAR data, with small baseline components, provide a solution to obtain reliable coherence from the forest. The PolInSAR data acquisition, with a small baseline, gives a very high altitude of ambiguity that is not suitable for the forest stand height of 30–40 m tall trees. To overcome this problem, coherence amplitude inversion (CAI) and three-stage inversion (TSI) were performed with a vertical wavenumber file that was generated by the altitude of ambiguity equal to twice the forest height. The main objective of this present study is to retrieve forest tree height with the help of multifrequency spaceborne PolInSAR data. PolInSAR-based vegetation height estimation has been studied by many using various algorithms and models [51–55]. The potential of spaceborne PolInSAR data is evaluated in this study using TSI and CAI modelling approaches. Three-stage inversion is extensively used in the PolInSAR model, which uses a random volume over the ground (RVoG) model for height estimation of forest vegetation [56,57]. This study utilizes monostatic multifrequency (L-, C-, and X-band) spaceborne PolInSAR data for inversion modelling to retrieve forest height.

2. Study Area and Dataset

The two forest ranges (Barkot and Thano) in Doon Valley, Uttarakhand, India were selected for spaceborne multifrequency-based inversion modeling. These forest ranges lie between the Shivalik Hills and the Himalayan Mountains. Figure 1 shows the standard false-color composite (FCC) of the Sentinel-2B optical multispectral data that was acquired on 16 December 2018. The three spectral bands (842 nm, 665 nm, and 560 nm) of 10 m spatial resolution were used to generate the FCC (Figure 1), which enhances the forest and vegetation cover in red. The vegetation cover of both the forest ranges of the Doon Valley is visible in the red color as shown in Figure 1. Nearby features, such as the agricultural area, the urban settlement of Rishikesh City, and the waters of the Ganges River, are clearly visible in various bands of the FCC according to their surface reflectance.

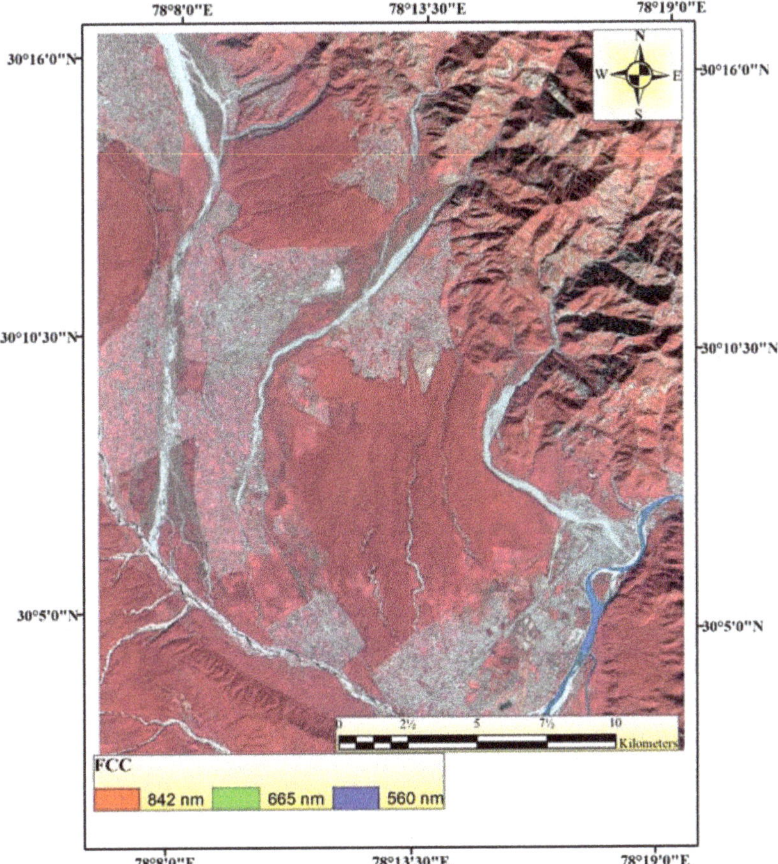

Figure 1. Sentinel-2 False Color Composite (FCC) image of the study area.

The vegetation type map of the two forest ranges is shown in Figure 2a. The topography of the terrain for these study sites is flat, which is advantageous to retrieve actual PolSAR-based parameters without rigor geometric correction. Sal (*Shorea robusta*) trees (Figure 2b,c)) are the main species of both the forest ranges. The other vegetation species are Teak (*Tectona grandis*) plantation, Sisham (*Dalbergia sissoo*), and Khair (*Acacia catechu*).

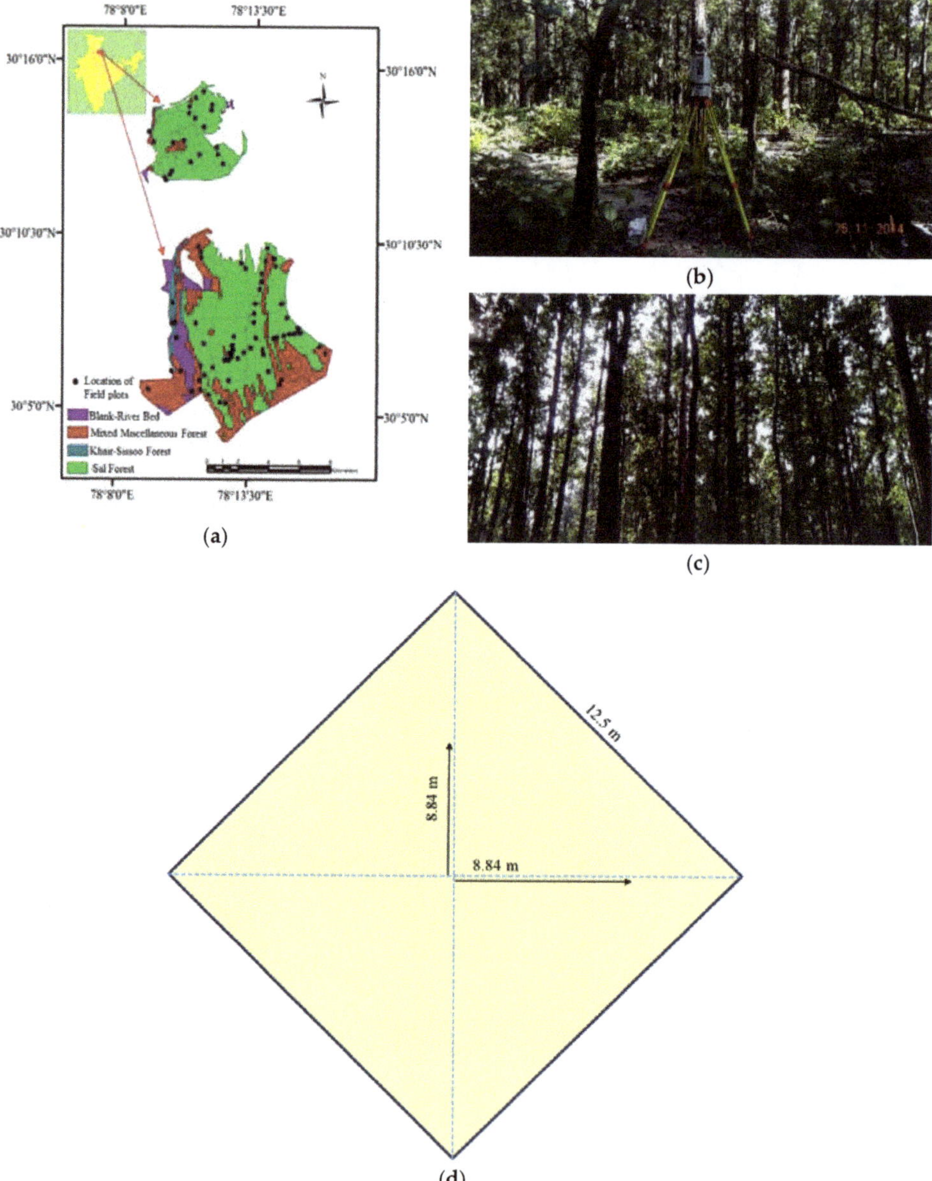

Figure 2. (**a**) Vegetation type map of the forest ranges, (**b**) Terrestrial Lase Scanner (TLS)-based forest height measurement, (**c**) Sal trees, and (**d**) sample plot design.

Figure 2a shows the location of the field-measurement in the Barkot and Thano forest ranges of Doon Valley. The Barkot and Thano Forests are covered by dense Sal (*Shorea robusta*), Khair-Sissoo/Sisham (*Acacia catechu-Dalbergia sissoo*), and mixed miscellaneous forests. Field-collected forest tree heights from 100 locations in the Barkot and Thano Forests [48,58] were used for validation of the PolInSAR-based modelled vegetation height map. Square plots of 12.5 m^2 size were made for the field-based forest height measurement. The field data collection for tree height measurement was done using the Nikon Forestry Pro Laser Range Finder and Laser Dendrometer (Criterion RD1000). The position of the plot locations was measured with the Trimble Juno handheld GPS. The field data were collected in the accessible region of the forest ranges and the distribution of plots is shown in Figure 2a. Figure 2b shows tree height measurements using the Terrestrial Lase Scanner (TLS) in the forest plot. In a separate research work, TLS-based measurement for forest parameter retrieval were carried out between November 2014 to February 2015 for limited plot locations [59]. It was found that the maximum difference between field-based forest height using handheld equipment and TLS data-based estimates was 0.81 m [59]. Figure 2d shows a sample plot design that was made in the field for tree height measurement. The forest ranges of Doon Valley predominantly dominate Sal trees, which can also be seen for both these forest ranges as shown in Figure 2a. Considering that most of the area is covered by Sal forest, we planned to collect a good number of data from this vegetation type, according to which plots were made in the field. Table 1 shows the number of plots of the field-collected data for forest height measurement in different vegetation species of the forest ranges.

Table 1. Number of plots of the field-collected data for forest height measurement.

Forest Vegetation Type	No. of Plots
Khair-Sisham Forest	5
Mixed Miscellaneous Forest	18
Sal Forest	77

Plot-wise average forest height of the field-measured data for the 100 locations in the Barkot and Thano forest ranges are shown in Figure 3. Error bars with a standard deviation of average forest height for all the 100 field-measured forest plots are shown in Figure 3. It is observed that, for most of the locations, the average forest height was more than 20 m.

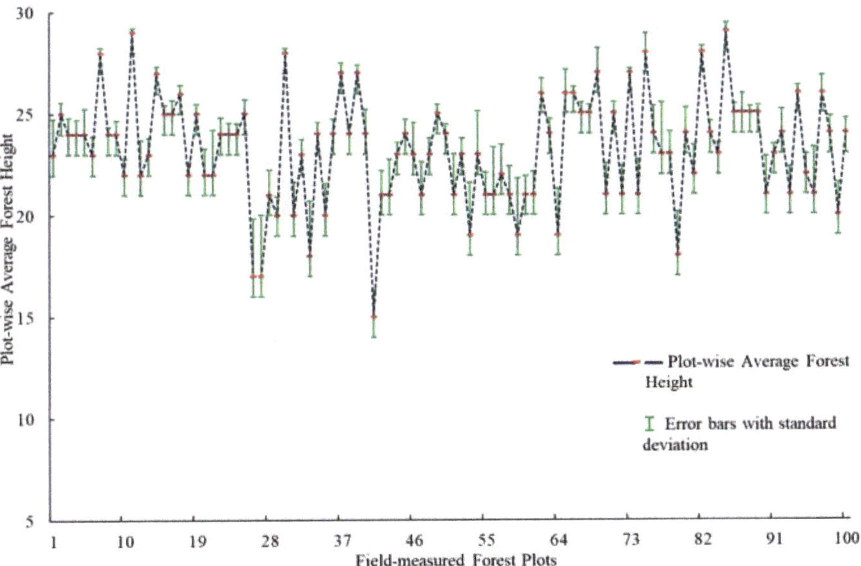

Figure 3. Field-measured plot-wise average forest height with error bars.

2.1. Global Forest Canopy Height Map

The global forest canopy height map was developed by integrating the lidar-based structural measurement for forest vegetation of the Global Ecosystem Dynamics Investigation (GEDI) data and Landsat multitemporal surface reflectance data [60,61]. The product id Forest_height_2019_NASIA of 30 m spatial resolution global forest canopy height map was used to prepare a forest height map of the Uttarakhand state of India. Most of the forest cover shows vegetation height with a difference of less than 35 m from the forest height values. Three more values, 101, 102, and 103, were also obtained from the data, representing water, snow/ice, and no data, respectively [60]. The Barkot and Thano forest ranges of Doon Valley, Uttarakhand, India, are shown in the red circle in Figure 4. A subset was taken from the Uttarakhand forest height map, which is shown in Figure 5 to show the vegetation height in the two forest ranges.

Figure 5 shows a subset of 30 m spatial resolution of the global forest canopy height that was generated by the integration of GEDI-based forest height metrics with Landsat multitemporal data.

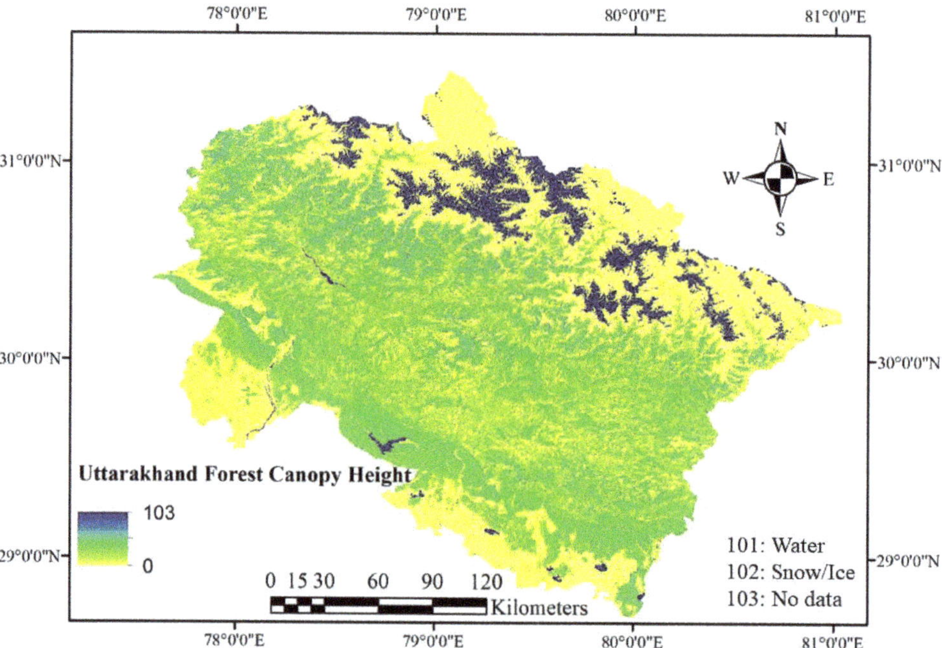

Figure 4. The 30 m spatial resolution global forest canopy height map of Uttarakhand, India.

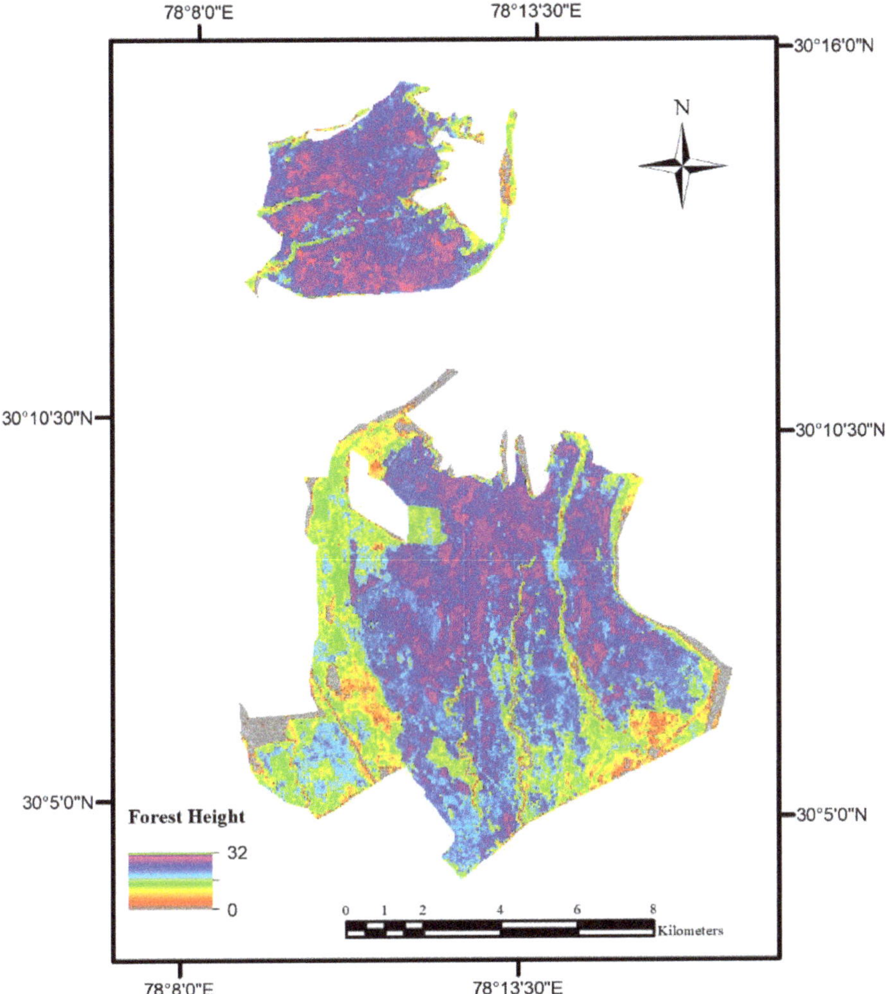

Figure 5. A subset of global forest canopy height map for the Barkot and Thano forest ranges.

2.2. Polarimetric SAR Interferometry (PolInSAR) Data

Polarimetric SAR Interferometry (PolInSAR) combines interferograms obtained at different polarizations [62]. Generally, all of the spaceborne SAR are monostatic SAR systems in which a single antenna is used to transmit and receive EM waves. Figure 6 shows the PolInSAR geometry in which PolSAR data were acquired by maintaining a temporal gap between two acquisitions to generate an interferometric pair. The two acquisitions provide two scattering matrices (Figure 6) represented by Equation (1).

The scattering matrix of the first acquisition is represented by S^1 and the second acquisition or slave data is represented by S^2. The matrix elements are represented as polarimetric combinations of horizontal and vertical polarizations of the electric field vector of the EM wave. To get an interferometric pass, the platform positions between two acquisitions must be separated by a baseline.

The multifrequency data used in this study were acquired from spaceborne SAR systems, ALOS-2 PALSAR-2, RADARSAT-2, and TerraSAR-X. A detailed description of the multifrequency polarimetric interferometric data is given in Table 2.

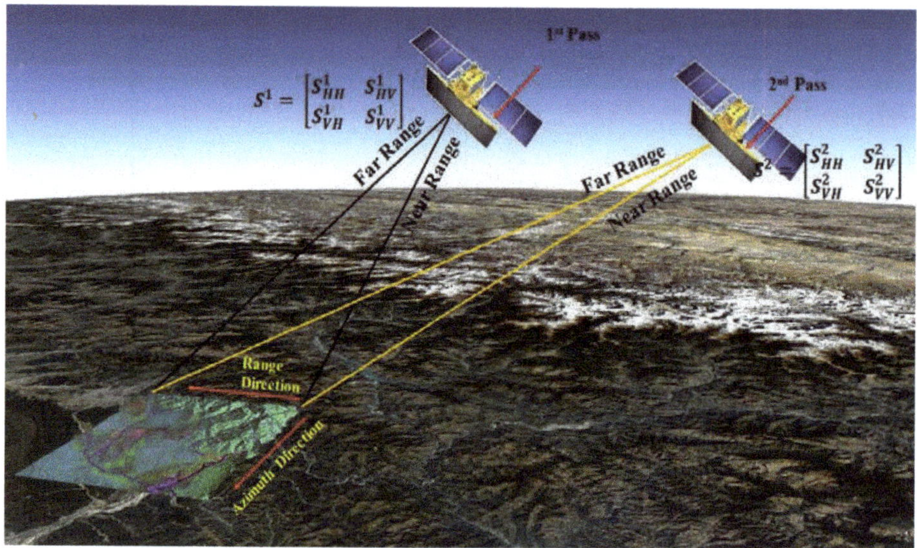

Figure 6. Polarimetric synthetic-aperture radar interferometry (PolInSAR) geometry.

Table 2. Description of the multifrequency polarimetric interferometric data.

PolInSAR Data	TerraSAR-X		RADARSAT-2		ALOS-2 PALSAR-2	
Acquisition	Reference	Secondary	Reference	Secondary	Reference	Secondary
Date of acquisition	21 January 2015	12 February 2015	27 January 2014	20 February 2014	9 August 2015	23 August 2015
Polarisation			Quad-pol (HH+HV+VH+VV)			
Wavelength (cm)/frequency (GHz)	3.10/9.64		5.55/5.4		24.25/1.236	
Resolution (m), range & azimuth	1.36 & 2.86		4.7 & 9.5		2.86 & 3.236	
Absolute orbit	42159	42493	32291	31948	6545	6752
Near range incidence angle	24.59	24.54	33.45	33.45	21.56	21.55
Far range incidence angle	26.73	26.79	35.07	35.07	25.93	25.93
Perpendicular baseline (m)	105		67.92		84.37	
Temporal baseline (days)	22		24		14	
Altitude of ambiguity (m)	35.85		215.05		455.16	

Figure 7 shows the color composite of model-based polarimetric decomposition of L-band, C-band, and X-band SAR data. Yamaguchi's four components decomposition model was used to generate scattering elements from the polarization orientation angle (POA)-compensated PolSAR coherency matrix [63]. Forest vegetation is a dominant source of volume scattering of the incident EM wave. In the polarimetric decomposition-based output of the multifrequency SAR data, this can be easily seen in green. The double-bounce scatterers and single-bounce scatterers are represented in red and blue, respectively. It is visible in the PolSAR data that the dry river channels of the Doon Valley showed surface scattering (blue color). The urban structures showed the dominance of double-bounce scattering, and these features are highlighted in red/pink in the false color composite image of the SAR data. The polarimetric decomposed image of L-band ALOS-2 PALSAR-2 showed high volume scattering in comparison to the TerraSAR-X and RADARSAT-2 C-band datasets. The long-wavelength of L-band ALOS-2 can penetrate the top canopy surface to provide scattering information contributed due to the entire volume of the tree. Due to the deep penetration of microwave signals into the forest vegetation and multiple scattering from the entire structure of the vegetation cover, the degree of volume scattering in L-band observations was higher than C- and X-band SAR data.

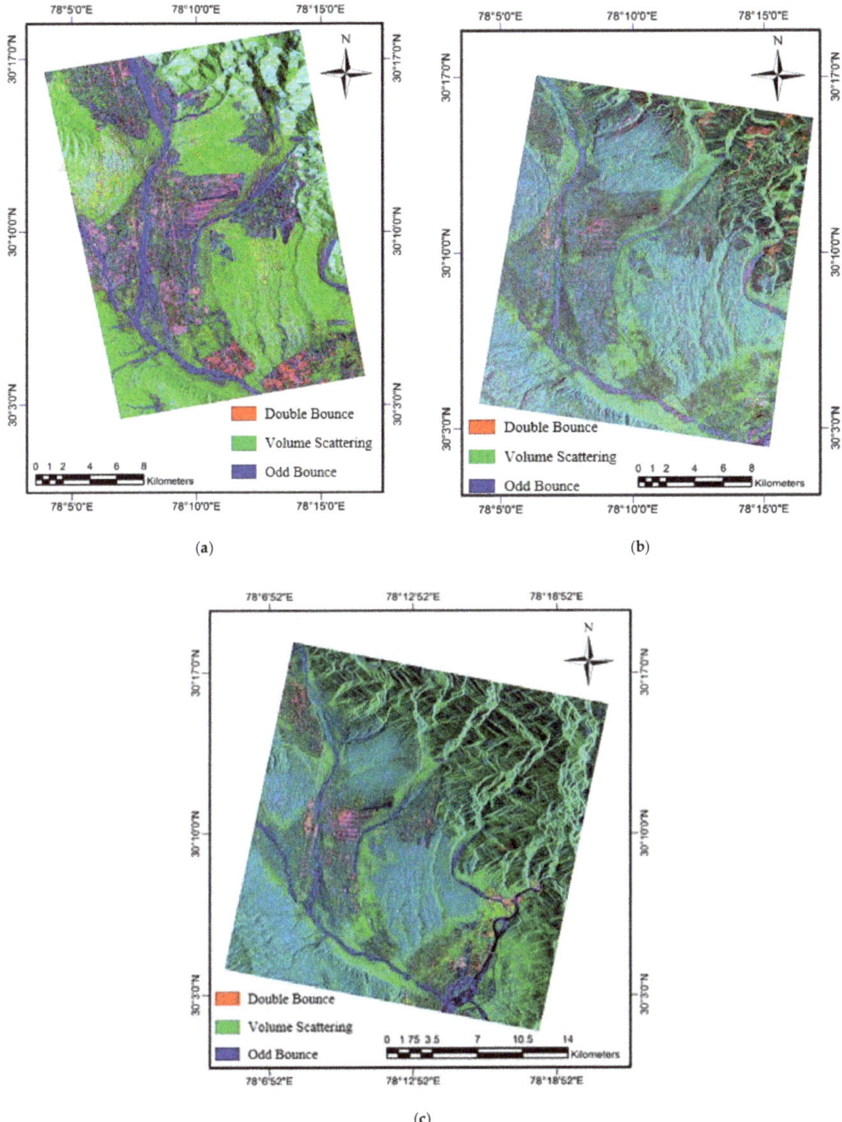

Figure 7. Polarimetric decomposition-based scattering representation of (**a**) ALOS-2 PALSAR-2 L-band, (**b**) TerraSAR-X band, and (**c**) RADARSAT-2 C-band SAR data.

3. Methodology

The present work utilized L-, C-, and X-band PolInSAR data-based TSI modelling for forest stand height estimation. The methodological steps are shown in Figure 8. The single-look complex data was initially corrected from slant range ambiguity to provide actual ground information in the range direction. Radiometric calibration was performed to calculate scattering matrix on multifrequency polarimetric interferometric pairs of SAR data. The scattering matrix was generated with four polarimetric combinations of first and second acquisitions of interferometric pairs of complex SAR data.

$$[S_1] = \begin{bmatrix} S_{HH}^1 & S_{HV}^1 \\ S_{VH}^1 & S_{VV}^1 \end{bmatrix} \text{ and } [S_2] = \begin{bmatrix} S_{HH}^2 & S_{HV}^2 \\ S_{VH}^2 & S_{VV}^2 \end{bmatrix}. \tag{1}$$

The scattering matrices were coregistered with subpixel accuracy for accurate phase estimation at interferometric processing. The coregistered PolInSAR pairs were orthorectified with the help of the Range Doppler Terrain Correction tool in SNAP v 6.0 [64].

The Pauli feature vector [65], which is a first-order polarimetric representation of the scattering matrix, was generated from the coregistered orthorectified product for all of the SAR data. The Pauli feature vectors of reference and secondary images were multiplied with their transconjugate to generate a 6×6 coherency matrix as shown in Equation (2) [41].

$$[T_6] := \left\langle \begin{bmatrix} \underline{k}_1 \\ \underline{k}_2 \end{bmatrix} \begin{bmatrix} \underline{k}_1^{*T} & \underline{k}_2^{*T} \end{bmatrix} \right\rangle = \begin{bmatrix} [T_{11}] & [\Omega_{12}] \\ [\Omega_{12}]^{*T} & [T_{22}] \end{bmatrix}$$

$$\underline{k}_1 = \tfrac{1}{\sqrt{2}} \begin{bmatrix} S_{HH}^1 + S_{VV}^1 & S_{HH}^1 - S_{VV}^1 & 2S_{HV}^1 \end{bmatrix}^T \text{ and} \tag{2}$$

$$\underline{k}_2 = \tfrac{1}{\sqrt{2}} \begin{bmatrix} S_{HH}^2 + S_{VV}^2 & S_{HH}^2 - S_{VV}^2 & 2S_{HV}^2 \end{bmatrix}^T.$$

where $[T_{11}]$ and $[T_{22}]$ are 3×3 coherency matrices of reference and secondary images and these matrices follow the characteristics of Hermitian matrices, while $[\Omega_{12}]$ is derived from the interferometric Hermitian product of Pauli feature vectors $\underline{k}_1$ and $\underline{k}_2$ of PolInSAR pairs.

The Interferometric coherence of PolSAR data was estimated with the help of a 6×6 coherency matrix that represents the complex correlation coefficients of a feature in the SAR data during the two acquisitions. The interferometric acquisition of PolSAR data is capable to retrieve coherence due to different scatterers in a SAR resolution cell.

When an EM wave hits the target, the coherence of an object changes with polarization due to different scattering behavior in different polarimetry channels. The PolInSAR coherence (γ) is represented in Equation (3) with the help of elements of the 6×6 coherency matrix [36,41].

$$\gamma = \frac{\left\langle \omega_1^{*T} [\Omega_{12}] \omega_2 \right\rangle}{\sqrt{\left\langle \omega_1^{*T} [T_{11}] \omega_1 \right\rangle \left\langle \omega_2^{*T} [T_{22}] \omega_2 \right\rangle}} \tag{3}$$

where ω_1 and ω_2 represent the two scattering mechanisms and * indicates the complex conjugate. The spatial averaging is also shown in Equation (3) as an angular bracket and 6×6 coherency matrix elements are calculated using Equations (4)–(6).

$$[T_{11}] = \begin{bmatrix} \left\langle |S_{HH}^1 + S_{VV}^1|^2 \right\rangle & \left\langle (S_{HH}^1 + S_{VV}^1)(S_{HH}^1 - S_{VV}^1)^* \right\rangle & 2\left\langle (S_{HH}^1 + S_{VV}^1)S_{HV}^{1*} \right\rangle \\ \left\langle (S_{HH}^1 - S_{VV}^1)(S_{HH}^1 + S_{VV}^1)^* \right\rangle & \left\langle |S_{HH}^1 - S_{VV}^1|^2 \right\rangle & 2\left\langle (S_{HH}^1 - S_{VV}^1)S_{HV}^{1*} \right\rangle \\ 2\left\langle S_{HV}^1 (S_{HH}^1 + S_{VV}^1)^* \right\rangle & 2\left\langle S_{HV}^1 (S_{HH}^1 - S_{VV}^1)^* \right\rangle & \left\langle 4|S_{HV}^1|^2 \right\rangle \end{bmatrix} \tag{4}$$

$$[T_{22}] = \begin{bmatrix} \langle |S^2_{HH} + S^2_{VV}|^2 \rangle & \langle (S^2_{HH} + S^2_{VV})(S^2_{HH} - S^2_{VV})^* \rangle & 2\langle (S^2_{HH} + S^2_{VV})S^{2*}_{HV} \rangle \\ \langle (S^2_{HH} + S^2_{VV})(S^2_{HH} - S^2_{VV})^* \rangle & \langle |S^2_{HH} - S^2_{VV}|^2 \rangle & 2\langle (S^2_{HH} - S^2_{VV})S^{2*}_{HV} \rangle \\ 2\langle S^2_{HV}(S^2_{HH} + S^2_{VV})^* \rangle & 2\langle S^2_{HV}(S^2_{HH} - S^2_{VV})^* \rangle & \langle 4|S^2_{HV}|^2 \rangle \end{bmatrix} \quad (5)$$

$$[\Omega_{12}] = \begin{bmatrix} \langle (S^1_{HH} + S^1_{VV})(S^{2*}_{HH} + S^{2*}_{VV}) \rangle & \langle (S^1_{HH} + S^1_{VV})(S^{2*}_{HH} - S^{2*}_{VV}) \rangle & 2\langle (S^1_{HH} + S^1_{VV})S^{2*}_{HV} \rangle \\ \langle (S^1_{HH} - S^1_{VV})(S^{2*}_{HH} + S^{2*}_{VV}) \rangle & \langle (S^1_{HH} - S^1_{VV})(S^{2*}_{HH} - S^{2*}_{VV}) \rangle & 2\langle (S^1_{HH} - S^1_{VV})S^{2*}_{HV} \rangle \\ 2\langle S^1_{HV}(S^{2*}_{HH} + S^{2*}_{VV}) \rangle & 2\langle S^1_{HV}(S^{2*}_{HH} - S^{2*}_{VV}) \rangle & \langle 4S^1_{HV}S^{2*}_{HV} \rangle \end{bmatrix} \quad (6)$$

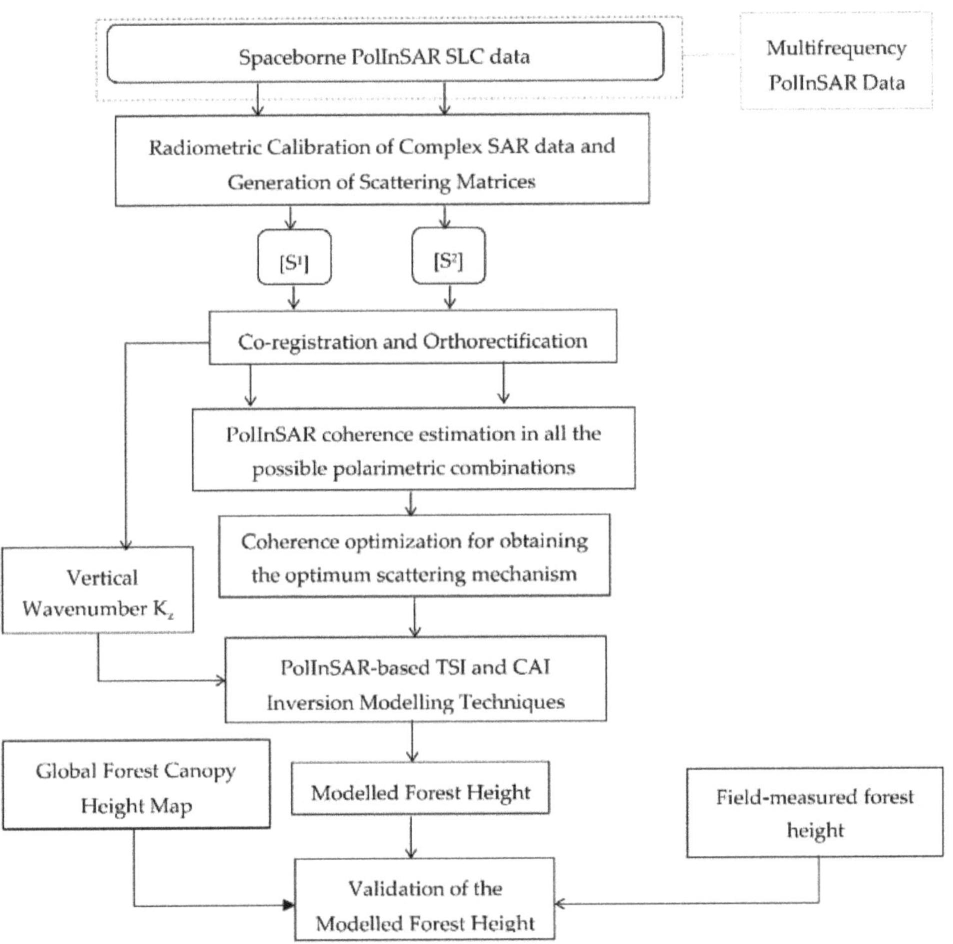

Figure 8. Methodology for PolInSAR inversion modelling for forest height retrieval.

3.1. Three-Stage Inversion (TSI) Modelling

The TSI model is an extension of the fully polarimetric SAR data-based expression of a two-layer model [66] to estimate the vertical structure of tree species [41]. The PolInSAR-based TSI model is very popular for forest height retrieval [49,58,67–70]. The TSI model includes the identification of the ground phase and volumetric coherence of the effective scattering center of forest vegetation for inversion-based forest height retrieval. PolInSAR-based TSI uses the complex coherences of all the polarimetric combinations to identify the ground phase and volumetric coherence with minimum ambiguity. A best-fit line is determined in the complex plane to find the optimum ground phase and volumetric coherence. The point on the best-fit line with the lowest value of ground to volume scattering ratio will be the most suitable input for volumetric coherence [41,71,72].

The complex coherence, γ, of the PolInSAR data for forest vegetation is shown in Equation (7) [41].

$$\gamma = \frac{\omega^{*T}\left(e^{i\phi_1} I_2^V + e^{\frac{-2\sigma h_v}{\cos\theta_0}} T_g e^{i\phi_1}\right)\omega}{\omega^{*T}\left(I_1^V + e^{\frac{-2\sigma h_v}{\cos\theta_0}} T_g\right)\omega} \tag{7}$$

where

$$I_1^V = e^{\frac{-2\sigma h_v}{\cos\theta}} \int_0^{h_v} e^{\frac{2\sigma z'}{\cos\theta_0}} T_V dz' \text{ and } I_2^V = e^{\frac{-2\sigma h_v}{\cos\theta}} \int_0^{h_v} e^{\frac{2\sigma z'}{\cos\theta_0}} e^{ik_z z'} T_V dz'$$

$$T_V = m_v \begin{bmatrix} 1 & 0 & 0 \\ 0 & \mu & 0 \\ 0 & 0 & \mu \end{bmatrix} 0 \le \mu \le 0.5$$

$$T_g = m_g \begin{bmatrix} 1 & t_{12} & 0 \\ t_{12}^* & t_{22} & 0 \\ 0 & 0 & t_{33} \end{bmatrix}.$$

The complex coherence is polarization-dependent and can be represented as a straight line in the complex plane of PolInSAR coherence:

$$\gamma(\omega) = e^{i\phi_1}(\gamma_v + L(\omega)(1 - \gamma_v))\ 0 \le L(\omega) \le 1 \tag{8}$$

where

$$L(\omega) = \frac{\mu(\omega)}{1 + \mu(\omega)} \text{ and } \mu(\omega) = \frac{2\sigma}{\cos\theta_0\left(e^{\frac{2\sigma h_v}{\cos\theta_0}} - 1\right)} \frac{\omega^{*T} T_g \omega}{\omega^{*T} T_V \omega} \tag{9}$$

$$\gamma_v = \frac{2\sigma}{\cos\theta_0\left(e^{\frac{2\sigma h_v}{\cos\theta_0}} - 1\right)} \int_0^{h_v} e^{ik_z z'} e^{2\sigma z' / \cos\theta_0} dz' \tag{10}$$

where, ω is a unitary complex vector, γ, is interferometric complex coherence that is dependent on the choice of polarization of the SAR data, φ is the interferometric phase of the ground topography, $\mu(\omega)$ shows ground to volume scattering ratio, θ_0 is the angle of incidence of the incident EM waves of the SAR system, and attenuation of the EM waves through vegetation canopy is represented by σ. In Equations (8) and (10) γ_v is polarization independent and is expressed as an integral of vertical wavenumber (k_z), angle of incidence, and wave extinction coefficient. The effective vertical interferometric wavenumber (k_z) is determined by the SAR imaging parameters. Baseline during the interferometric acquisition of SAR data and spaceborne platform to target range distance is used to derive k_z, as in

$$k_z = \frac{4\pi B_n}{\lambda R sin\theta}.$$ (11)

where B_n is the normal component of the baseline, λ is the wavelength of the SAR system, and R is the platform to the target distance. The value of R is retrieved from metadata by multiplying the speed of the electromagnetic wave and half of the total time taken by the wave to track the target. The interferometric wavenumber (k_z) depends on the SAR parameter and it is a crucial parameter that affects the PolInSAR inversion-based forest height [50]. The normal baseline component (B_n) plays an important role in deriving the vertical wavenumber [49]. This work utilized the vertical interferometric wavenumber that was generated from the spaceborne SAR parameters of multifrequency data and a derived normal baseline component. The normal baseline component was calculated according to the altitude of ambiguity greater than (approximately double) the forest height of the study area.

3.2. Coherence Amplitude Inversion (CAI) Modelling

Electromagnetic waves transmitted by a SAR system mainly interact with the top canopy surface of the forest vegetation. The top canopy surface contributes an amount of the backscattered signal, and twigs and branches of the forest vegetation contribute to the remaining portion. The top canopy surface is an unstable structure in nature, which shows very low interferometric coherence. If the SAR waves are able to interact with stable structures of the forest vegetation like big branches, then the interferometric coherence will become high for these features. Generally, in SAR interferometry, the height of an object is measured from the difference of the interferometric phases of the top and bottom of the surface. It is very difficult to retrieve interferometric phases of ground and forest vegetation for the locations having low interferometric coherence [44]. To overcome the problem of the interferometric phase coherence, an amplitude-based approach was proposed in which the ground is identified with polarimetric channels that have a low surface-to-volume ratio [71] The PolInSAR coherence amplitude-based inversion model for forest height retrieval is shown in Equation (12):

$$\min_{h_v}\left\{F = |||\widetilde{\gamma}_{w_v}| - |\frac{p}{p_1}\frac{e^{p_1 h_v}-1}{e^{ph_v}-1}|||\right\} \text{ where } \left\{\begin{array}{l} p = \frac{2\sigma}{cos\theta} \\ p_1 = p + ik_z \end{array}\right. .$$ (12)

Here, the function F is minimized for the inversion-based modelling [72], h_v represents the height of the vegetation, and σ represents the attenuation of an electromagnetic wave due to vegetation and volume only coherence is represented by $\widetilde{\gamma}_{w_v}$.

4. Results

This section describes the results of PolInSAR-based inversion modelling for forest height estimation using multifrequency spaceborne SAR data.

Multifrequency PolInSAR Data for Forest Height Estimation

Polarimetric SAR interferometry-based inversion modelling was implemented to generate the forest height. Forest height retrieval was performed with all the possible coherence combinations and interferometric vertical wavenumber file. The CAI technique assumes that volume scattering is at the top of the canopy and generally, cross-polarimetric channels are selected as input in the modelling approach for inversion-based height retrieval.

The PolInSAR data has the capability to generate complex coherence in several polarimetric combinations, but the combination having the lowest ground-to-volume ratio is the most suitable for inversion-based forest height retrieval. Figure 9a,b show the TSI-based and CAI-based forest height maps that were generated from the PolInSAR pair of ALOS-2 PALSAR-2 data. Forest heights generated from RADARSAT-2 and TerraSAR-X are shown in Figure 10a,b and Figure 11a,b, respectively. The modelled outputs of the forest heights were validated with field-measured values using the coefficient of determination (R^2), root mean square error (RMSE), standard error of the estimate (SE), and *p*-value. To validate the modelled output of the forest height of PolInSAR inversion-based modelling, field data were used for 100 locations in the study area.

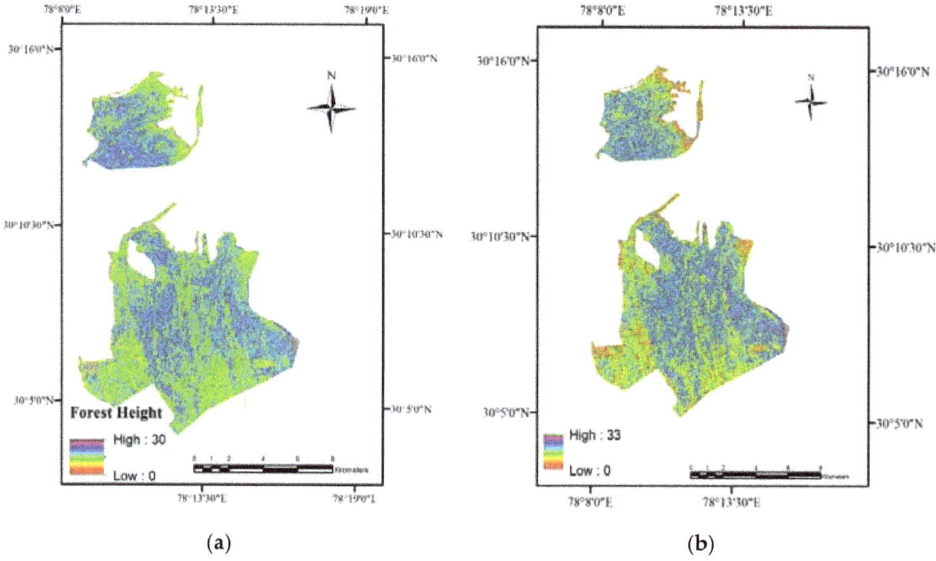

(a)　　　　　　　　　　　(b)

Figure 9. ALOS-2 PALSAR-2 inversion-based forest height for the Barkot and Thano forest ranges using (**a**) the TSI model and (**b**) the CAI model.

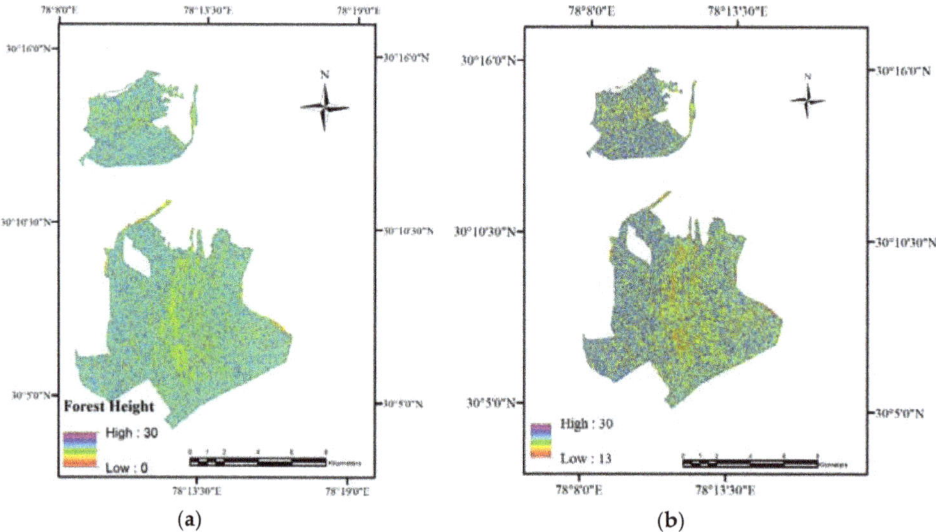

(a)　　　　　　　　　　　(b)

Figure 10. RADARSAT-2 inversion-based forest height for the Barkot and Thano forest ranges using (**a**) the TSI model and (**b**) the CAI model.

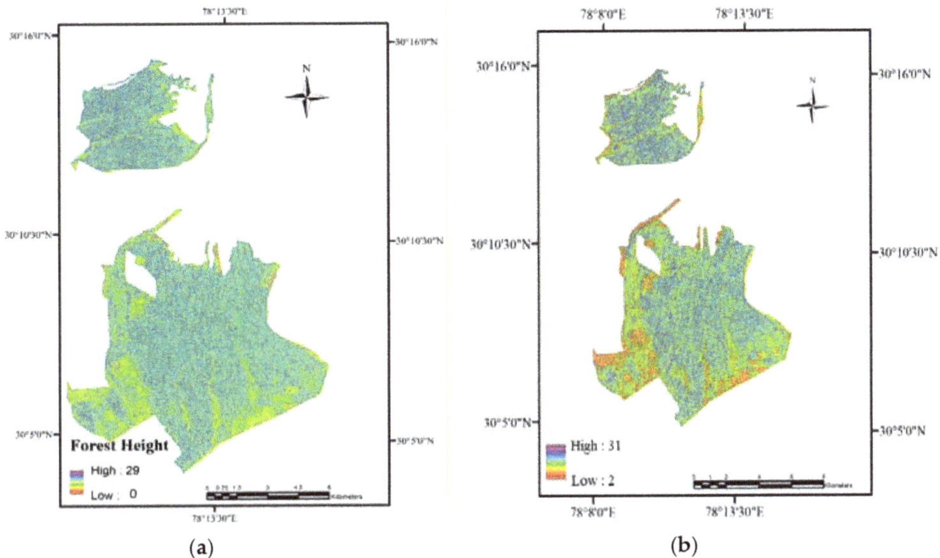

Figure 11. TerraSAR-X inversion-based forest height for the Barkot and Thano forest ranges using (**a**) the TSI model and (**b**) the CAI model.

Figure 12a shows the TSI-based forest height generated from the PolInSAR pair of ALOS-2 PALSAR-2 data. Forest heights generated from RADARSAT-2 and TerraSAR-X are shown in Figure 12b,c, respectively. In addition to the best-fit regression, the line confidence interval (CI) and prediction interval (PI) were also drawn. The CI is marked with a broken line in yellow, the PI is marked with a solid line in black, and the best-fit regression line is shown in blue. The CI with a 95% confidence interval ensures that it has a 95% probability of being the best-fit line between the upper and lower limits that are marked as the broken line in yellow in Figure 12. The CI with 95% still has a probability of 5% in which the true best-fit linear regression line may fall outside the confidence interval limits/boundaries. It is evident in Figure 12 that the best-fit regression line for all the modelled output for forest height retrieval did not go outside of the upper and lower limits of the CI.

The 95% prediction interval (PI) indicates the area where 95% of the datapoints will fall and predicts a range in which a future observation will fall. Figure 12 shows that one datapoint fell outside the prediction interval boundary for the linear regression with ALOS-2 PALSAR-2 and RADARSAT-2 data. The linear regression with TerraSAR-X data shows that four datapoints fell outside the boundary of the prediction interval.

The linear regression between the field data and modelled output of multifrequency PolInSAR data (Figure 12) shows that the coefficients of determination (R^2) of TSI-based forest height for ALOS-2 PALSAR-2, RADARSAT-2, and TerraSAR-X were 0.53, 0.32, and 0.43, respectively, with RMSE values of 2.87 m, 3.74 m, and 4.53 m. The standard error of the estimate (SE) for TSI-based approach for L-band, C-band, and X-band data were 1.56 m, 1.73 m, and 2.17 m, respectively (Table 3). The relationship between field-measured forest height and TSI-based modelled height shows the highest R^2 from ALOS-2 PALSAR-2 data. The lowest R^2 (0.32) was obtained from RADARSAT-2 data. RMSE obtained from the multifrequency SAR data shows that the error value of PolInSAR inversion-based height retrieval was 2.87 m for ALOS-2 PALSAR-2, 3.74 m for RADARSAT-2, and 4.53 m for TerraSAR-X data. From Figure 12 and Table 3, it is obvious that the highest R^2 and lowest RMSE was obtained from the long-wavelength L-band PolInSAR pair of ALOS-2 PALSAR-2. The reason for the accuracy of L-band likely lies in the canopy penetration capability of L-band SAR for the detection of ground coherence, phase, and identification of the volume scattering center at the top of the canopy. A moderate accuracy in forest height retrieval was obtained from X-band SAR data due to the high spatial resolution of TerraSAR-X, which enables the retrieval of ground information through canopy gaps.

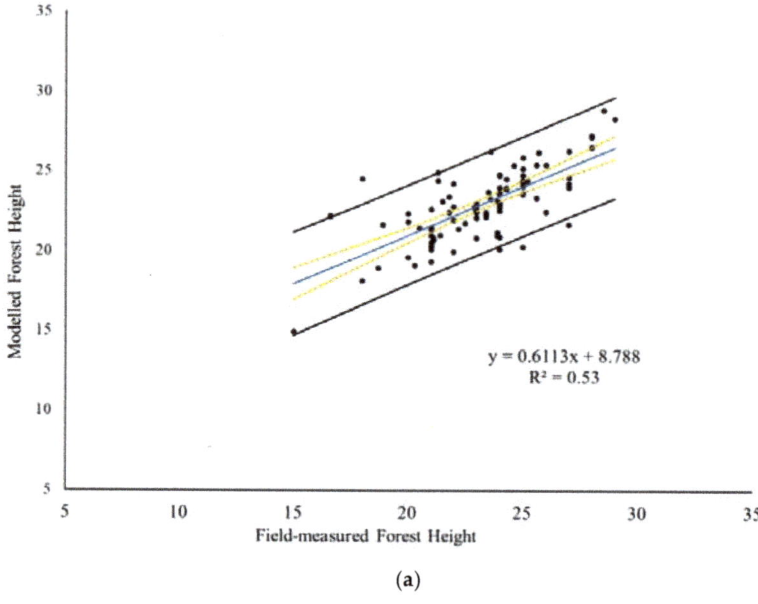

(a)

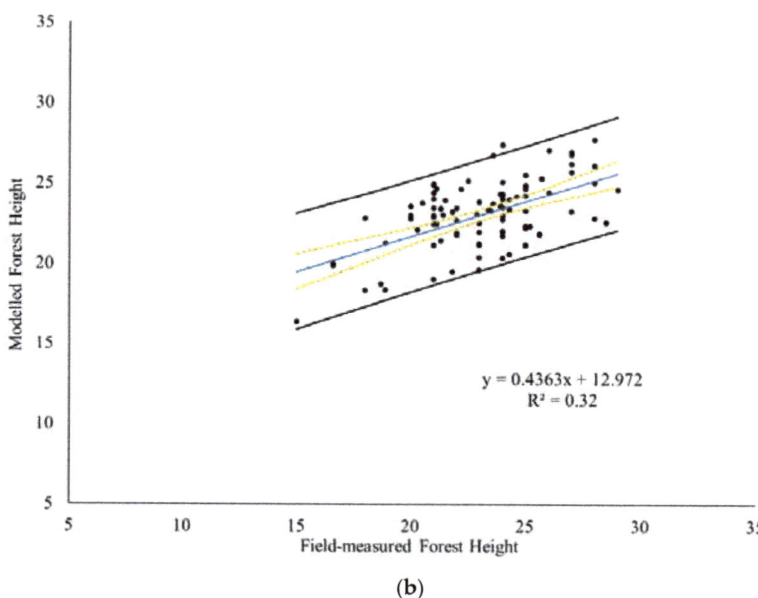

(b)

Figure 12. *Cont.*

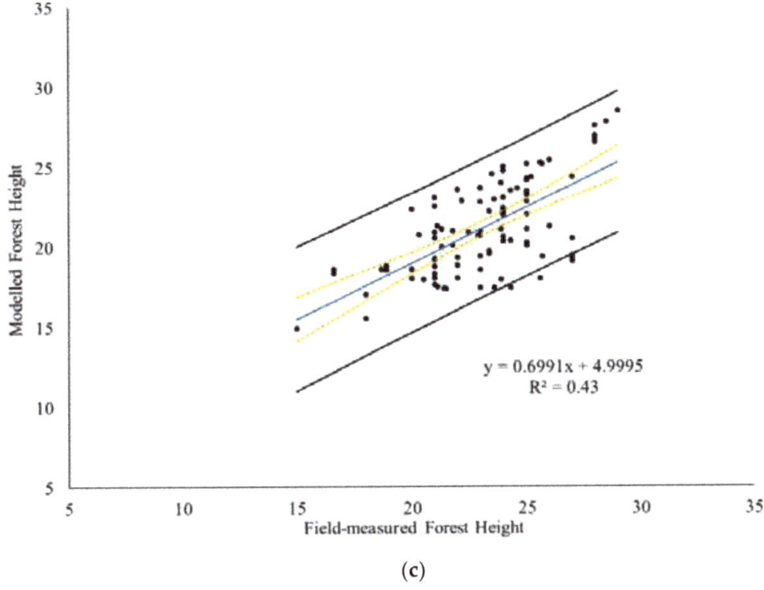

(c)

Figure 12. Linear regression with 95% confidence interval for TSI-based inversion modelling of (**a**) L-band SAR data, (**b**) C-band SAR data, and (**c**) X-band SAR data.

CAI-based forest height maps of multifrequency PolInSAR data are shown in Figure 13. The forest height values of modelled output varied from 0 m to 30 m in the TSI-based inversion. CAI-based PolInSAR vegetation height values ranged from 0 m to 33 m. The best-fit regression line of CAI-based forest height for ALOS-2 PALSAR-2, RADARSAT-2, and TerraSAR-X data fell between the upper and lower limits of the 95% confidence interval. Four datapoints of the linear regression with CAI modelling for ALOS-2 PALSAR-2 fell outside of the prediction interval boundary with the 95% confidence interval. Two datapoints for RADARSAT-2 and three for TerraSAR-X fell outside of the prediction interval boundary.

Figure 13 shows the relation between the field-based forest height and CAI-based modelled height. A very low coefficient of determination was obtained from CAI-based height retrieval for all the PolInSAR data. CAI-based PolInSAR inversion for forest height retrieval from ALOS-2 PALSAR-2 showed an R^2 of 0.04. No correlation was obtained between modelled forest height of RADARSAT-2 data and field-measured forest height. The relationship between field-measured forest height and CAI-based forest height of TerraSAR-X data showed an R^2 of 0.05. A comparison of CAI-based PolInSAR inversion with field-measured forest height gave an RMSE of 4.48 m for ALOS-2 PALSAR-2, 3.88 m for RADARSAT-2 and 4.90 m for TerraSAR-X data. The standard error of the estimate (SE) were 2.53 m for ALOS-2 PALSAR-2, 2.09 m for RADARSAT-2, and 3.91 m for TerraSAR-X data.

Global forest canopy height map-based vegetation height values for the field-measured plot locations were retrieved to make a comparison as shown in Figure 14. For the 100 plot locations, a comparison of vegetation height values of field-measured forest heights, global forest canopy height-based measurements, and PolInSAR-based height values of TSI and CAI models for multifrequency spaceborne SAR is shown in Figure 15. Figure 14 shows the best-fit regression line for global forest canopy height map-based vegetation height was between the 95% confidence interval boundaries, and three points fell outside of the prediction interval boundaries. Figure 15 shows that the vegetation height for plot numbers 45, 71, and 94 were underestimated in the global forest canopy height map. The field-based measurements for the forest height for plots 45, 71, and 94 were 24 m, 25 m, and 26 m, respectively, and their corresponding values in the global forest canopy height map were 0 m, 5 m, and 8 m. The linear regression between field-measured and global forest canopy height-based forest height shows a very low coefficient of determination (0.0022) with 5.82 m RMSE and 5.33 m SE.

Table 3. Comparison of forest height values obtained from different spaceborne earth observation missions.

Mission	Approach	Coefficient of Determination (R^2)	Root Mean Square Error (RMSE)	Standard Error (SE)	*p*-Value with 95% Confidence Level
ALOS-2	TSI	0.53	2.87 m	1.56 m	1.73×10^{-17}
PALSAR-2	CAI	0.04	4.48 m	2.53 m	0.040
RADARSAT-2	TSI	0.32	3.74 m	1.73 m	1.22×10^{-9}
	CAI	0.00	3.88 m	2.09 m	0.906
TerraSAR-X	TSI	0.43	4.53 m	2.17 m	2.08×10^{-13}
	CAI	0.05	4.90 m	3.91 m	0.019
GEDI	Integration of GEDI-derived canopy height with Landsat timeseries data	0.0022	5.82 m	5.33 m	0.644

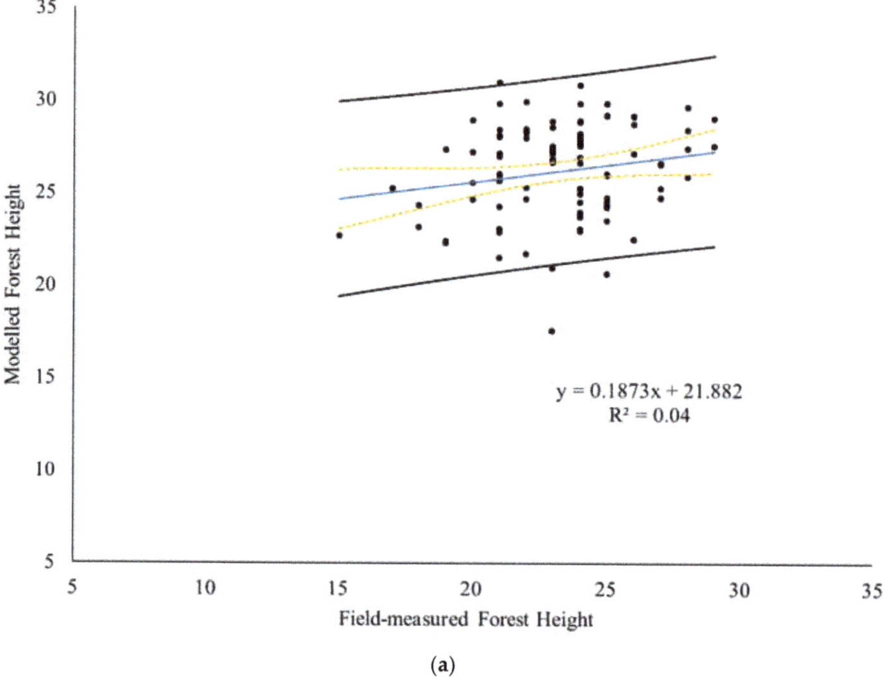

(**a**)

Figure 13. *Cont.*

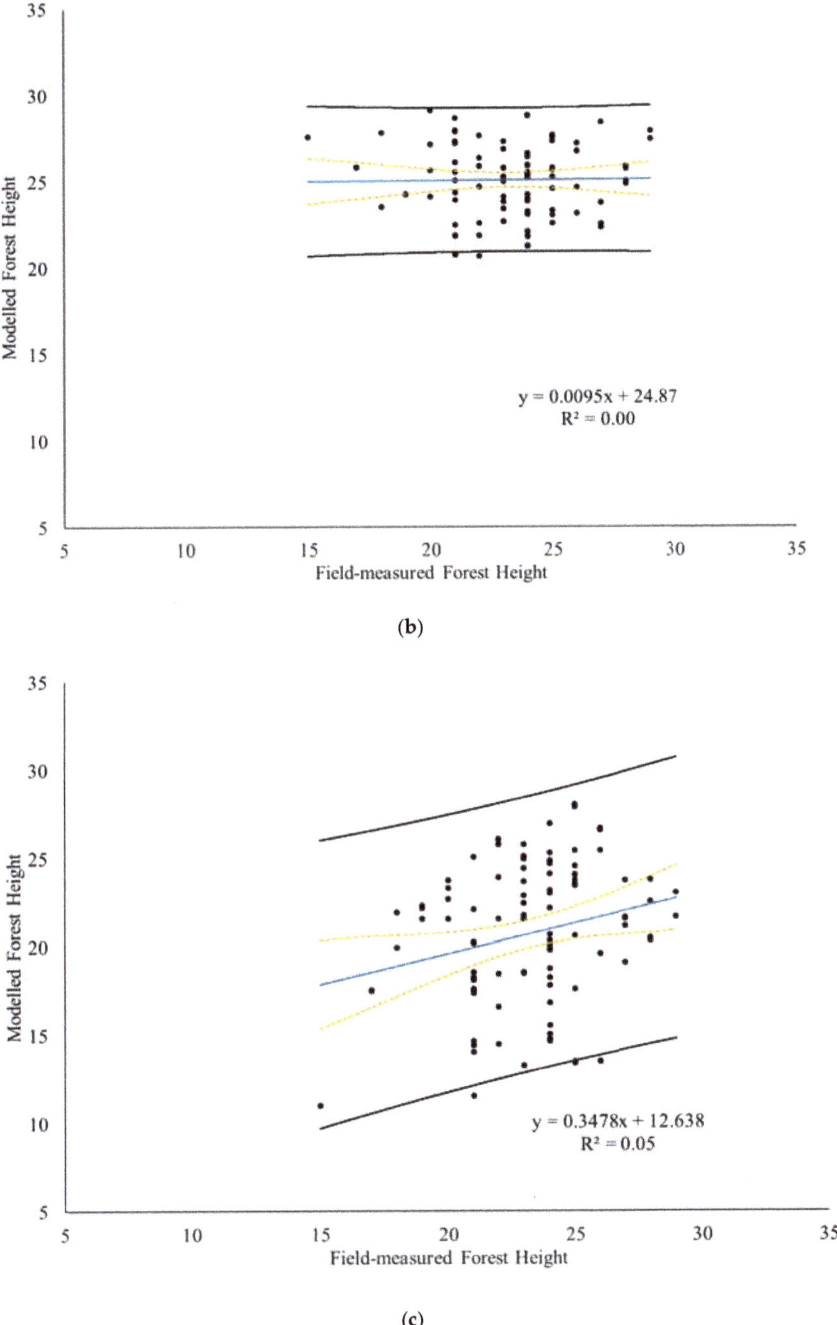

(b)

(c)

Figure 13. Linear regression with 95% confidence interval for CAI-based inversion modelling of (a) L-band SAR data, (b) C-band SAR data, and (c) X-band SAR data.

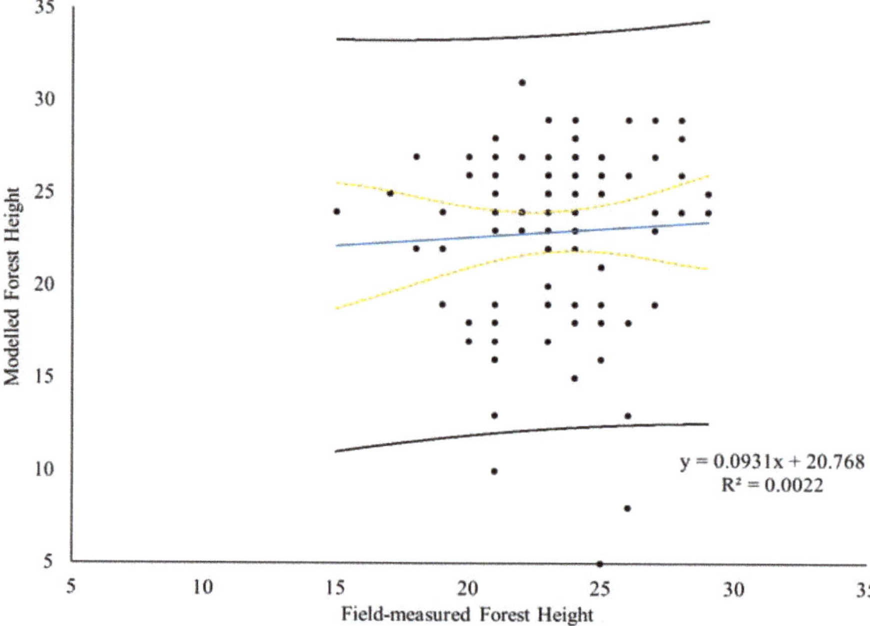

Figure 14. Linear regression with 95% confidence interval between field-measured and global forest canopy height map-based forest height.

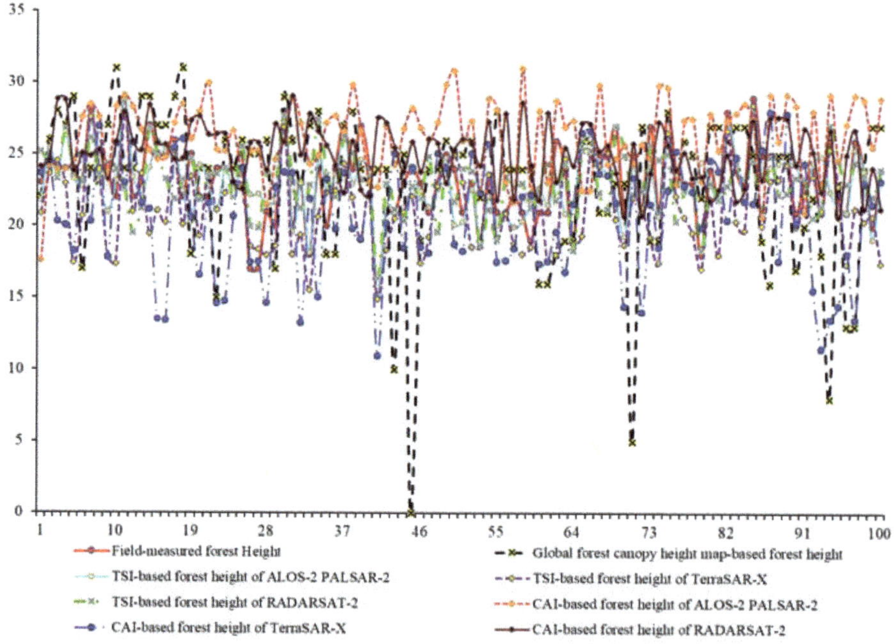

Figure 15. A comparison of forest height values of field-measured, PolInSAR-derived, and global canopy height map.

A comparison of forest height estimates of different spaceborne missions is shown in Table 3. The model-estimated forest heights were statistically tested for their significance using p-values, which indicate the probability of a result being observed. For a statistically significant observation, the p-value should be smaller than the significance level. The smaller p-value shows that there was a significant relation between field-measured and model-estimated forest height values. Higher p-values (near to 1) indicate observations by chance [73]. In Table 3, a very low p-value was shown by most of the forest height observations except for the GEDI mission and RADARSAT-2-based CAI model. The p-value with a 95% confidence level for the GEDI mission and RADARSAT-2-based CAI model was greater than the significance level (0.05). It is visible in Figure 13 and Table 3 that the R^2 between CAI-based forest height and field-measured forest height was very low for L-, C-, and X-band PolInSAR data. The RMSE of ALOS-2 PALSAR-2, RADARSAT-2, and TerraSAR-X were 4.48 m, 3.88 m, and 4.90 m, respectively. The standard errors of CAI model estimates with L-, C-, and X- band data were 2.53 m, 2.09 m, and 3.91 m, respectively. It is evident from Table 3, Figure 12, and Figure 13 that TSI-based PolInSAR inversion provided better R^2, RMSE, SE, and p-values in forest height retrieval in comparison to the CAI-based modelling approach

5. Discussion

Nowadays, polarimetric SAR data acquisition in interferometric mode shows a great potential in modelling to estimate forest parameters [74–76]. Several studies have been carried out for SAR image-based forest height estimation and most successfully demonstrated airborne systems [42,77,78]. Very few studies have studied spaceborne SAR for forest height retrieval compared to aerial SAR systems [48,55,58]. An automated framework was proposed by Lie et al. (2021) to retrieve the vegetation height of a tropical forest using spaceborne TanDEM-X SAR interferometry with 2–3 m accuracy [79]. Lee and Fatoyinbo (2015) performed the PolInSAR inversion technique and retrieved mangrove height using X-band TanDEM-X data with a good correlation coefficient (0.851–0.919) and low RMSE values (1.069–1.727 m) [80]. An investigation was done by Kugler et al. (2014) to find the suitability of X-band TanDEM-X data for PolInSAR inversion-based vegetation height retrieval of a boreal, a temperate, and a tropical site [81]. In the investigation, it was found that the modelled vegetation height for boreal forest showed a very good result with 0.91 coefficient of determination and a low RMSE (1.58 m). Linear regression between the lidar-based and PolInSAR-model-based vegetation height for the temperate and tropical forest showed a 0.89 coefficient of determination with 2.3 m RMSE and a 0.98 coefficient of determination with 2.1 m RMSE [81]. Several other studies were also conducted in which highly accurate results were reported by the researchers [68,82–85].

In this study, field data were collected for 100 plot locations and used to validate the forest height values of multifrequency spaceborne SAR data and GEDI-derived canopy height. The field data acquisition was done in 2014 and 2015. Spaceborne SAR-system-based L-band ALOS-2 PALSAR-2 and X-band TerraSAR-X data were acquired in 2015 and RADARSAT-2 data were acquired in 2014. Due to the deep penetration capacity of L-band SAR, it was easily possible to detect PolInSAR complex coherences contributed by ground and volume scatterers. The lowest RMSE and SE and highest coefficient of determination were obtained from TSI model-based forest height of ALOS-2 PALSAR-2 data. RADARSAT-2 data also showed reasonable results with its limitation of less penetration capability of the SAR system. The reason behind the reasonably good result of the C-band RADARSAT-2 is the interferometric coherence, which was preserved for the study area due to interferometric data acquisition with a very short baseline. The very short baseline data may be good to preserve interferometric coherence, but the altitude of ambiguity was very high. To resolve this issue, the vertical wavenumber was simulated with an appropriate altitude of ambiguity according to the maximum forest height of the study area. Due to precise complex coherence and simulated vertical baseline, it became possible to obtain reasonably good results from the C-band SAR data. A comparison of results of all the sensors shows that a high RMSE and SE and a reasonably good coefficient of determination with a significant p-value was obtained from the X-band TerraSAR-X SAR system. The reason behind the reasonably acceptable results for the X-band SAR is its spatial resolution, which was highest among all the spaceborne SAR sensors used in this study. The PolInSAR TerraSAR-X data was acquired with a range resolution of 1.36 m and Azimuth resolution of 2.86 m, which enables the sensor to detect ground information through canopy gaps. The analysis of forest height values of spaceborne missions suggests that vegetation height values from the TSI-based model have greater significance for this forest area than other approaches. The complex coherence optimization process of the TSI approach to find ground and top of the vegetation coherence is more rigorous and authentic in comparison to the CAI method as shown in Equations (10) and (11).

The integrated GEDI-derived canopy height with Landsat timeseries data showed a very low coefficient of determination (0.0022) with a high RMSE (5.81 m) and SE (5.33 m). The *p*-value was also very high, which does not indicate a significant relationship between the vegetation height values of field-measured and GEDI-derived estimates. The main reason behind this may be a large temporal gap (approximately 4 years) between the field data collection and GEDI product. The field data was acquired in 2014 and the GEDI-derived canopy height map was generated in 2019. The error was highest at three plot locations, which were plot numbers 45, 71, and 94. The field-measured values for these locations were 24 m for plot number 45, 25 m for plot number 71, and 26 m for plot number 94, and their respective values in GEDI-derived product were 0 m, 5 m, and 8 m. To understand the reason why the error was high, high-resolution timeseries images of Google Earth were used to visualize the point locations. It is visible in the Google Earth image (Figure 16) of the study area that plot number 45, for which the vegetation height value in GEDI-derived global forest map was assigned as zero, lies in the dense forest with a height of 24 m in the field-measured data.

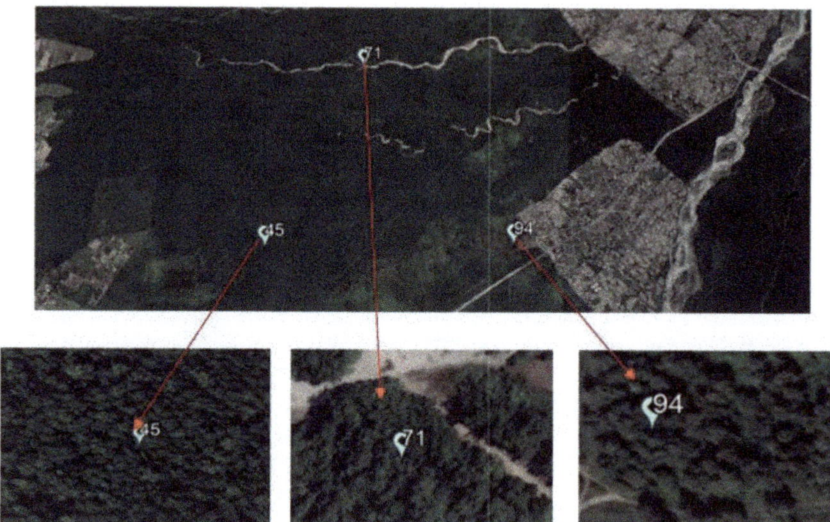

Figure 16. Google Earth view of plot numbers 45, 71, and 94.

6. Conclusions

The prime focus of the study was to evaluate the PolInSAR data for forest height retrieval. ALOS-2 PALSAR-2, RADARSAT-2, and TerraSAR-X acquired multifrequency spaceborne PolInSAR data in L-, C-, and X-band. It was found that the PolInSAR coherence has the capability to characterize the features based on the coherence contribution due to different types of scatters in the acquired area. The Polarimetric Interferometric SAR (PolInSAR) technique provides interferometric dataset pairs with multiple polarizations, which have the potential to accurately estimate the forest height. In contrast to the normal interferometric SAR datasets, the availability of multiple polarizations in the PolInSAR datasets helps to optimize the interferometric coherence using the combination of different polarization channels for accurate scattering mechanism retrieval from the ground targets. The optimization of the interferometric coherence maximizes the quality of interferograms and phase information required for accurate forest height retrieval. The multifrequency PolInSAR datasets were analyzed in this research to assess the potential of these datasets for accurate forest height retrieval. GEDI-derived global forest canopy map-based vegetation height values were also tested against the field-measured height values. The vegetation height values of GEDI-derived product showed mismatches with field measurement at several locations, and consequently, a high RMSE (5.82 m) and SE (5.33 m) and a very low R^2 (0.0022) were obtained through the linear regression. In the field-measured vegetation height and GEDI-derived estimates, there was a gap of approximately 4–5 years. Due to this large temporal gap and the mismatch in height values at several locations, the *p*-value, with a 95% confidence level, had not shown a significant relation with the field data. The TSI and CAI modelling approaches were implemented on monostatic PolInSAR data to generate forest height maps. The statistical analysis of the modelled vegetation height showed the supremacy of the TSI approach over other techniques. By performing the Three-Stage Inversion (TSI) technique for forest height estimation using the L-band, C-band, and X-band PolInSAR datasets, it was found that the ALOS-2 PALSAR-2 L-band dataset was able to provide a better tree height estimate with an R^2 value of 0.53 for the linear regression between the modelled tree height and field data. The R^2 values obtained for the TSI model with C-band and X-band datasets were 0.32 and 0.43, respectively. The mean error analysis of the tree height estimated from the TSI modelling and field data showed that, for the L-band ALOS-2 PALSAR-2 dataset, the error was 2.87 m; for the C-band RADARSAT-2 dataset, the error was 3.74 m; and for the X-band TerraSAR-X dataset, the mean error was 4.53 m. The L-band TSI model-based forest height showed the most reliable result with low RMSE (2.87 m), relatively higher R^2 (0.53), and a significant *p*-value (1.73×10^{-17}). The CAI-based PolInSAR inversion showed a very low coefficient of determination between model-derived forest height and field-measured forest height. The *p*-value analysis of the CAI model-based vegetation height showed marginal significant relation between field-measured and model-estimated forest height values for L- and X-band SAR data. No significant relationships between field-measured and CAI model-estimated forest height were obtained for C-band SAR data. The *p*-value of CAI model-based vegetation height for RADARSAT-2 data was greater than the significance level. This trend indicates that the TSI model is best suited to correctly assess forest height. Whereas the study area of the Doon Valley forest ranges was mainly dominated by saline trees, future research could include a variety of vegetation and forest types for PolInSAR-model-based forest elevation retrieval under different geographic and climatic conditions.

Author Contributions: Conceptualization, S.K. and S.P.S.K.; methodology, S.K.; software, S.K. and S.P.S.K.; validation, S.K., and H.G.; formal analysis, S.K.; H.G. investigation, writing—original draft preparation, S.K.; writing—review and editing, S.K., S.P.S.K., H.G., P.K.S., and P.K.T., satellite data acquisition, S.K., P.K.T., and field data S.K., S.P.S.K.; visualization, S.K., H.G.; funding acquisition, P.K.S., H.G. All authors have read and agreed to the published version of the manuscript.

Funding: This research received no external funding.

Acknowledgments: The authors would like to express their sincere gratitude to the whole research team of ESA for providing SNAP 6.0 and PolSARPro 4.2 tools for polarimetric processing of the SAR data. The authors are thankful to the German Aerospace Center (DLR) Oberpfaffenhofen for providing TerraSAR-X/TanDEM-X data under the Project Id -NTI_POLI6635 on PolInSAR Tomography for aboveground biomass estimation and the Japan Aerospace Exploration Agency (JAXA) for providing the L-band ALOS-2 PALSAR-2 datasets under the proposal number 1408 of RA4 with the title Hydrological parameter retrieval and glacier dynamics study with L-band SAR data. The Technology Development Programmes (TDP) of the Indian Space Research Organisation (ISRO) supported this research. The RADARSAT-2 data were purchased from the MDA corporation under the ISRO TDP. The research was carried under a TDP code: Z1Z12022E301-303 on the development of semi-empirical/numerical models for biophysical characterization of tropical forests using spaceborne PolSAR and PolInSAR techniques. The authors are thankful to the Indian Institute of Remote Sensing (IIRS), ISRO for providing all the support to carry out this research work. Valuable suggestions from Shree Kamal Pandey, Scientist, Geoweb services, IT & Distance Learning Department, IIRS, and Shree Bhaskar R. Nikam, Scientist, Water Resources Department, IIRS, helped in analysing the modelling output.

Conflicts of Interest: The authors declare no conflict of interest.

References

1. Chirici, G.; Chiesi, M.; Corona, P.; Salvati, R.; Papale, D.; Fibbi, L.; Sirca, C.; Spano, D.; Duce, P.; Marras, S.; et al. Estimating daily forest carbon fluxes using a combination of ground and remotely sensed data. *J. Geophys. Res. Biogeosci.* **2016**, *121*, 266–279. [CrossRef]
2. Adloff, M.; Reick, C.H.; Claussen, M. Earth system model simulations show different feedback strengths of the terrestrial carbon cycle under glacial and interglacial conditions. *Earth Syst. Dyn.* **2018**, *9*, 413–425. [CrossRef]
3. Zhang, S.; Pang, B.; Zhang, Z. Carbon footprint analysis of two different types of hydropower schemes: Comparing earth-rockfill dams and concrete gravity dams using hybrid life cycle assessment. *J. Clean. Prod.* **2015**, *103*, 854–862. [CrossRef]
4. Shao, P.; Zeng, X.; Sakaguchi, K.; Monson, R.K.; Zeng, X. Terrestrial carbon cycle: Climate relations in eight CMIP5 earth system models. *J. Clim.* **2013**, *26*, 8744–8764. [CrossRef]
5. Falkowski, P.; Scholes, R.J.; Boyle, E.; Canadell, J.; Canfield, D.; Elser, J.; Gruber, N.; Hibbard, K.; Hogberg, P.; Linder, S.; et al. The global carbon cycle: A test of our knowledge of earth as a system. *Science* **2000**, *290*, 291–296. [CrossRef]
6. Jones, A.D.; Calvin, K.V.; Shi, X.; Di Vittorio, A.V.; Bond-Lamberty, B.; Thornton, P.E.; Collins, W.D. Quantifying Human-Mediated Carbon Cycle Feedbacks. *Geophys. Res. Lett.* **2018**, *45*, 11370–11379. [CrossRef]
7. Chen, H.; Zhu, Q.; Peng, C.; Wu, N.; Wang, Y.; Fang, X.; Gao, Y.; Zhu, D.; Yang, G.; Tian, J.; et al. The impacts of climate change and human activities on biogeochemical cycles on the Qinghai-Tibetan Plateau. *Glob. Chang. Biol.* **2013**, *19*, 2940–2955. [CrossRef]
8. Raupach, M.R. Ecosystem Services and the Global Carbon Cycle. In *Ecosystem Services and Carbon Sequestration in the Biosphere*; Lal, R., Lorenz, K., Hüttl, R.F., Schneider, B.U., von Braun, J., Eds.; Springer: Dordrecht, The Netherlands, 2013; pp. 155–181. ISBN 978-94-007-6455-2.
9. Van Leeuwen, M.; Nieuwenhuis, M. Retrieval of forest structural parameters using LiDAR remote sensing. *Eur. J. For. Res.* **2010**, *129*, 749–770. [CrossRef]
10. Kangas, A.; Astrup, R.; Breidenbach, J.; Fridman, J.; Gobakken, T.; Korhonen, K.T.; Maltamo, M.; Nilsson, M.; Nord-Larsen, T.; Næsset, E.; et al. Remote sensing and forest inventories in Nordic countries–roadmap for the future. *Scand. J. For. Res.* **2018**, *33*, 397–412. [CrossRef]
11. Gleason, C.; Im, J. A review of remote sensing of forest biomass and biofuel: Options for small-area applications. *GIScience Remote Sens.* **2011**, *48*, 141–170. [CrossRef]
12. Chen, X.; An, S.; Chen, J.; Liu, Y.; Xu, C.; Yang, H. A review on forest ecosystem biophysical parameter retrieval from remotely sensed data. *Chin. J. Ecol.* **2005**, *24*, 1074–1079.
13. Zhao, J.; Li, J.; Liu, Q. Review of forest vertical structure parameter inversion based on remote sensing technology. *Yaogan Xuebao/J. Remote Sens.* **2013**, *17*, 697–716.
14. Lutz, D.A.; Washington-Allen, R.A.; Shugart, H.H. Remote sensing of boreal forest biophysical and inventory parameters: A review. *Can. J. Remote Sens.* **2008**, *34*, S286–S313. [CrossRef]
15. Story, M.; Polcyn, F.; Imhoff, M.; Vermillion, C.; Khan, F. Forest Canopy Characterization and Vegetation Penetration Assessment with Space-Borne Radar. *IEEE Trans. Geosci. Remote Sens.* **1986**, *GE-24*, 535–542.
16. Beaudoin, A.; Le Toan, T.; Goze, S.; Nezry, E.; Lopes, A.; Mougin, E.; Hsu, C.C.; Han, H.C.; Kong, J.A.; Shin, R.T. Retrieval of forest biomass from SAR data. *Int. J. Remote Sens.* **1994**, *15*, 2777–2796. [CrossRef]
17. Luckman, A.; Baker, J.; Kuplich, T.M.; Corina da Costa, F.Y.; Alejandro, C.F. A study of the relationship between radar backscatter and regenerating tropical forest biomass for spaceborne SAR instruments. *Remote Sens. Environ.* **1997**, *60*, 1–13. [CrossRef]
18. Luckman, A.J. Correction of SAR imagery for variation in pixel scattering area caused by topography. *IEEE Trans. Geosci. Remote Sens.* **1998**, *36*, 344–350. [CrossRef]
19. Kumar, S.; Pandey, U.; Kushwaha, S.P.; Chatterjee, R.S.; Bijker, W. Aboveground biomass estimation of tropical forest from Envisat advanced synthetic aperture radar data using modeling approach. *J. Appl. Remote Sens.* **2012**, *6*, 063588. [CrossRef]
20. Main, R.; Mathieu, R.; Kleynhans, W.; Wessels, K.; Naidoo, L.; Asner, G.P. Hyper-temporal C-band SAR for baseline woody structural assessments in deciduous savannas. *Remote Sens.* **2016**, *8*, 661. [CrossRef]

21. Vaglio, G.L.; Pirotti, F.; Callegari, M.; Chen, Q.; Cuozzo, G.; Lingua, E.; Notarnicola, C.; Papale, D. Potential of ALOS2 and NDVI to estimate forest above-ground biomass, and comparison with lidar-derived estimates. *Remote Sens.* **2017**, *9*, 18. [CrossRef]

22. Ningthoujam, R.K.; Balzter, H.; Tansey, K.; Morrison, K.; Johnson, S.C.M.; Gerard, F.; George, C.; Malhi, Y.; Burbidge, G.; Doody, S.; et al. Airborne S-band SAR for forest biophysical retrieval in temperate mixed forests of the UK. *Remote Sens.* **2016**, *8*, 609. [CrossRef]

23. Bharadwaj, P.S.; Kumar, S.; Kushwaha, S.P.S.; Bijker, W. Polarimetric scattering model for estimation of above ground biomass of multilayer vegetation using ALOS-PALSAR quad-pol data. *Phys. Chem. Earth Parts A/B/C* **2015**, *83–84*, 187–195. [CrossRef]

24. Solberg, S.; Næsset, E.; Gobakken, T.; Bollandsås, O.-M. Forest biomass change estimated from height change in interferometric SAR height models. *Carbon Balance Manag.* **2014**, *9*, 1–12. [CrossRef] [PubMed]

25. Lefsky, M.A.; Harding, D.J.; Keller, M.; Cohen, W.B.; Carabajal, C.C.; Bom Espirio-Santo, F.; Hunter, M.O.; Oliveira, R. Estimates of forest canopy height and aboveground biomass using ICESat. *Geophys. Res. Lett.* **2005**, *32*. [CrossRef]

26. Wang, M.; Sun, R.; Xiao, Z. Estimation of forest canopy height and aboveground biomass from spaceborne LiDAR and Landsat imageries in Maryland. *Remote Sens.* **2018**, *10*, 344. [CrossRef]

27. Wang, X.; Ouyang, S.; Sun, O.J.; Fang, J. Forest biomass patterns across northeast China are strongly shaped by forest height. *For. Ecol. Manage.* **2013**, *293*, 149–160. [CrossRef]

28. Treuhaft, R.N.; Madsen, S.N.; Moghaddam, M.; van Zyl, J.J. Vegetation characteristics and underlying topography from interferometric radar. *Radio Sci.* **1996**, *31*, 1449–1485. [CrossRef]

29. Balzter, H.; Rowland, C.S.; Saich, P. Forest canopy height and carbon estimation at Monks Wood National Nature Reserve, UK, using dual-wavelength {SAR} interferometry. *Remote Sens. Environ.* **2007**, *108*, 224–239. [CrossRef]

30. Askne, J.I.H.; Dammert, P.B.G.; Ulander, L.M.H.; Smith, G.S. C-Band Repeat-Pass Interferometric SAR Observations of the Forest. *IEEE Trans. Geosci. Remote Sens.* **1997**, *35*, 25–35. [CrossRef]

31. Tebaldini, S. Single and multipolarimetric SAR tomography of forested areas: A parametric approach. *IEEE Trans. Geosci. Remote Sens.* **2010**, *48*, 2375–2387. [CrossRef]

32. Aguilera, E.; Nannini, M.; Reigber, A. Wavelet-based compressed sensing for SAR tomography of forested areas. *IEEE Trans. Geosci. Remote Sens.* **2013**, *51*, 5283–5295. [CrossRef]

33. Nannini, M.; Scheiber, R.; Horn, R.; Moreira, A. First 3-D reconstructions of targets hidden beneath foliage by means of polarimetric SAR tomography. *IEEE Geosci. Remote Sens. Lett.* **2012**, *9*, 60–64. [CrossRef]

34. Kumar, S.; Joshi, S.K. SAR Tomography for forest structure investigation. In Proceedings of the Asia Pacific Microwave Conference 2016, New Delhi, India, 5–9 December 2016; pp. 1–4.

35. Xue, F.; Wang, X.; Xu, F.; Wang, Y. Polarimetric SAR Interferometry: A Tutorial for Analyzing System Parameters. *IEEE Geosci. Remote Sens. Mag.* **2020**, *8*, 83–107. [CrossRef]

36. Cloude, S.R.; Papathanassiou, K.P. Polarimetric SAR interferometry. *IEEE Trans. Geosci. Remote Sens.* **1998**, *36*, 1551–1565. [CrossRef]

37. Kumar, S. *PolInSAR and PolTomSAR based Modelling for Characterization of Forest Parameters*; Indian Institute of Technology: Roorkee, India, 2019.

38. Garestier, F.; Dubois-Fernandez, P.; Champion, I.; Toan, T. Le Pine forest investigation using high resolution P-band Pol-InSAR data. *Remote Sens. Environ.* **2011**, *115*, 2897–2905. [CrossRef]

39. Le Toan, T.; Quegan, S.; Davidson, M.W.J.; Balzter, H.; Paillou, P.; Papathanassiou, K.; Plummer, S.; Rocca, F.; Saatchi, S.; Shugart, H.; et al. The BIOMASS mission: Mapping global forest biomass to better understand the terrestrial carbon cycle. *Remote Sens. Environ.* **2011**, *115*, 2850–2860. [CrossRef]

40. Oveisgharan, S.; Saatchi, S.S.; Hensley, S. Sensitivity of Pol-InSAR Measurements to Vegetation Parameters. *IEEE Trans. Geosci. Remote Sens.* **2015**, *53*, 6561–6572. [CrossRef]

41. Cloude, S.R.; Papathanassiou, K.P. Three-stage inversion process for polarimetric SAR interferometry. *IEEE Proc. Radar Sonar Navig.* **2003**, *150*, 125–134. [CrossRef]

42. Neumann, M.; Ferro-Famil, L.; Reigber, A. Estimation of forest structure, ground, and canopy layer characteristics from multibaseline polarimetric interferometric SAR data. *IEEE Trans. Geosci. Remote Sens.* **2010**, *48*, 1086–1104. [CrossRef]

43. Asopa, U.; Kumar, S. UAVSAR Tomography for Vertical Profile Generation of Tropical Forest of Mondah National Park, Gabon. *Earth Sp. Sci.* **2020**, *7*, e2020EA001230. [CrossRef]

44. Papathanassiou, K.P.; Cloude, S.R. Single-baseline polarimetric SAR interferometry. *IEEE Trans. Geosci. Remote Sens.* **2001**, *39*, 2352–2363. [CrossRef]

45. Garestier, F.; Dubois-Fernandez, P.C.; Papathanassiou, K.P. Pine forest height inversion using single-pass X-band PolInSAR data. *IEEE Trans. Geosci. Remote Sens.* **2008**, *46*, 59–68. [CrossRef]

46. Neumann, M.; Saatchi, S.S.; Ulander, L.M.H.; Fransson, J.E.S. Assessing Performance of L- and P-Band Polarimetric Interferometric SAR Data in Estimating Boreal Forest Above-Ground Biomass. *IEEE Trans. Geosci. Remote Sens.* **2012**, *50*, 714–726. [CrossRef]

47. Shiroma, G.H.X.; de Macedo, K.A.C.; Wimmer, C.; Moreira, J.R.; Fernandes, D. The Dual-Band PolInSAR Method for Forest Parametrization. *IEEE J. Sel. Top. Appl. Earth Obs. Remote Sens.* **2016**, *9*, 3189–3201. [CrossRef]

48. Kumar, S.; Khati, U.G.; Chandola, S.; Agrawal, S.; Kushwaha, S.P.S. Polarimetric SAR Interferometry based modeling for tree height and aboveground biomass retrieval in a tropical deciduous forest. *Adv. Sp. Res.* **2017**, *60*, 571–586. [CrossRef]

49. Liao, Z.; He, B.; van Dijk, A.I.J.M.; Bai, X.; Quan, X. The impacts of spatial baseline on forest canopy height model and digital terrain model retrieval using P-band PolInSAR data. *Remote Sens. Environ.* **2018**, *210*, 403–421. [CrossRef]

50. Kugler, F.; Lee, S.-K.; Hajnsek, I.; Papathanassiou, K.P. Forest Height Estimation by Means of Pol-InSAR Data Inversion: The Role of the Vertical Wavenumber. *IEEE Trans. Geosci. Remote Sens.* **2015**, *53*, 5294–5311. [CrossRef]

51. Erten, E.; Lopez-Sanchez, J.M.; Yuzugullu, O.; Hajnsek, I. Retrieval of agricultural crop height from space: A comparison of SAR techniques. *Remote Sens. Environ.* **2016**, *187*, 130–144. [CrossRef]

52. López-Martínez, C.; Fàbregas, X.; Pipia, L. Forest parameter estimation in the Pol-InSAR context employing the multiplicative–additive speckle noise model. *ISPRS J. Photogramm. Remote Sens.* **2011**, *66*, 597–607. [CrossRef]

53. Lopez-Sanchez, J.M.; Vicente-Guijalba, F.; Erten, E.; Campos-Taberner, M.; Garcia-Haro, F.J. Retrieval of vegetation height in rice fields using polarimetric SAR interferometry with TanDEM-X data. *Remote Sens. Environ.* **2017**, *192*, 30–44. [CrossRef]

54. Stebler, O.; Meier, E.; Nüesch, D. Multi-baseline polarimetric {SAR} interferometry—First experimental spaceborne and airborne results. *ISPRS J. Photogramm. Remote Sens.* **2002**, *56*, 149–166. [CrossRef]

55. Zhang, Y.; He, C.; Xu, X.; Chen, D. Forest vertical parameter estimation using PolInSAR imagery based on radiometric correction. *ISPRS Int. J. Geo-Inf.* **2016**, *5*, 186. [CrossRef]

56. Bao, Z.; Guo, R.; Suo, Z.; Lu, H. S-RVoG model for forest parameters inversion over underlying topography. *Electron. Lett.* **2013**, *49*, 618–620.

57. Wang, C.; Wang, L.; Fu, H.; Xie, Q.; Zhu, J. The Impact of Forest Density on Forest Height Inversion Modeling from Polarimetric InSAR Data. *Remote Sens.* **2016**, *8*, 291. [CrossRef]

58. Kumar, S.; Garg, R.D.; Kushwaha, S.P.S.; Jayawardhana, W.G.N.N.; Agarwal, S. Bistatic PolInSAR Inversion Modelling for Plant Height Retrieval in a Tropical Forest. *Proc. Natl. Acad. Sci. India Sect. A Phys. Sci.* **2017**, *87*, 817–826. [CrossRef]

59. Kumar, S.; Sara, R.; Singh, J.; Agrawal, S.; Kushwaha, S.P.S. Spaceborne PolInSAR and ground-based TLS data modeling for characterization of forest structural and biophysical parameters. *Remote Sens. Appl. Soc. Environ.* **2018**, *11*, 241–253. [CrossRef]

60. Hansen, M.; Kommareddy, I. Global Forest Canopy Height. 2019. Available online: https://glad.umd.edu/dataset/gedi (accessed on 22 November 2020).

61. Potapov, P.; Li, X.; Hernandez-Serna, A.; Tyukavina, A.; Hansen, M.C.; Kommareddy, A.; Pickens, A.; Turubanova, S.; Tang, H.; Silva, C.E.; et al. Mapping global forest canopy height through integration of GEDI and Landsat data. *Remote Sens. Environ.* **2020**, 112165. [CrossRef]

62. Moreira, A.; Prats-iraola, P.; Younis, M.; Krieger, G.; Hajnsek, I.; Papathanassiou, K.P. A tutorial on synthetic aperture radar. *IEEE Geosci. Remote Sens. Mag.* **2013**, *1*, 6–43. [CrossRef]

63. Yamaguchi, Y.; Sato, A.; Boerner, W.M.; Sato, R.; Yamada, H. Four-Component Scattering Power Decomposition With Rotation of Coherency Matrix. *IEEE Trans. Geosci. Remote Sens.* **2011**, *49*, 2251–2258. [CrossRef]

64. European Space Agency Sentinel Application Platform (SNAP) V 6.0. Available online: https://step.esa.int/main/toolboxes/snap/ (accessed on 14 September 2018).

65. Cloude, S.R.; Pottier, E. A Review of Target Decomposition Theorems in Radar Polarimetry. *IEEE Trans. Geosci. Remote Sens.* **1996**, *34*, 498–518. [CrossRef]

66. Treuhaft, R.N.; Siqueira, P.R. Vertical structure of vegetated land surfaces from interferometric and polarimetric radar. *Radio Sci.* **2000**, *35*, 141–177. [CrossRef]

67. Xie, Q.; Zhu, J.; Wang, C.; Fu, H.; Lopez-Sanchez, J.M.; Ballester-Berman, J.D. A modified dual-baseline PolInSAR method for forest height estimation. *Remote Sens.* **2017**, *9*, 819. [CrossRef]

68. Liao, Z.; He, B.; Bai, X.; Quan, X. Improving Forest Height Retrieval by Reducing the Ambiguity of Volume-Only Coherence Using Multi-Baseline PolInSAR Data. *IEEE Trans. Geosci. Remote Sens.* **2019**, *57*, 8856–8866. [CrossRef]

69. Wenxue, F.; Huadong, G.; Xinwu, L.; Bangsen, T.; Zhongchang, S. Extended Three-Stage Polarimetric SAR Interferometry Algorithm by Dual-Polarization Data. *IEEE Trans. Geosci. Remote Sens.* **2016**, *54*, 2792–2802. [CrossRef]

70. Khati, U.; Singh, G.; Kumar, S. Potential of Space-Borne PolInSAR for Forest Canopy Height Estimation Over India-A Case Study Using Fully Polarimetric L-, C-, and X-Band SAR Data. *IEEE J. Sel. Top. Appl. Earth Obs. Remote Sens.* **2018**, *11*, 2406–2416. [CrossRef]

71. Cloude, S.R. Pol-InSAR Training Course. Available online: https://earth.esa.int/documents/653194/656796/Pol-InSAR_Training_Course.pdf (accessed on 14 September 2020).

72. Cloude, S.R. Polarization coherence tomography. *Radio Sci.* **2006**, *41*, 1–27. [CrossRef]

73. Chekanov, S.V. Probability and Statistics. In *Numeric Computation and Statistical Data Analysis on the Java Platform*; Springer International Publishing: Cham, Switzerland, 2016; pp. 351–397. ISBN 978-3-319-28531-3.

74. Liao, Z.; He, B.; Quan, X.; van Dijk, A.I.J.M.; Qiu, S.; Yin, C. Biomass estimation in dense tropical forest using multiple information from single-baseline P-band PolInSAR data. *Remote Sens. Environ.* **2019**, *221*, 489–507. [CrossRef]

75. Denbina, M.; Simard, M.; Hawkins, B. Forest Height Estimation Using Multibaseline PolInSAR and Sparse Lidar Data Fusion. *IEEE J. Sel. Top. Appl. Earth Obs. Remote Sens.* **2018**, *11*, 3415–3433. [CrossRef]

76. Aghabalaei, A.; Ebadi, H.; Maghsoudi, Y. Forest height estimation based on the RVoG inversion model and the PolInSAR decomposition technique. *Int. J. Remote Sens.* **2020**, *41*, 2684–2703. [CrossRef]

77. Lee, S.-K.; Kugler, F.; Papathanassiou, K.P.; Hajnsek, I. Quantification of temporal decorrelation effects at L-band for polarimetric SAR interferometry applications. *IEEE J. Sel. Top. Appl. Earth Obs. Remote Sens.* **2013**, *6*, 1351–1367. [CrossRef]

78. Simard, M.; Denbina, M. An assessment of temporal decorrelation compensation methods for forest canopy height estimation using airborne L-band same-day repeat-pass polarimetric SAR interferometry. *IEEE J. Sel. Top. Appl. Earth Obs. Remote Sens.* **2018**, *11*, 95–111. [CrossRef]

79. Lei, Y.; Treuhaft, R.; Gonçalves, F. Automated estimation of forest height and underlying topography over a Brazilian tropical forest with single-baseline single-polarization TanDEM-X SAR interferometry. *Remote Sens. Environ.* **2021**, *252*, 112132. [CrossRef]

80. Lee, S.-K.; Fatoyinbo, T.E. TanDEM-X Pol-InSAR Inversion for Mangrove Canopy Height Estimation. *IEEE J. Sel. Top. Appl. Earth Obs. Remote Sens.* **2015**, *8*, 3608–3618. [CrossRef]

81. Kugler, F.; Schulze, D.; Hajnsek, I.; Pretzsch, H.; Papathanassiou, K.P. TanDEM-X Pol-InSAR performance for forest height estimation. *IEEE Trans. Geosci. Remote Sens.* **2014**, *52*, 6404–6422. [CrossRef]

82. Wang, X.; Xu, F. A PolinSAR Inversion Error Model on Polarimetric System Parameters for Forest Height Mapping. *IEEE Trans. Geosci. Remote Sens.* **2019**, *57*, 5669–5685. [CrossRef]

83. Sun, X.; Wang, B.; Xiang, M.; Jiang, S.; Fu, X. Forest height estimation based on constrained Gaussian Vertical Backscatter model using multi-baseline P-band Pol-InSAR data. *Remote Sens.* **2019**, *11*, 42. [CrossRef]

84. Brigot, G.; Simard, M.; Colin-Koeniguer, E.; Boulch, A. Retrieval of forest vertical structure from PolInSAR Data by machine learning using LIDAR-Derived features. *Remote Sens.* **2019**, *11*, 381. [CrossRef]

85. Berninger, A.; Lohberger, S.; Zhang, D.; Siegert, F. Canopy height and above-ground biomass retrieval in tropical forests using multi-pass X-and C-band Pol-InSAR data. *Remote Sens.* **2019**, *11*, 2105. [CrossRef]

Publisher's Note: MDPI stays neutral with regard to jurisdictional claims in published maps and institutional affiliations.

Article

Potential Lidar Height, Intensity, and Ratio Parameters for Plot Dominant Species Discrimination and Volume Estimation

Taejin Park [1,2]

[1] Bay Area Environmental Research Institute, Moffett Field, CA 94035, USA; tpark@baeri.org; Tel.: +1-617-893-1988

[2] NASA Ames Research Center, NASA, Moffett Field, CA 94035, USA

Received: 21 September 2020; Accepted: 7 October 2020; Published: 8 October 2020

Abstract: Precise stand species classification and volume estimation are key research topics for automated forest inventory. This study aims to explore the feasibility of light detection and ranging (lidar) height, intensity, and ratio parameters for discriminating dominant species (*Pinus densiflora*, *Larix kaempferi*, and *Quercus* spp.) and estimating volume at plot scale. To achieve these objectives, multiple linear discriminant and regression analyses were utilized after a separate selection of explanatory variables from extracted 38 lidar height, intensity, and ratio parameters. A kappa accuracy of 0.75 was achieved in discriminating the plot-dominant species from three different species by adopting a combination of nine selected explanatory variables. Further investigation found that dispersion and mean of lidar intensity within a plot are key classifiers of identifying three species. Species-specific optimal plot volume models for *Pinus densiflora, Larix kaempferi,* and *Quercus* spp. were evaluated by coefficients of determination of 0.71, 0.74, and 0.56, respectively. Compared to species classification, height-related lidar variables play a key role in modeling forest plot volume. Several explanatory variables for each modeling practice were correlated to canopy vertical and horizontal structures and were enough to represent species-specific characteristics in both approaches for species classification and plot volume estimation. Additionally, observed different variable combinations for two important applications imply that future studies should use proper variable combinations for each purpose.

Keywords: lidar; remote sensing; forest structure; stand volume; stand dominant species

---

1. Introduction

Precise and quantitative information of forests is essential for forest management and planning [1–3]. Forest stand volume is one of the most important structural variables characterizing economic and environmental value of the forest stand. As it can estimate stand biomass and carbon content, accurate stand volume estimation is critical for understanding forest carbon dynamics. Field surveying can provide accurate and extensive forest inventories nationwide; however, it is not only labor-intensive and time-consuming, but also difficult to seamlessly cover large forested area. Together with increasing needs of high quality and large-scale forest information, in this context, remote sensing has become a more powerful tool in forest management [4–7].

Laser surveying techniques known as light detection and ranging (lidar) have been widely used to characterize a large-scale three-dimensional forest structure and its dynamics [1,7–9]. Studies have proven the potential of lidar for estimating forest stand volume [10–12]. Approaches for estimating stand volume using lidar data can be categorized into (a) individual tree detection-based approaches [1,9,13] and (b) height distributional approaches at the stand or plot level [13,14]. The individual tree detection based approaches for estimating forest stand volume, tree height, and diameter at breast height (DBH) are required for calculating volume using an allometric function [1]. However, direct estimation of DBH is problematic due to stem concealment by obstructive upper canopy structures. To obtain more accurate DBH estimates, several studies have suggested ways to apply the statistical relationship between measured DBH and crown width derived from lidar data [13]. Although stand volume could be estimated by using height and DBH derived from lidar data, modeling procedures for DBH estimates and the necessity of accurately isolating individual trees tend to increase the uncertainty of stand volume estimates.

To avoid the complications and limitations of approaches based on individual tree detection, stand volume has been estimated directly using height distribution parameters of large- and small-footprint lidar systems at the stand or plot level [10,15,16]. This approach assumes that stand volume is closely related to actual vertical and horizontal forest structures and that lidar height distributions can represent the forest structures [15]. In other words, the forest structure from tree top to ground can be explained by structurally arranged large- and small-footprint lidar data [17]. In practice, large-footprint full-waveform lidar can provide height distribution data depending on the time-varying intensity of the returned energy within the laser pulse; however, this system is not appropriate for demonstrating finer scale forest attributes, including volume and biomass [18]. Small-footprint discrete lidar can produce a height distribution of forest vertical structures by accumulating all returns per sampling unit. This distribution then can be employed to estimate fine-scale and stand-level forest volume [15,19].

Tree species information is also critical for correctly valuing forests in terms of economic, ecological, and technical perspectives. Inaccurate species identification can result in prominent bias in the estimates of stem volume and biomass as allometric dependencies are species-specific [1,20]. The approach using height distribution data for volume estimation has been used to discriminate tree species at the individual tree and plot or stand level [21–23]. These species discrimination analyses have been successfully used to differentiate forest species between coniferous and broadleaf forests [24]. Both height and intensity data from lidar clearly show the potential to classify individual or stand dominant species by characterizing spectral, vertical, and horizontal profiles of canopy structure.

Reviewing the literature for related research reveals an opportunity to develop a sequential analytic framework that identify stand-dominant species and estimate stand volume. This study thus aims to achieve these two main objectives using height, intensity, and ratio parameters derived from airborne lidar data. Both objectives focused on the relationship between forest attributes and various lidar distributional characteristics at the plot scale, thus this study can provide insight of how different sets of lidar height, intensity, and ratio variables play a role in species classification and plot volume estimation. The detailed first objective is to identify explanatory parameters and their combination for classifying plot-dominant species between homogeneously distributed Korean red pine (*Pinus densiflora*), Japanese Larch (*Larix kaempferi*), and oaks (*Quercus* spp.) with determining statistical criteria. The second objective is to examine the appropriate parameters for estimating species-specific plot volume based on classification results under critical statistical criteria and select an optimal volume model for each plot-dominant species. Following results and possible limitations are compared and discussed.

2. Materials

2.1. Study Area

The study sites are located in Obin-ri (Coordinates: upper left 127°27′04.08″E, 37°31′15.80″N and lower right 127°29′22.89″E, 37°30′15.45″N) and at the Korea University Yangpyeong Experimental Forest (upper left 127°40′45.77″E, 37°30′56.22″N and lower right 127°43′19.11″E, 37°29′36.76″N), Yangpyeong-gun, central South Korea (Figure 1). The forests of the study region range from 21 m to 220 m above sea level. The main tree species in these sites are *Pinus densiflora*, *Larix kaempferi*, and *Quercus* spp. For this study, a total of 90 plots (30 plots for each species) were surveyed and investigated. Each 30 plots for each species were split into 20 training and 10 testing sites. The training sites were used for developing stand dominant species classification and volume estimation model, then the testing sites were used for verifying the developed models.

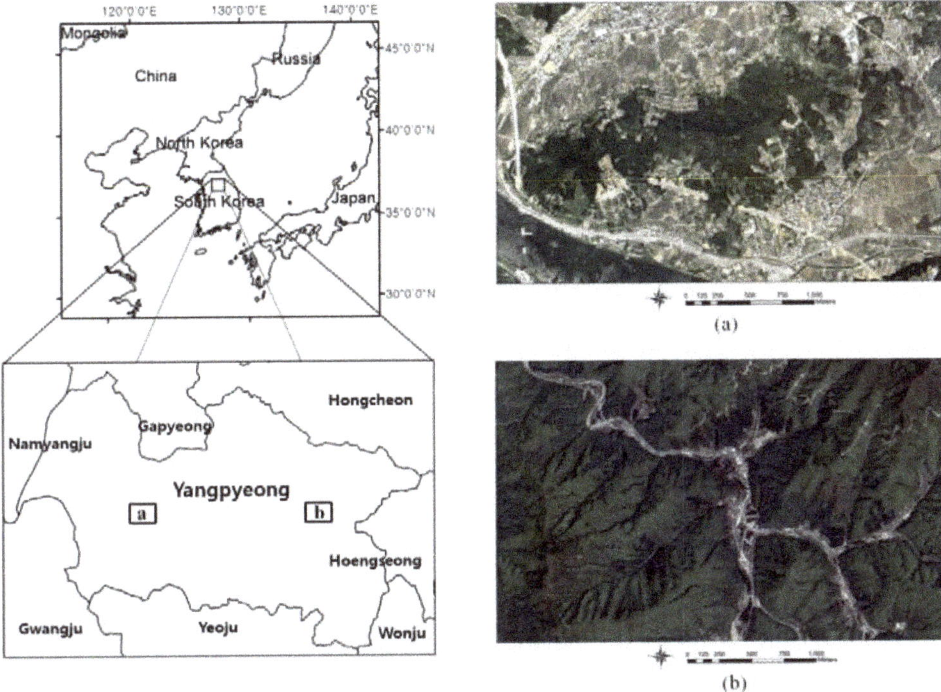

Figure 1. Geographical location of study area with digital aerial photographs. (**a**) Obin-ri; (**b**) Korea University Yangpyeong Experimental Forest.

2.2. Lidar Data Acquisition

A small-footprint lidar system (ALTM 3070 developed by Optech, Inc.) was used to acquire the lidar data. The flight took place on 4 May 2009, and LiDAR data acquisition was performed at an altitude of 1400 m and with a point density of 5–8 points per square meter. Each point return provides the x, y, z position and intensity information derived from a near-infrared (1064 nm) laser pulse. The radiometric resolution, scan frequency, and scan width of the recorded lidar data were 12 bits, 70 Hz, and ±20°, respectively. This study only used lidar data within a ±10° scanning angle to reduce the scan angle effect on tree height and volume estimation [25]. Prior to extracting parameters from the lidar data, every lidar return was classified by the automatic procedure of the TerraScam Program [26]. The lidar returns were first classified into two groups, ground returns and above-ground returns. Ground returns were determined to be reflected from the ground within the plots, and above-ground returns included all returns other than ground returns. Then, every return of the lidar data recorded as height above sea level was converted to local height measure using a digital terrain model (DTM) generated from ground returns.

2.3. Ground Data Acquisition

A field survey was performed on 25–27 September 2010 across two study regions, i.e., the Obin-ri and the Korea University Yangpyeong Experimental Forest. Training and testing plots are 20 and 10 square plots (0.04 ha or 400 m^2, 20 m sides) of each plot-dominant species, respectively [27]. For the survey, species, DBH, height, and age of every individual trees > 5 cm DBH were measured and tree density within each plot was recorded (Table S1). The dominant tree species, which accounts for more than 75% of the total tree species within the plot, was determined by counting every individual tree species [28]. The volume of each individual tree was calculated using species-specific allometric equations from the field-measured DBH and tree height; thereafter, plot volume was summed from every individual trees within each plot [28]. The species-specific allometric equations for volume estimation were developed by Korea Forest Research Institute [29]. The survey followed a standard procedure guided by the 5th National Forest Inventory [28].

The position of the center of each sample and test plot was acquired using the GPS Pathfinder Pro XR, which is manufactured by Trimble. This position information was used to spatially join and geo-match field surveying data with lidar data. The acquired GPS data were processed by a differential correction method using supplementary information received from a continuous GPS station near the study area. The corrected positions of the geo-referenced plots were obtained with position errors of less than 1 m. This process aimed to correct some errors that could be related to differences between the satellites and receivers, the atmosphere, satellite orbits, and reflective surfaces near the receiver when surveying with a single GPS receiver [30].

3. Methods

This study utilized multiple linear discriminant and regression analysis for plot-level species classification and volume estimation, respectively. Each process used the LiDAR height, intensity, ratio parameters, and field measurements of the plot-dominant species and volume as independent and dependent variables, respectively (Figure 2). Multiple linear discriminant analysis was adopted to classify plot-dominant species using the independent dataset, and the results were then verified by a cross-validation procedure with separate testing data. Thereafter, multiple linear regression analysis was performed on the LiDAR dataset, with the results of each plot-dominant species then evaluated using a corrected Akaike's Information Criterion (AIC_c) [31] for selecting the optimal equation candidates. The candidate models showing the best performance were then verified and selected as an optimal volume model using testing datasets.

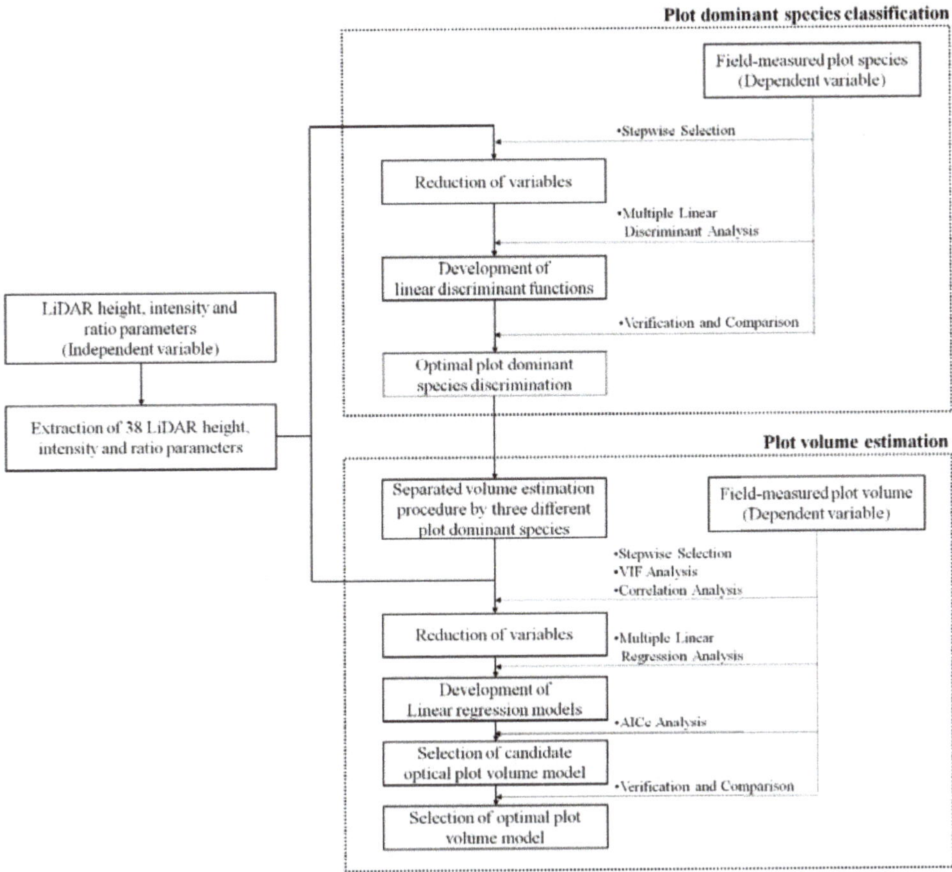

Figure 2. Flowchart of plot dominant species classification and plot volume estimation.

3.1. Extraction of Lidar Height, Intensity, and Ratio Parameters

I extracted lidar height, intensity, and ratio parameters for constructing independent variables from two separate LiDAR return types, i.e., canopy and total returns. Canopy LiDAR returns are defined as returns that are higher than 1.0 m normalized height from DTM, while total returns include every return falling in the target plot [8,16]. According to Chen et al. [13], the height threshold for discriminating canopy returns from total returns might vary due to various forest stand conditions. This study set the threshold at 1.0 m as it effectively differentiates canopy crowns (i.e., lower than minimum crown base heights) and ground (Table S1). Then, 38 variables for the modeling process were extracted based on a dataset classified into canopy and total returns.

The lidar variables include percentile, mean, maximum, minimum, median, mode, standard deviation, coefficient of variation, kurtosis, skewness, and range of height and intensity data as those are closely related to volume and species information [32] (Table 1). These height and intensity variables were extracted from only canopy returns. The intensity variables can be significant descriptors of tree specifications, however, the intensity values might be affected by variations in laser path length and target reflectivity [33]. Thus, for intensity variables, I calibrated and normalized it by following an approach suggested by Kwak et al. [34].

Table 1. Definition of lidar height, intensity, and ratio metrics.

Independent Variables		
Height Parameters Based on Canopy Returns	**Intensity Parameters Based on Canopy Returns**	**Ratio Parameters Based on Integrated Canopy and Ground Returns**
$HEI_{,i}$, $I = 10, 20, \dots , 100$ percentile height	$INT_{,mean}$, mean of intensity	Num_T, number of total returns
$HEI_{,mean}$, mean of height	$INT_{,max}$, maximum of intensity	Num_C, number of canopy returns
$HEI_{,max}$, maximum of height	$INT_{,min}$, minimum of intensity	CRR, canopy return ratio
$HEI_{,min}$, minimum of height	$INT_{,med}$, median of intensity	$INT_{,TSum}$, sum of total intensity
$HEI_{,med}$, median of height	$INT_{,mode}$, mode of intensity	$INT_{,CSum}$, sum of canopy intensity
$HEI_{,mode}$, mode of height	$INT_{,std}$, standard deviation of intensity	CIR, canopy intensity ratio
$HEI_{,std}$, standard deviation of height	$INT_{,cv}$, coefficient of variation of intensity	
$HEI_{,cv}$, coefficient of variation of height	$INT_{,se}$, standard error of mean of intensity	
$HEI_{,se}$, standard error of mean of height	$INT_{,kurt}$, kurtosis of intensity distribution	
$HEI_{,kurt}$, kurtosis of height distribution	$INT_{,skew}$, skewness of intensity distribution	
$HEI_{,skew}$, skewness of height distribution	$INT_{,range}$, range of intensity	
$HEI_{,range}$, range of height		

To consider return transmission from the canopy to ground, this study added not only the number of returns from the canopy and total, but also the canopy return ratio (CRR, Equation (1)) and canopy intensity ratio (CIR, Equation (2)) (Table 1).

$$Canopy\ Return\ Ratio = \frac{number\ of\ canopy\ returns}{number\ of\ total\ returns} \tag{1}$$

$$Canopy\ Intensity\ Ratio = \frac{sum\ of\ intensities\ of\ canopy\ returns}{um\ of\ intensities\ of\ total\ returns} \tag{2}$$

3.2. Plot-Dominant Species Classification

3.2.1. Selection of Explanatory Variables

Multiple linear discriminant analysis is a multivariate technique for separating distinct sets of observations and allocating new observations into predefined classes, i.e., plot-dominant species in this study [35]. This analysis is fundamentally based on minimizing the expected misclassification cost. Fisher's linear discriminant analysis, a widely used discriminant analysis function [35], was used to classify three plot-dominant species (*Pinus densiflora*, *Larix kaempferi*, and *Quercus* spp.) in this study.

Thirty-eight independent variables were first extracted to discriminate plot-dominant species. However, the use of all of the candidate variables to separate plot-dominant species would be inefficient due to the need for intensive, time-consuming collection and management of all of the data. In addition, the use of too many variables is referred to as overfitting, a condition in which the results are dependent on sampling errors [35]. Therefore, a reduced discriminant analysis, with essential explanatory variables, need to be performed using a stepwise selection method based on the Wilks' λ criteria approach (0.05 significance level used in this study). The Wilks' λ, also known as the ratio of generalized variance, is a test statistic used in multivariate analysis of variance to test whether differences exist between the means of specified groups, i.e., plot-dominant-species groups in this study [35]. Further, reduced parameters from the stepwise technique were also evaluated for their discriminant ability by using both box-whisker plots and Tukey's honestly significant difference (HSD) test. From the 38 lidar height, intensity, and ratio parameters, several parameters were selected as explanatory variables for plot-dominant species discrimination by stepwise selection. The discriminating power of each variable was then evaluated.

3.2.2. Development and Assessment of Linear Discriminant Analysis

Fisher's linear discriminant analysis is a linear dimensionality reduction technique using canonical discriminant functions [36]. The procedure constructs a discriminant function by maximizing the ratio between the groups' and within the groups' variances [35]. Fisher's method yields a linear function that divides the variable space into two dimensions by developing a canonical discriminant function. This canonical discriminant function is commonly written as Equation (3):

$$D = b_1 X_1 + b_2 X_2 + \cdots + b_n X_n \tag{3}$$

where D is the discriminant score, b_i is the discriminant coefficient of the *ith* independent variable, and X_i is the *ith* independent variable. Canonical discriminant functions can be used to calculate the discriminant score of each plot for determining the centroid of scores related with species group separation. The centroid of each species group was calculated by averaging discriminant scores derived from these functions. These canonical functions might be evaluated by explanation degree for discriminant score variations referred from canonical correlation coefficients [35]. In addition, standardized canonical discriminant function coefficients are used to evaluate which variable has higher discriminating power in each developed canonical discriminant function [35].

A cross-validation approach (leave-one-out method) was used to assess the accuracy of the discriminant analysis [37]. The classification result was evaluated based on an originally grouped—and cross-validated—accuracy assessment process by hit ratio explaining the correctly corresponding discrimination performance. Additionally, Cohen's kappa coefficients were calculated to measure the agreement between the classifications of the best performance combination case. The Kappa class value was used to rate the agreement as poor (0.40), fair (0.40–0.55), good (0.55–0.70), very good (0.70–0.85), and excellent (>0.85) [38]. Among manifold variable combinations, the variable combination showing the best performance was applied to determine plot-dominant species and this information was sequentially used to develop the species-specific volume model. In addition to such evaluation procedures, this study verified the variable combinations showing the best performance in the training plots by applying these to the 30 separate testing plots.

3.3. Plot Volume Estimation

3.3.1. Selection of Explanatory Variables

To estimate the plot volume dominated by different species, multiple linear regression modeling was separately performed. The 38 independent variables from the lidar height, intensity, and ratio metrics were first utilized for regression modeling of the plot volume. The use of the fully developed model using all candidate variables would be inefficient due to the same reason as in the discriminant analysis; thus, this study reduced the model using stepwise selection methods at the 0.05 significance level [13]. However, such selected variables under stepwise selection might have linear dependency relationships, i.e., a problem referred to as multicollinearity. The multicollinearity can disturb the estimation of a least square estimator in the regression procedure, so that the estimated value may be unreliable due to increased variation. O'Brien [39] suggested a variance inflation factor (VIF) that can evaluate the multicollinearity between selected independent variables. Moreover, Kutner et al. [40] mentioned that a VIF below 10 is suitable for selecting independent variables, but values above 10 indicate a multicollinearity problem. Therefore, I only selected independent variables that have VIF lower than 10. Then, highly correlated independent variables were further eliminated (Pearson's correlation coefficients over 0.5) [18].

3.3.2. Development and Assessment of Linear Volume Models

In this study, multiple linear regression analysis was performed using combinations of selected variables for each species with 10 or fewer regressed functions selected, as shown in Equation (4):

$$y = \alpha + \beta_1 x_1 + \beta_2 x_2 + \beta_3 x_3 + \ldots + \beta_n x_n + \varepsilon \tag{4}$$

where y is the plot volume measured in the field survey; $x_1, x_2, x_3, \ldots, x_n$ are the selected variables derived from lidar height, intensity, and ratio parameters; $\alpha, \beta_1, \beta_2, \beta_3, \ldots, \beta_n$ are the estimated regression parameters; and ε is its residuals.

This study utilized Akaike's information criterion (AIC) to select an optimal plot volume model for each dominant species. To remedy potential bias introduced by the size of sample, AIC was substituted into the corrected AIC (AIC_c) in this study (Equation (5)),

$$AIC_c = n\left[\ln\left(\frac{\sum (y - \hat{y})^2}{n}\right)\right] + 2k + \frac{2k(k+1)}{n-k-1} \tag{5}$$

where k is the number of parameters in the model and n the number of observations. y and $\hat{y}$ stand for the plot volume and its estimate from the model. The smaller AIC_c is, the more closely the model approaches reality. When comparing different regression models, the estimated AIC_c values are generally normalized by subtracting the minimum AIC_c values, according to Equation (6):

$$\Delta_i = AIC_{ci} - AIC_{cmin} \tag{6}$$

where AIC_{ci} is the AIC_c value of the ith model and AIC_{cmin} is the minimum AIC_{ci} value. The results of this transformation allow the following criteria. Regression models with $\Delta_i \geq 2$ were rejected and models with $\Delta_i \leq 2$ were selected because only models with $\Delta_i \leq 2$ are accepted as providing substantial support. With these AIC criteria, other supplementary statistics including R^2, adjusted R^2, RMSE, and SSE of the models were also considered and compared when selecting the optimal plot volume model.

4. Results

4.1. Plot-Dominant Species

4.1.1. Explanatory Variables for Plot-Dominant Species Classification

Explanatory variables of the LiDAR height, intensity, and ratio parameters for plot-dominant species classification were chosen through the stepwise selection method based on *Wilks'* λ criteria at a significance level of 0.05. The selected metrics were the 80th and 90th percentiles and the standard deviation of height ($HEI_{,80}$, $HEI_{,90}$, and $HEI_{,std}$), the mean, mode, standard deviation, coefficient of variation, and skewness of intensity ($INT_{,mean}$, $INT_{,mode}$, $INT_{,std}$, $INT_{,cv}$, and $INT_{,skew}$), and the canopy return ratio (CRR). In order to determine the statistical significance of the differences in the three plot-dominant species, the *Wilks'* λ statistic, F-value, and Tukey's HSD test were examined (Table 2).

The *Wilks'* λ and F-value are generalized tests for determining the probability level of equality of population centroids, assuming equality of dispersion. The variables including lower λ statistics and higher F statistics at a high level of significance indicated high discrimination ability in plot-dominant-species classification. According to these statistics, $HEI_{,80}$, $HEI_{,90}$, and $INT_{,std}$ were the most effective discrimination parameters among the nine selected explanatory variables. The results of multiple species comparisons by Tukey's HSD test showed which parameter was the distinctively meaningful factor for differentiating between two particular species. Tukey's HSD tests revealed significant differences between *Larix kaempferi* and other species for three variables ($HEI_{,80}$, $HEI_{,90}$, and $INT_{,mode}$). In the case of *Pinus densiflora* and *Quercus* spp., four ($HEI_{,80}$, $HEI_{,90}$, $INT_{,mode}$, and $INT_{,cv}$) and two ($HEI_{,80}$ and $HEI_{,90}$) variables showed differences with other species under critical statistical criteria ($p < 0.001$). Similar patterns can be found in the box–whisker plots (Figure S1).

Table 2. Results of stepwise selection and Tukey's HSD test for plot dominant species discrimination (both under 0.05 significant level).

| Selected Variable | Significantly Different Species by Tukey's HSD Test | | | Wilks λ | F | *p*-Value |
	Larix kaempferi (LK)	*Pinus densiflors* (PD)	*Quercus* spp. (Qs)			
$HEI_{,80}$	PD, Qs	LK, Qs	LK, PD	0.467	32.56	<0.001
$HEI_{,90}$	PD, Qs	LK, Qs	LK, PD	0.409	41.10	<0.001
$HEI_{,std}$	Qs	·	LK	0.893	3.42	<0.039
$INT_{,mean}$	Qs	LK	LK	0.892	3.46	<0.038
$INT_{,mode}$	PD, Qs	LK	LK	0.715	11.34	<0.001
$INT_{,std}$	PD	LK, Qs	PD	0.534	24.87	<0.001
$INT_{,cv}$	PD	LK, Qs	PD	0.665	14.34	<0.001
$INT_{,skew}$	PD	LK	·	0.724	10.86	<0.001
CRR	Qs	·	·	0.836	5.58	<0.006

4.1.2. Evaluation of Plot-Dominant-Species Discrimination

A set of explanatory variables was used to differentiate three plot-dominant species using the linear discriminant analysis. The 9 selected lidar height, intensity, and ratio parameters generated 511 combinations ($2^9 - 1$) of independent variables for discriminant analysis. Every possible combination classified plot-dominant species of the 60 sampled plots and was evaluated by original grouped and cross-validated accuracy assessments. The best performance in species discrimination was obtained from a combination of all explanatory variables, including $HEI_{,80}$, $HEI_{,90}$, $HEI_{,std}$, $INT_{,mean}$, $INT_{,mode}$, $INT_{,std}$, $INT_{,cv}$, $INT_{,skew}$, and CRR. Original grouped and cross-validated accuracy of this model were 95.0 and 93.3%, respectively. On the other hand, using only one variable, $INT_{,mean}$, produced the lowest classification accuracies of 46.7% (original grouped accuracy) and 45.0% (cross-validated accuracy). Among the 511 possible combinations, Table 3 shows 10 variable combinations in order of cross-validated results. The cross-validated results of the 10 combinations range from 88.3 to 93.3% and most combinations have more than six variables showing that a general tendency between number of variables and model accuracy (Figure S2).

The all of the combinations generated canonical discriminant functions to calculate discriminant scores and classify plot-dominant species. In particular, the highest accuracy was achieved by the combination with all variables (93.3%). This combination had first and second canonical discriminant functions that could explain 86.3 and 67.4% of the discriminant score variance, respectively, refers from canonical correlation coefficient. According to the result of the x^2 test, both functions had a *p*-value that was lower than 0.05, which meant that this function significantly discriminated *Pinus densiflora*, *Larix kaempferi*, and *Quercus* spp. groups. The discriminant score centroid and distribution calculated by canonical discriminant functions is shown in Figure S3. The classification agreement was excellent as indicated by a kappa value greater than 0.90 [38] (Table S3).

Table 3. Accuracy assessment results of multiple linear discriminant analysis by various combinations (upper ranked 10 combinations).

No. (AE1, AE2)	Variable Combination (RD1, RD2)									Number of Variables	Accuracy (%) Original Grouped	Accuracy (%) Cross Validated
1 (86.3, 67.4)	$HEI_{,80}$ (4, 4)	$HEI_{,90}$ (3, 7)	$HEI_{,std}$ (6, 8)	$INT_{,mean}$ (2, 2)	$INT_{,mode}$ (9, 6)	$INT_{,std}$ (5, 3)	$INT_{,cv}$ (1, 1)	$INT_{,skew}$ (7, 5)	CRR (8, 9)	9	95.0	93.3
2 (81.7, 53.3)	$HEI_{,80}$ (4, 3)	$HEI_{,90}$ (1, 6)	$HEI_{,std}$ (5, 7)	$INT_{,mean}$ (3, 5)	$INT_{,mode}$ (6, 4)	$INT_{,cv}$ (2, 1)	$INT_{,skew}$ (7, 2)			7	93.3	90.0
3 (85.4, 60.1)	$HEI_{,80}$ (5, 5)	$HEI_{,90}$ (3, 7)	$HEI_{,std}$ (4, 8)	$INT_{,mean}$ (1, 2)	$INT_{,mode}$ (8, 4)	$INT_{,std}$ (6, 3)	$INT_{,cv}$ (2, 1)	CRR (7, 6)		8	93.3	90.0
4 (84.8, 64.2)	$HEI_{,90}$ (3, 5)	$HEI_{,std}$ (4, 8)	$INT_{,mean}$ (1, 2)	$INT_{,mode}$ (7, 6)	$INT_{,std}$ (5, 3)	$INT_{,cv}$ (2, 1)	$INT_{,skew}$ (8, 4)	CRR (6, 7)		8	93.3	90.0
5 (76.2, 50.8)	$HEI_{,90}$ (1, 4)	$INT_{,mode}$ (4, 3)	$INT_{,std}$ (2, 6)	$INT_{,cv}$ (3, 1)	$INT_{,skew}$ (6, 2)	CRR (5, 5)				6	91.7	90.0
6 (81.7, 66.7)	$HEI_{,80}$ (3, 5)	$HEI_{,std}$ (4, 7)	$INT_{,mean}$ (1, 2)	$INT_{,mode}$ (6, 6)	$INT_{,std}$ (7, 3)	$INT_{,cv}$ (2, 1)	$INT_{,skew}$ (8, 4)	CRR (5, 8)		8	91.7	90.0
7 (84.8, 59.6)	$HEI_{,90}$ (3, 5)	$HEI_{,std}$ (4, 7)	$INT_{,mean}$ (1, 2)	$INT_{,mode}$ (7, 4)	$INT_{,std}$ (5, 3)	$INT_{,cv}$ (2, 1)	CRR (6, 6)			7	90.0	90.0
8 (83.2, 67.1)	$HEI_{,80}$ (4, 4)	$HEI_{,90}$ (3, 6)	$HEI_{,std}$ (6, 8)	$INT_{,mean}$ (2, 2)	$INT_{,mode}$ (8, 7)	$INT_{,std}$ (5, 3)	$INT_{,cv}$ (1, 1)	$INT_{,skew}$ (7, 5)		8	96.7	88.3
9 (81.2, 62.4)	$HEI_{,90}$ (3, 5)	$HEI_{,std}$ (4, 7)	$INT_{,mean}$ (1, 2)	$INT_{,mode}$ (5, 6)	$INT_{,std}$ (6, 3)	$INT_{,cv}$ (2, 1)	$INT_{,skew}$ (7, 4)			7	93.3	88.3
10 (80.8, 51.8)	$HEI_{,90}$ (3, 4)	$HEI_{,std}$ (4, 6)	$INT_{,mean}$ (1, 3)	$INT_{,mode}$ (5, 5)	$INT_{,cv}$ (2, 1)	$INT_{,skew}$ (6, 2)				6	91.7	88.3

AE1 & 2 (%): Ability of explainable variation by developed canonical discriminant function 1 and 2, respectively. RD1 & 2 (%): Rank of discriminant power in developed canonical discriminant function 1and 2, respectively.

4.2. Plot Volume

4.2.1. Explanatory Variables for Plot Volume

In the case of the plots dominated by *Larix kaempferi*, five variables were selected using the stepwise selection method at a significance level of 0.05: 90th percentile height ($HEI_{,90}$), standard deviation of height (HEI_{std}), mode of the intensity ($INT_{,mode}$), standard error of the mean of the intensity ($INT_{,se}$), and the sum of the intensity ($INT_{,TSum}$). These variables were shown to have low multicollinearity by a VIF of approximately 1 (Table 4). Then, the selected variables were examined by comparing Pearson's correlation coefficients derived from correlation analysis between candidate variables. From the results, HEI_{std} was highly correlated with $HEI_{,90}$ with a coefficient higher than 0.5, so this variable was eliminated to reduce multicollinearity (Table 5). Therefore, the explanatory variables to regress the plot volume model for *Larix kaempferi* were $HEI_{,90}$, $INT_{,mode}$, $INT_{,se}$, and $INT_{,TSum}$.

Table 4. Results of variable selection to each species by variance inflation.

Species	Variable	DF	Parameter Estimate	Standard Error	t Value	pr > |t|	Variance Inflation
	Intercept	1	−11.2321	1.91114	−5.88	<0.0001	0.0000
	$HEI_{,90}$	1	1.44559	0.28551	5.06	0.0002	4.07915
Larix	$HEI_{,std}$	1	−2.24028	0.62636	−3.58	0.0030	4.03682
kaempferi	$INT_{,mode}$	1	−0.65035	0.19885	−3.27	0.0056	1.00719
	$INT_{,se}$	1	63.19259	20.10955	3.14	0.0072	1.48727
	$INT_{,TSum}$	1	0.00112	0.000192	5.83	<0.0001	1.37517
	Intercept	1	1.19921	2.42319	0.49	0.6279	0.00000
	$HEI_{,mean}$	1	0.38271	0.11826	3.24	0.0055	1.48927
P. densiflora	$HEI_{,mode}$	1	0.07408	0.04048	1.83	0.0872	1.20480
	$INT_{,std}$	1	2.54207	1.37647	1.85	0.0846	1.52052
	$INT_{,range}$	1	−0.85953	0.24021	−3.58	0.0027	1.86565
	Intercept	1	−0.55562	1.09140	−0.51	0.6181	0.00000
	$HEI_{,80}$	1	−0.47026	0.19574	−2.40	0.0297	8.50554
Quercus	$HEI_{,90}$	1	0.72066	0.18835	3.83	0.0017	8.57314
spp.	$INT_{,mode}$	1	0.06646	0.02167	3.07	0.0078	1.24328
	$INT_{,kurt}$	1	0.84446	0.42291	2.00	0.0643	1.22767

Table 5. Results of variable selection to each species by correlation coefficient.

Species	Variables	$HEI_{,90}$	$HEI_{,std}$	$INT_{,mode}$	$INT_{,se}$	$INT_{,TSum}$
	$HEI_{,90}$	1.00000	0.98055	−0.0341	0.18118	−0.15654
Larix	$HEI_{,std}$	0.98055	1.00000	−0.03256	0.26466	−0.07958
kaempferi	$INT_{,mode}$	−0.0341	−0.03256	1.00000	−0.04156	−0.03433
	$INT_{,se}$	0.18118	0.26466	−0.04156	1.00000	−0.14369
	$INT_{,TSum}$	−0.15654	−0.07958	−0.03433	−0.14369	1.00000

Species	Variables	$HEI_{,mean}$	$HEI_{,mode}$	$INT_{,std}$	$INT_{,range}$
	$HEI_{,mean}$	1.00000	0.20974	−0.47106	−0.43012
Pinus	$HEI_{,mode}$	0.20974	1.00000	−0.20509	−0.41167
densiflora	$INT_{,std}$	−0.47106	−0.20509	1.00000	0.44358
	$INT_{,range}$	−0.43012	−0.41167	0.44358	1.00000

Species	Variables	$HEI_{,80}$	$HEI_{,90}$	$INT_{,mode}$	$INT_{,kurt}$
	$HEI_{,80}$	1.00000	0.93215	0.15998	−0.07099
Quercus spp.	$HEI_{,90}$	0.93215	1.00000	0.25939	0.14843
	$INT_{,mode}$	0.15998	0.25939	1.00000	0.45776
	$INT_{,kurt}$	−0.07099	0.14843	0.45776	1.00000

In the case of *Pinus densiflora*, the mean of the height ($HEI_{,mean}$), the mode of the height ($HEI_{,mode}$), the standard deviation of intensity ($INT_{,std}$), and the range of intensity ($INT_{,range}$) were selected. The multicollinearity between the selected variables was low compared to their VIFs (below 10), as shown in Table 4. An assessment of the one-to-one correlations between candidate variables was completed; however, all four variables were included in the regression analysis due their relatively low correlation coefficients (below 0.5) (Table 5).

To extract explanatory variables, acquired LiDAR parameters for *Quercus* spp. plots were reduced by the same procedure. Consequently, the candidate independent variables, shown in Table 4, are the 80th percentile of height ($HEI_{,80}$), the 90th percentile of height ($HEI_{,90}$), the mode of the intensity ($INT_{,mode}$), and the kurtosis of the intensity distribution ($INT_{,kurt}$). The multicollinearity between the selected variables was weak, as each VIF had a value below 10 (Table 4). Furthermore, as a result of the correlation analysis between the selected variables, $HEI_{,80}$ and $HEI_{,90}$ were highly correlated ($r > 0.5$) (Table 5). Therefore, $HEI_{,90}$ was rejected as an explanatory independent variable for regressing the plot volume model because both the probability value from a *t*-test and the VIF of $HEI_{,80}$ were lower than those for $HEI_{,90}$. Eventually, $HEI_{,80}$, $INT_{,mode}$, and $INT_{,kurt}$ were adopted for the multiple linear regression analysis for predicting the plot volume.

4.2.2. Evaluation of Plot-Volume Models

The four independent variables selected for *Larix kaempferi* were used in developing regression models. The predictable equation was estimated by adopting the optimal regression model for estimating the plot volume represented by ΔAIC_c less than 2 (Table S4). The two optimal regression models found with $\Delta AIC_c \leq 2$ were model no. 1, which estimates the plot volume using $HEI_{,90}$, $INT_{,mode}$, and I_T, and model no. 2, in which the explanatory variables were $HEI_{,90}$, $INT_{,mode}$, $INT_{,se}$, and INT, when AIC_c was employed as the first criterion for selecting the best model. In addition, models 3 to 7 were rejected because their ΔAIC_c values were higher than 2, and the significance of their statistics was reduced with increasing ΔAIC_c. Among these two models, I chose the model no. 2 as it shows a better performance in RMSE, SEE, R^2, and adjusted R^2.

For *Pinus densiflora* plots, the combinable regression model was estimated with optimal regression models having ΔAIC_c less than 2 (Table S5). However, the best model and the only one with $\Delta AIC_c \leq 2$ was model no. 1, by which the plot volume could be estimated using $HEI_{,mean}$, $HEI_{,mode}$, $INT_{,std}$, and $INT_{,range}$. Models 2 to 7 were eliminated as candidate regression models because their ΔAIC_c was higher than 2 and because they lacked statistical significance with increasing ΔAIC_c.

In the case of *Quercus* spp. plots, three explanatory variables ($HEI_{,80}$, $HEI_{,mode}$, and $INT_{,kurt}$) were selected and applied to the regression analysis. As a result of the regression procedure, one regression model, no. 1, with $\Delta AIC_c \leq 2$ was recommended. Models 2 to 7 were eliminated as candidate regression models for estimating plot volume of *Quercus* spp. stands because their ΔAIC_c was greater than 2 and because they lacked statistical significance with increasing ΔAIC_c (Table S6).

The developed plot volume models for each species were evaluated by implementing the models to independent 30 testing plots (10 plots for each dominant species) (Table 6; Figure S4). The model for *Larix kaempferi* shows high performance ($R^2 = 0.71$, RMSE $= 2.8$ m^3). In the case of *Pinus densiflora*, validation results showed that R^2 and RMSE values of the model were 0.74 and 2.59 m^3, respectively. When compared with plot volume predictions for *Larix kaempferi*, the best model of *Pinus densiflora* showed slightly better validation performance. Prediction by the model for *Pinus densiflora* was slightly overestimated, while estimation by the model for *Larix kaempferi* was underestimated. For *Quercus* spp., the developed model predicted plot volume with relatively a lower accuracy of R^2 (0.56) and RMSE (3.01 m^3) than those of other species (Figure S4d). The best plot-volume model of *Quercus* spp. produced overestimated plot volume predictions. The significance of the *t*-test for the developed and selected optimal plot volume models for each dominant species was statistically satisfactory.

Table 6. Selected optimal regression models for each species.

Species	Optimal Plot Volume Equation	R^2	RMSE (m^3)	t-test ($\alpha = 0.05$)
				Pr > | t |
Larix kaempferi	$PV = 0.43730 \cdot HEI_{,90} - 0.68725 \cdot INT_{,mode} + 24.2152 \cdot INT_{,se} - 0.000782 \cdot INT_{,TSum} - 5.85002$	0.7075	2.772	0.966
Pinus densiflora	$PV = 0.38271 \cdot HEI_{,mean} + 0.07408 \cdot HEI_{,mode} + 2.54207 \cdot INT_{,std} - 0.85953 \cdot INT_{,range} + 1.19921$	0.7368	2.590	0.852
Quercus. spp.	$PV = 0.28685 \cdot HEI_{,80} + 0.07623 \cdot INT_{,mode} + 0.31517 \cdot INT_{,kurt} - 0.71001$	0.5641	3.010	0.925

5. Discussion

5.1. Plot-Dominant Species Classification

The stepwise technique selected nine explanatory variables: the 80th and 90th percentiles and standard deviation of height ($HEI_{,80}$, $HEI_{,90}$, and $HEI_{,std}$), the mean, mode, standard deviation, coefficient of variation, skewness of intensity ($INT_{,mean}$, $INT_{,mode}$, $INT_{,std}$, $INT_{,cv}$, and $INT_{,skew}$), and the canopy return ratio (CRR) (Table 2). For the three plot-dominant species, each species showed significant differences in height percentile parameters (such as $HEI_{,80}$ and $HEI_{,90}$) (Figure S5). Næsset and Bjerknes [41] and Holmgren and Persson [23] showed that the 90 percentile statistic could be used as a determining factor for species identification according to the close relationship between the height of the dominant trees in the plot and crown shape. This study determined that the $HEI_{,80}$ and $HEI_{,90}$ parameters could represent the dominant height of trees within a plot and showed high potential for being powerful discriminant variables. The height standard deviation of lidar returns which shows how much variation or dispersion exists from the average of returns was closely related with crown depth and corresponded to crown base height. The standard deviation of height was usually higher for *Larix kaempferi* than *Quercus* spp. and *Pinus densiflora* because *Larix kaempferi* usually has deeper tree crowns than those of the other species. This distinct aspect of standard deviation of height could be a potential indicator for differentiating plot-dominant species [23].

Descriptive statistics of lidar near infrared intensity returns provide useful discriminators for species identification through recording the different characteristics of the near-infrared radiation reflected from forest canopies (Figure S6). In this study, five intensity statistics were effectively used to classify plot-dominant species. The mean of intensity showed the highest value among the three dominant species for *Quercus* spp. because reflectivity characteristics of broadleaf and dense foliage produce higher intensity values than needle-like leaves and sparse foliage [42]. Coniferous trees generally show significantly lower reflection values than broadleaf trees [43], so our result supports his finding of species-specific intensity differences (Figure S6). In addition, due to densely covered leaves or other components, this study also found that the mean of intensity of *Pinus densiflora* was higher than that of *Larix kaempferi*. This is likely due to higher reflectivity of *Pinus densiflora* and denser canopy structure (Figure S1) [44].

The statistics related to dispersion of lidar intensity, such as standard deviation and coefficient of variation, were considered to be explanatory variables for identifying individual trees or dominant species in previous studies [20,24,42]. The coefficient of variation of intensity, a normalized measure of dispersion, was closely related to the standard deviation of intensity. Generally, intensity was affected not only by canopy closure, but also by specific reflectivity characteristics that depend on species. Consequently, dense forest canopies were associated with low lidar penetration rates and therefore such forests had low coefficients of variation and standard deviation [33]. When this study examined intensity dispersion and the influences of canopy closure, *Pinus densiflora* showed the lowest standard deviation and coefficient of variation. These lowest values originated from its permeability, which is determined by the densely covered canopy. The intensity dispersion of *Quercus* spp. showed higher standard deviation and higher variability of normalized standard deviation. Considering the variability of field measured tree density, the intensity dispersions of *Quercus* spp. might be crucially influenced by the intermingled effects of tree density and degree of canopy closure.

The mode of intensity might be expected to describe the concentrativeness of the returned intensity distribution. The mode could be strongly influenced by the shape and structure of the canopy and the degree of canopy closure. Species-specific reflective characteristics were shown in three plot-dominant species; however, the dense canopy closure of *Pinus densiflora* and variability of tree density of *Quercus* spp. affected those distributions. Highly dense needle-like leaves increased the returned intensity value, and the severe variation of tree density in *Quercus* spp. dispersed its mode of the intensity distribution. The skewness of intensity is related to the asymmetry of the recorded intensity distribution. Positive skewness of intensity would be associated with a minimally skewed distribution and negative skewness would be relevant to a highly skewed distribution. $INT_{,skew}$ shown in box-whisker plots indicated that larger laser pulses were reflected from branches or bark in plots of *Larix kaempferi*, while highly recorded intensities were from leaves in plots of *Pinus densiflora* considering the structural and canopy closure characteristics of each species. Because Roberts et al. [45] found that intensities from branch and bark had lower spectral reflectance than that of leaves, this skewness of intensity suggested that the asymmetry of intensity might be an explainable indicator of reflective objects.

The canopy return ratio was generally used to measure the degree of laser penetration (i.e., gap probability) against canopy components such as leaves, branches, and stem. The canopy return ratio showed a different pattern compared to other lidar height and intensity parameters revealing its potential for describing canopy structures (Figures S5 and S6). This is because the canopy return ratio was derived from calculating the ratio between the number of total returns and the number canopy returns while the other parameters were calculated from only canopy returns. Higher ratio value means that laser pulses are dominantly interrupted by dense canopy components. To the contrary, lower canopy return ratios can be considered as open canopy forests. Comparing these ratios between species, *Quecus* spp. showed the densest canopy cover with the highest canopy return ratio. This ratio based variable can be an important explanatory indicator for differentiating stand species.

5.2. Plot-Volume Estimation

For the plot volume model, this study identified different sets of explanatory variables for different plot-dominant species. This is likely because of different canopy structural conditions such as canopy cover ratio, crown depth, tree density, and others of the plot-dominant species. Across all cases, $HEI_{,90}$, $HEI_{,80}$, $HEI_{,mean}$, $HEI_{,mode}$, and $HEI_{,kurt}$ variables were chosen for plot volume estimation. These key structural parameters were corroborated by findings of previous studies that were able to estimate plot or stand volume using only canopy height distributional parameters [13,14]. The characteristics of these parameters were closely related to volumetric canopy structures [14,19]. The higher percentiles of height (80th and 90th) and mean of height were highly correlated with actual canopy height which has been used to calculate volume based on allometric relationships. In addition, the mode of height and kurtosis of the height distribution might help explain crown geometric volume [9,13]. Because crown geometric volume correlated with stem volume and was derived from 3-dimensional crown structures, these crown structures might be represented by variables based on their meanings. Therefore, these two variables also have a meaningful explanatory ability for plot volume estimation through expanding the linear relationship [9,13].

Intensity data were also included in plot volume estimation. The intensity parameters used here were $INT_{,mode}$, $INT_{,se}$, $INT_{,TSum}$, $INT_{,std}$, $INT_{,range}$, $INT_{,mode}$, and $INT_{,kurt}$. According to van Aardt et al. [18], various statistics of lidar intensity data, such as median, standard deviation, minimum, and others, were adopted to predict stand volume under specified species information. As described in Section 5.1, lidar intensity at the near-infrared region contains canopy structure and reflectivity information which may improve the volume estimation model [46].

In multiple linear regression analysis, this study could distinguish the explanatory ability of each variable for plot volume estimation. For the case of *Larix kaempferi*, $HEI_{,90}$ and $INT_{,TSum}$ acted as key explanatory variables at the highest significance. The plot volume dominated by *Pinus densiflora* took $HEI_{,mean}$ and $INT_{,range}$ as highly significant variables among the selected four variables. In the case of the *Quercus* spp., $INT_{,mode}$ largely explained plot volume of this species along with the $HEI_{,80}$ and $INT_{,kurt}$ parameters. Different key explanatory variables could be attributed by the characteristics of canopy structures of each species. In general, *Larix kaempferi* showed a fat corn shape crown with deeper crown depth, whereas the other two species had a horizontally flattened and rounded shape in the study area. Particularly, the canopy depth of *Quercus* spp. was relatively deeper than that of *Pinus densiflora*.

Canopy structural characteristics of each species may determine performance of lidar based volume estimation model. Several studies have reported that plot- or stand-volume estimation using lidar data is highly feasible but most of them examined over coniferous forests. For instance, Næsset noted a range of the coefficient of determination of 0.80–0.93 for stand volume estimation in coniferous forest and age and site-quality classes. Means et al. [16] conducted volume modeling for coniferous forest with R^2 values of 0.95 (including mature plots) and 0.97 (not including mature plots). By including coniferous, mixed and broadleaf forests, van Aardt et al. [18] attempted to develop a species-specific stand volume model and showed a limited performance ranging from R^2 of 0.40–0.70. The performance of this study is relatively better, at 0.71, 0.74, and 0.56 for *Pinus densiflora*, *L. kaempferi*, and *Quercus* spp., respectively. Most studies including this study confirms that lidar based volume estimation performs better in coniferous forest than broadleaf or mixed forests. In this study, the plots dominated by *Quercus* spp. had especially large variability from the forest condition, so its plot volume model suffered. These accuracy shortcomings can be also attributed to unexplainable variability in the forest condition, such as stand density, species mixed ratio, etc. Therefore, further study is required to consider various forest conditions (age, density, ratio of mixture, site quality, etc.) and to precisely survey unbiased samples.

6. Conclusions

This study shows that lidar height, intensity, and ratio parameters are applicable for discriminating plot-dominant species (*Pinus densiflora*, *Larix kaempferi*, and *Quercus* spp.) and for estimating plot volume sequentially. A kappa accuracy of 0.75 was achieved in plot-dominant species classification, and species-specific optimal plot volume models were developed and evaluated by coefficients of determination of 0.71, 0.74, and 0.56, respectively. Further investigation found that dispersion and mean of lidar intensity within a plot are key classifiers of identifying three species while height related lidar variables play a key role in modeling forest plot volume. Selected explanatory variables are closely correlated to vertical and horizontal canopy structures and are enough to represent species-specific characteristics in both discriminative analysis and volume estimation. Additionally, observed different variable combinations for two important applications imply that future studies should use proper variable combinations for each purpose. This study only investigated over homogeneous forest stands without considering characteristics of mixed forest stands, such as species mixture, age class mixture, etc. Considering the characteristics of mixed forest stands can help provide an unbiased implementation for discriminating species and estimating volume.

Supplementary Materials: The following are available online at http://www.mdpi.com/2072-4292/12/19/3266/s1, Figure S1. Box-whisker plots for visualizing the distributional characteristics of selected parameters by three plot dominant species. Figure S2. Original grouped- and Cross validated-accuracy of discriminant analysis by number of selected variables in combinations. Figure S3. Distribution of discriminant score and its centroid by first and second canonical discriminant function. Figure S4. Evaluation of the developed plot volume models using independent testing data. Figure S5. Three dimensional view of the lidar height distribution of each species. X and Y axes are a spatial coordinate of the plot (meter scale). Figure S6. Three dimensional view of the lidar intensity distribution of each species. X and Y axes are a spatial coordinate of the plot (meter scale). Table S1. Descriptive statistics of the field measurements. Table S2. Sorted the highest accuracy results of linear discriminant analysis by number of variables. Table S3. Error matrix of plot dominant species classification results by discriminant analysis (case on the highest performance of 93.3%). Table S4. Result of plot volume parameters estimated multiple regression analysis to *Larix kaempferi*. Table S5. Result of plot volume parameters estimated multiple regression analysis to *Pinus densiflora*. Table S6. Result of plot volume parameters estimated multiple regression analysis to *Quercus* spp.

Funding: This study is partially supported by NASA's Carbon Monitoring System program (80NSSC18K0173-CMS).

Conflicts of Interest: The author declares no conflict of interest.

References

1. Jung, S.-E.; Kwak, D.-A.; Park, T.; Lee, W.-K.; Yoo, S. Estimating crown variables of individual trees using airborne and terrestrial laser scanners. *Remote Sens.* **2011**, *3*, 2346–2363. [CrossRef]
2. Matasci, G.; Hermosilla, T.; Wulder, M.A.; White, J.C.; Coops, N.C.; Hobart, G.W.; Zald, H.S. Large-area mapping of Canadian boreal forest cover, height, biomass and other structural attributes using Landsat composites and lidar plots. *Remote Sens. Environ.* **2018**, *209*, 90–106. [CrossRef]
3. Næsset, E.; McRoberts, R.E.; Pekkarinen, A.; Saatchi, S.; Santoro, M.; Trier, Ø.D.; Zahabu, E.; Gobakken, T. Use of local and global maps of forest canopy height and aboveground biomass to enhance local estimates of biomass in miombo woodlands in Tanzania. *Int. J. Appl. Earth Obs. Geoinf.* **2020**. [CrossRef]
4. Choi, S.; Kempes, C.P.; Park, T.; Ganguly, S.; Wang, W.; Xu, L.; Basu, S.; Dungan, J.L.; Simard, M.; Saatchi, S.S. Application of the metabolic scaling theory and water–energy balance equation to model large-scale patterns of maximum forest canopy height. *Glob. Ecol. Biogeogr.* **2016**, *25*, 1428–1442. [CrossRef]
5. Ni, X.; Park, T.; Choi, S.; Shi, Y.; Cao, C.; Wang, X.; Lefsky, M.A.; Simard, M.; Myneni, R.B. Allometric scaling and resource limitations model of tree heights: Part 3. Model optimization and testing over continental China. *Remote Sens.* **2014**, *6*, 3533–3553. [CrossRef]
6. Saatchi, S.S.; Harris, N.L.; Brown, S.; Lefsky, M.; Mitchard, E.T.; Salas, W.; Zutta, B.R.; Buermann, W.; Lewis, S.L.; Hagen, S. Benchmark map of forest carbon stocks in tropical regions across three continents. *Proc. Natl. Acad. Sci. USA* **2011**, *108*, 9899–9904. [CrossRef]
7. Park, T.; Lee, W.-K.; Lee, J.-Y.; Byun, W.-H.; Kwak, D.-A.; Cui, G.; Kim, M.-I.; Jung, R.; Pujiono, E.; Oh, S. Forest plot volume estimation using national forest inventory, forest type map and airborne LiDAR data. *For. Sci. Technol.* **2012**, *8*, 89–98. [CrossRef]
8. Lefsky, M.A.; Cohen, W.; Acker, S.; Parker, G.G.; Spies, T.; Harding, D. Lidar remote sensing of the canopy structure and biophysical properties of Douglas-fir western hemlock forests. *Remote Sens. Environ.* **1999**, *70*, 339–361. [CrossRef]
9. Kwak, D.-A.; Lee, W.-K.; Cho, H.-K.; Lee, S.-H.; Son, Y.; Kafatos, M.; Kim, S.-R. Estimating stem volume and biomass of Pinus koraiensis using LiDAR data. *J. Plant Res.* **2010**, *123*, 421–432. [CrossRef]
10. Harding, D.; Lefsky, M.; Parker, G.; Blair, J. Laser altimeter canopy height profiles: Methods and validation for closed-canopy, broadleaf forests. *Remote Sens. Environ.* **2001**, *76*, 283–297. [CrossRef]
11. Parker, G.A.; Smith, J.M. Optimality theory in evolutionary biology. *Nature* **1990**, *348*, 27–33. [CrossRef]
12. Park, T.; Kennedy, R.E.; Choi, S.; Wu, J.; Lefsky, M.A.; Bi, J.; Mantooth, J.A.; Myneni, R.B.; Knyazikhin, Y. Application of physically-based slope correction for maximum forest canopy height estimation using waveform lidar across different footprint sizes and locations: Tests on LVIS and GLAS. *Remote Sens.* **2014**, *6*, 6566–6586. [CrossRef]
13. Chen, Q.; Gong, P.; Baldocchi, D.; Tian, Y.Q. Estimating basal area and stem volume for individual trees from lidar data. *Photogramm. Eng. Remote Sens.* **2007**, *73*, 1355–1365. [CrossRef]
14. Maltamo, M.; Suvanto, A.; Packalén, P. Comparison of basal area and stem frequency diameter distribution modelling using airborne laser scanner data and calibration estimation. *For. Ecol. Manag.* **2007**, *247*, 26–34. [CrossRef]

15. Drake, J.B.; Dubayah, R.O.; Clark, D.B.; Knox, R.G.; Blair, J.B.; Hofton, M.A.; Chazdon, R.L.; Weishampel, J.F.; Prince, S. Estimation of tropical forest structural characteristics using large-footprint lidar. *Remote Sens. Environ.* **2002**, *79*, 305–319. [CrossRef]

16. Means, J.E.; Acker, S.A.; Fitt, B.J.; Renslow, M.; Emerson, L.; Hendrix, C.J. Predicting forest stand characteristics with airborne scanning lidar. *Photogramm. Eng. Remote Sens.* **2000**, *66*, 1367–1372.

17. Dubayah, R.; Prince, S.; JaJa, J.; Blair, J.; Bufton, J.L.; Knox, R.; Luthcke, S.B.; Clarke, D.B.; Weishampel, J. The vegetation canopy lidar mission. In *Land Satellite Information in the Next Decade II: Sources and Applications*; ASPRS: Washington, DC, USA, 1997.

18. Aardt, J.A.v.; Wynne, R.H.; Oderwald, R.G. Forest volume and biomass estimation using small-footprint lidar-distributional parameters on a per-segment basis. *For. Sci.* **2006**, *52*, 636–649.

19. Næsset, E. Predicting forest stand characteristics with airborne scanning laser using a practical two-stage procedure and field data. *Remote Sens. Environ.* **2002**, *80*, 88–99. [CrossRef]

20. Korpela, I.S.; Tokola, T.E. Potential of aerial image-based monoscopic and multiview single-tree forest inventory: A simulation approach. *For. Sci.* **2006**, *52*, 136–147.

21. Brandtberg, T. Classifying individual tree species under leaf-off and leaf-on conditions using airborne lidar. *ISPRS J. Photogramm. Remote Sens.* **2007**, *61*, 325–340. [CrossRef]

22. Brandtberg, T.; Warner, T.A.; Landenberger, R.E.; McGraw, J.B. Detection and analysis of individual leaf-off tree crowns in small footprint, high sampling density lidar data from the eastern deciduous forest in North America. *Remote Sens. Environ.* **2003**, *85*, 290–303. [CrossRef]

23. Holmgren, J.; Persson, Å. Identifying species of individual trees using airborne laser scanner. *Remote Sens. Environ.* **2004**, *90*, 415–423. [CrossRef]

24. Korpela, I.; Ørka, H.O.; Maltamo, M.; Tokola, T.; Hyyppä, J. Tree species classification using airborne LiDAR–effects of stand and tree parameters, downsizing of training set, intensity normalization, and sensor type. *Silva Fenn.* **2010**, *44*, 319–339. [CrossRef]

25. Naesset, E. Estimating timber volume of forest stands using airborne laser scanner data. *Remote Sens. Environ.* **1997**, *61*, 246–253. [CrossRef]

26. Soininen, A. *TerraScan User's Guide*; Terrasolid: Helsinki, Finland, 2004.

27. Gholamy, A.; Kreinovich, V.; Kosheleva, O. *Why 70/30 or 80/20 Relation Between Training and Testing Sets: A Pedagogical Explanation*; ScholarWorks@UTEP: Berkeley, CA, USA, 2018.

28. Korean Forest Research Institute. *Guideline for 5th National Forest Inventory*; KFRI: Seoul, Korea, 2008.

29. Korean Forest Research Institute. *Volume, Biomass, and Yield Table*; KFRI: Seoul, Korea, 2014.

30. Lee, J.-K.; Hwang, C.-S.; Jung, S.-H. Analysis of accuracy for the control points using the GPS continuous stations. *J. Korean Soc. Civ. Eng.* **2003**, *23*, 401–409.

31. Akaike, H. A new look at the statistical model identification. *IEEE Trans. Autom. Control.* **1974**, *19*, 716–723. [CrossRef]

32. Maltamo, M.; Mustonen, K.; Hyyppä, J.; Pitkänen, J.; Yu, X. The accuracy of estimating individual tree variables with airborne laser scanning in a boreal nature reserve. *Can. J. For. Res.* **2004**, *34*, 1791–1801. [CrossRef]

33. Donoghue, D.N.; Watt, P.J.; Cox, N.J.; Wilson, J. Remote sensing of species mixtures in conifer plantations using LiDAR height and intensity data. *Remote Sens. Environ.* **2007**, *110*, 509–522. [CrossRef]

34. Kwak, D.-A.; Cui, G.; Lee, W.-K.; Cho, H.-K.; Jeon, S.W.; Lee, S.-H. Estimating plot volume using LiDAR height and intensity distributional parameters. *Int. J. Remote Sens.* **2014**, *35*, 4601–4629. [CrossRef]

35. Johnson, R.A.; Wichern, D.W. *Applied Multivariate Statistical Analysis*; Prentice Hall: Upper Saddle River, NJ, USA, 2002; Volume 5.

36. Fisher, R.A. The use of multiple measurements in taxonomic problems. *Ann. Eugen.* **1936**, *7*, 179–188. [CrossRef]

37. Lachenbruch, P.A.; Mickey, M.R. Estimation of error rates in discriminant analysis. *Technometrics* **1968**, *10*, 1–11. [CrossRef]

38. Monserud, R.A.; Leemans, R. Comparing global vegetation maps with the Kappa statistic. *Ecol. Model.* **1992**, *62*, 275–293. [CrossRef]

39. O'brien, R.M. A caution regarding rules of thumb for variance inflation factors. *Qual. Quant.* **2007**, *41*, 673–690. [CrossRef]

40. Kutner, M.H.; Nachtsheim, C.J.; Neter, J.; Li, W. *Applied Linear Statistical Models*; McGraw-Hill Irwin: New York, NY, USA, 2005; Volume 5.
41. Næsset, E.; Bjerknes, K.-O. Estimating tree heights and number of stems in young forest stands using airborne laser scanner data. *Remote Sens. Environ.* **2001**, *78*, 328–340. [CrossRef]
42. Moffiet, T.; Mengersen, K.; Witte, C.; King, R.; Denham, R. Airborne laser scanning: Exploratory data analysis indicates potential variables for classification of individual trees or forest stands according to species. *ISPRS J. Photogramm. Remote Sens.* **2005**, *59*, 289–309. [CrossRef]
43. Schreier, H.; Lougheed, J.; Tucker, C.; Leckie, D. Automated measurements of terrain reflection and height variations using an airborne infrared laser system. *Int. J. Remote Sens.* **1985**, *6*, 101–113. [CrossRef]
44. Rautiainen, M.; Lukeš, P.; Homolova, L.; Hovi, A.; Pisek, J.; Mõttus, M. Spectral properties of coniferous forests: A review of in situ and laboratory measurements. *Remote Sens.* **2018**, *10*, 207. [CrossRef]
45. Roberts, D.A.; Ustin, S.L.; Ogunjemiyo, S.; Greenberg, J.; Dobrowski, S.Z.; Chen, J.; Hinckley, T.M. Spectral and structural measures of northwest forest vegetation at leaf to landscape scales. *Ecosystems* **2004**, *7*, 545–562. [CrossRef]
46. Knyazikhin, Y.; Schull, M.A.; Stenberg, P.; Mõttus, M.; Rautiainen, M.; Yang, Y.; Marshak, A.; Carmona, P.L.; Kaufmann, R.K.; Lewis, P. Hyperspectral remote sensing of foliar nitrogen content. *Proc. Natl. Acad. Sci. USA* **2013**, *110*, E185–E192. [CrossRef]

 *remote sensing*

Article

Assessing the Influence of UAV Altitude on Extracted Biophysical Parameters of Young Oil Palm

Ram Avtar [1,2,*], Stanley Anak Suab [2], Mohd Shahrizan Syukur [3], Alexius Korom [4], Deha Agus Umarhadi [2] and Ali P. Yunus [5]

1   Faculty of Environmental Earth Science, Hokkaido University, Sapporo 060-0810, Japan
2   Graduate School of Environmental Earth Science, Hokkaido University, Sapporo 060-0810, Japan; stan@eis.hokudai.ac.jp (S.A.S.); deha@eis.hokudai.ac.jp (D.A.U.)
3   Faculty of Plantation and Agrotechnology, Universiti Teknologi MARA (UiTM), Shah Alam, Selangor 40450, Malaysia; 2016321111@isiswa.uitm.edu.my
4   Faculty of Plantation and Agrotechnology, Universiti Teknologi MARA (UiTM) Sabah Branch, Kota Kinabalu, Sabah 88997, Malaysia; alexi502@uitm.edu.my
5   Center for Climate Change Adaptation, National Institute for Environmental Studies, Tsukuba, Ibaraki 305-8506, Japan; pulpadan.yunusali@nies.go.jp
*   Correspondence: ram@ees.hokudai.ac.jp; Tel.: +81-011-706-2261

Received: 27 August 2020; Accepted: 14 September 2020; Published: 17 September 2020

Abstract: The information on biophysical parameters—such as height, crown area, and vegetation indices such as the normalized difference vegetation index (NDVI) and normalized difference red edge index (NDRE)—are useful to monitor health conditions and the growth of oil palm trees in precision agriculture practices. The use of multispectral sensors mounted on unmanned aerial vehicles (UAV) provides high spatio-temporal resolution data to study plant health. However, the influence of UAV altitude when extracting biophysical parameters of oil palm from a multispectral sensor has not yet been well explored. Therefore, this study utilized the MicaSense RedEdge sensor mounted on a DJI Phantom–4 UAV platform for aerial photogrammetry. Three different close-range multispectral aerial images were acquired at a flight altitude of 20 m, 60 m, and 80 m above ground level (AGL) over the young oil palm plantation area in Malaysia. The images were processed using the structure from motion (SfM) technique in Pix4DMapper software and produced multispectral orthomosaic aerial images, digital surface model (DSM), and point clouds. Meanwhile, canopy height models (CHM) were generated by subtracting DSM and digital elevation models (DEM). Oil palm tree heights and crown projected area (CPA) were extracted from CHM and the orthomosaic. NDVI and NDRE were calculated using the red, red-edge, and near-infrared spectral bands of orthomosaic data. The accuracy of the extracted height and CPA were evaluated by assessing accuracy from a different altitude of UAV data with ground measured CPA and height. Correlations, root mean square deviation (RMSD), and central tendency were used to compare UAV extracted biophysical parameters with ground data. Based on our results, flying at an altitude of 60 m is the best and optimal flight altitude for estimating biophysical parameters followed by 80 m altitude. The 20 m UAV altitude showed a tendency of overestimation in biophysical parameters of young oil palm and is less consistent when extracting parameters among the others. The methodology and results are a step toward precision agriculture in the oil palm plantation area.

Keywords: UAV; different altitudes; multispectral; biophysical parameters; young oil palm

---

1. Introduction

The oil palm (*Elaeis guineensis*) is an important industrial cash crop for major producer countries such as Indonesia, Malaysia, and Thailand, which provide sizeable economic benefits both from employment and income through exports [1]. Malaysia is the second-largest producer of oil palm and employs more than 600,000 high- and low -skilled laborers. In the next few years, ove 66,000 new jobs are expected to be created through continued research and innovation [2]. According to the Department of Statistics in Malaysia, oil palm is a significant contributor to the gross domestic product (GDP) of the agriculture sector by 46% in 2017 [3]. Oil palm is the most significant source of vegetable oil because of its high yield and extended productivity with a lifespan up to 25 years [4]. Currently, 4.49 million hectares of land in Malaysia is planted with oil palm, which produced 17.73 million tons of oil palm [5]. More land areas for expansion of oil palm plantations are controversial and not sustainable; hence, there is a need to optimize and maximize oil palm yield and production [6]. Moreover, the yield of oil palm plantation depends mostly on plant health. In addition, the corresponding market price depends heavily on the quality of oil palm [1]. However, oil palm growth is susceptible to the effects of climate change through a range of expected biotic (e.g., pests, diseases, pollinators, associated crops) and abiotic (e.g., temperature, rainfall, soil moisture, soil pH) stresses [7].

Information on oil palm plantation health conditions provides valuable inputs for the oil palm companies for planning, decisions, and management strategies. Information technology plays a vital role in increasing the cost-effectiveness of agriculture practices in precision agriculture. Precision agriculture implements management activities, both spatially and temporally. These include pre-planting, planting, fertilizing, crop protection, harvesting, and irrigation [8]. Remote sensing is one of the main tools that supports precision agriculture as the spatial data provider with its spectral capability to detect some variables, including soil properties, plant health, and crop yields [9]. In the case of oil palm plantations, plant health detection at an early stage is crucial to curb future losses from underperforming trees. There could be several reasons for low yields, such as diseases, pest attacks, weak quality seedlings, fertilizers, climatic, and edaphic factors that require further investigations. Previous studies reported that early monitoring of oil palm health not only promotes appropriate and effective remedial measures but also extends oil palm lifespan and increases productivity [10]. The health of oil palm can be monitored by studying the biophysical parameters such as height, crown size, and vegetation vigor. Spectral reflectance-based vegetation indices are effective in monitoring vegetation vigor and phenological parameters [11]. Several biophysical parameters such as leaf area index (LAI), crown diameter, crown projection area (CPA), vigor, and tree height are positively correlated with the plant growth stage [12]. Real-time quantification of these parameters can be useful for detecting the health of a tree, which allows the selection of appropriate remedial measures such as the use of fertilizer, insecticides, and irrigation to improve tree health. Meng et al. studied real-time detection of ground objects using unmanned aerial vehicle (UAV) and deep learning methods in China [13].

The recent development in UAV techniques made it possible to apply low altitude photogrammetric techniques in precision agriculture due to their flexibility and low cost [14]. In the oil palm industry, UAV-based imaging provides low cost flexible data acquisition with less weather constraints and higher spatial/temporal resolution, as compared to high-resolution satellite data [15]. There are various applications of UAV, such as monitoring canopy structure and condition, mapping biomass, and precision agriculture [16,17]. There are ground-based sensors available for precision agriculture applications, but UAV-based monitoring is advantageous in generating smaller ground sample distances, instantaneous calibration to reflectance, and point cloud construction [18]. The structure from motion (SfM) technique is useful to characterize individual trees [17]. Díaz-Varela et al. and Zarco-Tejada et al. used the SfM technique to estimate olive tree height and crown diameter in Spain [19,20]. Previous studies showed that UAV-based SfM derived canopy cover of oil palm showed 20-50% overestimation as compared to ground-based measurement [21]. Usually, before conducting aerial surveys, several parameters need to be optimized. These include flight altitude, image overlap, speed, resolution, and area of coverage [22]. Logically, higher flight altitude captures a smaller number of images with lower ground sample distance (GSD) because of the broad field of view of the camera sensors onboard the UAV. High flight altitude can influence the accuracy of information derived for an object due to the decline of the image detail [23]. Hence, lower altitude UAV flight (15–30 m) can provide more accurate and detailed image information [22]. However, it has been reported that there is no significant difference in the normalized difference vegetation index (NDVI) value between two objects (weed and crop) in images taken from 60 m, 80 m, and 100 m above ground level [24]. Moreover, the reconstruction of 3D point clouds is sensitive to the movement of twigs and leaves induced by wind [22]. As the UAV flies lower, the camera captures more images so that the possibility of the object movement becomes greater. In addition, more images require more storage capacity and more computing power for processing.

Oil palm plantations generally cover a large area. However, UAV data collected at low flight altitudes can only cover a small area in a given time. There is a trade-off between flight altitude and area covered during the flight [22]. Some aerial surveys conducted over oil palm considered using 80 m, 100 m, and 150 m flight height [1,23,25]. See et al. found that 80 m altitude aerial images produced a fair amount of accuracy in individual tree identification and tree crown delineation in matured oil palm plantations [1]. A previous study on the comparison between fixed-wing and multi-rotor UAVs suggests that flying altitude below 150 m is suitable for environmental mapping for better representation of vegetation features. Multi-rotor UAV systems are more accurate and better suited for small areas than fixed-wing drones [22]. The flight altitude can directly influence the details and quality of the derived biophysical vegetation parameters. There is a lack of studies about the influence of low flight altitude on the extraction of biophysical parameters of young oil palm. Only a few studies noticed the impact of flight altitude on data acquisition and processing time [26]. Furthermore, Torres-Sanchez et al. investigated the influence of UAV collected image overlap on computation and DSM accuracy in olive orchards in Calancha, Spain [27]. Therefore, it is necessary to understand the influence of flight altitude on derived biophysical parameters of young oil palm for precision agriculture studies. This study attempts to compare the influence of different flight altitudes to derive biophysical parameters of young oil palm using the SfM technique.

2. Study Area

The study site lies between latitude 5°8′8.368″N to 5°8′4.852″N and longitude 118°24′26.299″E to 118°24′35.717″E in the Lahad Datu district of the Eastern coast of Sabah, Malaysia. The area of interest (AOI) covers 5.2 acres of young oil palm planted trees (Figure 1). In total, 241 young oil palm trees with the age of 3 to 4 years were present in the study area. All the oil palm trees were planted with a fixed tree spacing of approximately 8 × 8 m. The climate of the study area is tropical, with an average annual temperature of 26.9 °C and an average rainfall of 2063 mm [28]. Figure 1 shows the location of the study area and individual oil palm trees in the inset aerial images of the AOI.

Besides oil palm trees, the ground area is covered with *Mucuna Bracteata*, a type of land cover crop purposely planted to protect the soil from weeds. *Mucuna Bracteata* in oil palm plantations also helps to maintain soil moisture content, supply organic matter, and protect from soil erosion [29]. Some of the trees were affected by Rhinoceros beetle (*Oryctes rhinoceros* L.) and other pests in the study area. Hence, some of the trees were showing damaged fronts and dying leaves. Rhinoceros beetles (RB) destroy the young oil palm trees by burrowing into the shoots and young fronts. Figure 2a,b show a healthy and diseased oil palm tree, respectively. Figure 2c shows a variety of damages caused by RB and other pests. Overall, the affected oil palm trees show biological and physical damages such as dying leaves, stunted growth, and irregular crowns. Therefore, the detection of these diseased trees is essential to follow up treatment and control over the spreading to other healthy oil palm trees.

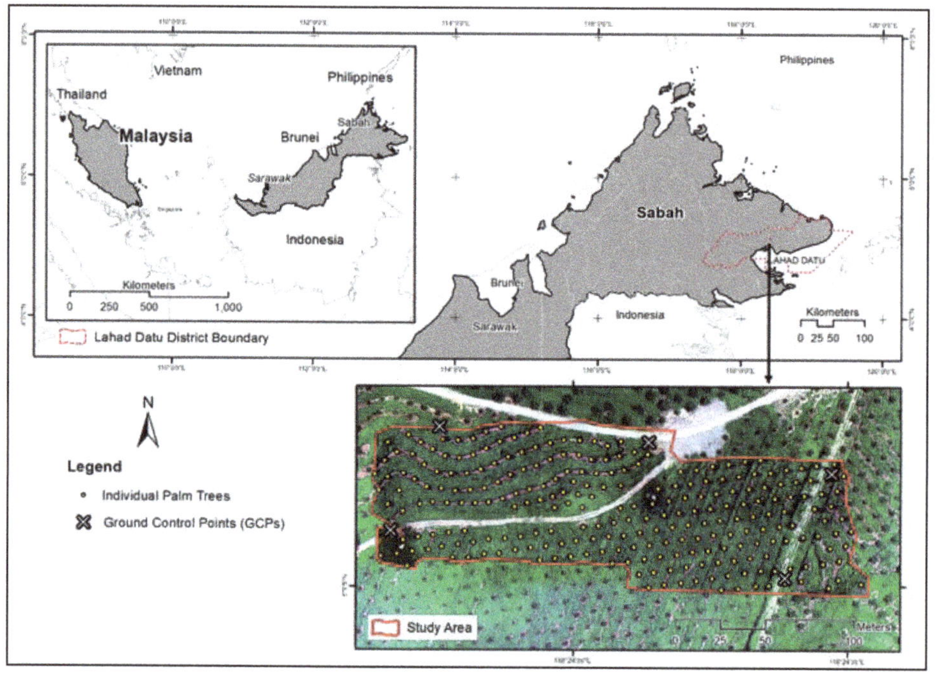

Figure 1. Location of the study area in Lahad Datu, Sabah, Malaysian Borneo.

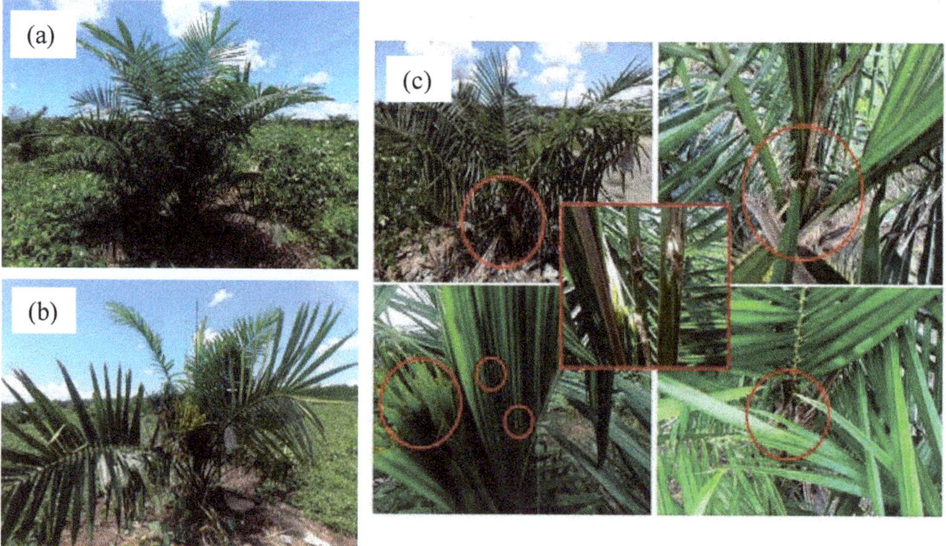

Figure 2. Field photographs of Oil palm: (**a**) Healthy tree, (**b**) diseased tree, and (**c**) damaged tree due to the rhinoceros beetle (*Oryctes Rhinoceros* L.) and other pests.

3. Materials and Methods

3.1. Aerial Imaging Tools and Data Collections

For the aerial surveys, DJI Phantom–4 mounted with a MicaSense RedEdge multispectral sensor at downward-facing/nadir was used. A multi-rotor UAV platform was chosen for this study because it is capable of capturing images at low altitudes for close-range photogrammetry. Phantom–4 is a quadcopter, which enables vertical take-off and landing, as well as slow flight speed to provide a stable platform for the multispectral camera. The MicaSense RedEdge camera was mounted with a GPS device to acquire geotagged images with 2–3 m accuracy [30]. It captures information in five spectral bands within the visible to red-edge and infrared spectrum. A downwelling light sensor (DLS) and calibrated reflectance panel were used to calibrate the images according to ambient light (Figure 3a). The MicaSense RedEdge Multispectral sensor was calibrated on-site before each flight using the reference panel for accurate ground reflectance calibration (Figure 3a). Tables 1 and 2 show the specifications of MicaSense RedEdge sensor and details about the spectral bands with wavelength and bandwidth, respectively. Ground control points (GCPs) were collected using Leica GS20 real-time differential GPS base and rover system with sub-meter accuracy (Figure 3b). Handheld Garmin GPSMAP 60CSx and GoPro Hero-6 Action Camera were used to record location points and capture images of oil palm tree conditions. Oil palm tree height and crown diameter samples were also measured at the ground using the measuring tape and a height stick.

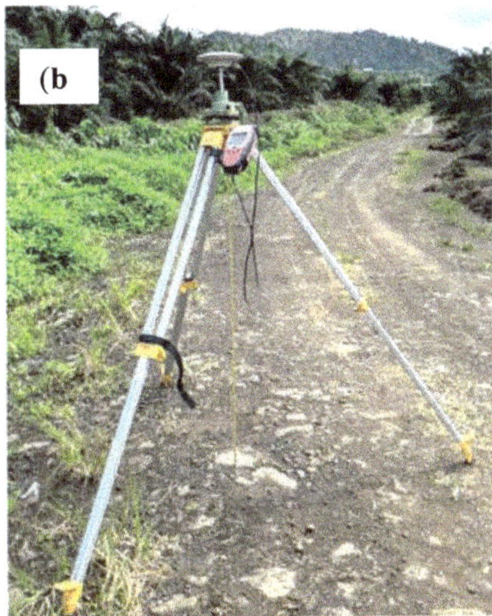

Figure 3. Equipment: (**a**) flight preparation and sensor calibration; and (**b**) ground control points (GCPs) collection.

Table 1. Specifications of MicaSense RedEdge sensor.

Parameter	Specification
Spectral bands	Blue, green, red, red edge, near-infrared
Ground sample distance	8.2 cm/Pixel (per band) at 120 m above ground level
Capture speed	Programmable by seconds interval for all bands
Format	RAW 12–bit camera
Foal length / field of view (FOV)	5.5 cm/47.2 degrees (FOV)
Image resolution	1280 × 960 pixels

Table 2. MicaSense RedEdge spectral bands with respective wavelength and bandwidth values.

Band	Center Wavelength (nm)	Bandwidth (nm)
Blue (B)	475	32
Green (G)	560	27
Red (R)	668	16
Red edge (R–Edge)	717	12
Near-infrared (NIR)	842	57

The UAV surveys were conducted on 29 and 30 August 2018 between 10–12 a.m. The weather conditions during the data acquisitions were adequate with enough solar illumination, calm wind with a slight breeze, and no clouds. The flight missions were planned using the DJI flight planner and executed by Pix4DCapture apps. A single grid type flight plan was deployed in an automatic mode at three different flight altitudes of 20 m, 60 m, and 80 m above ground level (AGL). The UAV flight speed of each flight altitude was at 5 m/s. The MicaSense RedEdge camera was programmed using Bluetooth connection to capture every 2 seconds and preview the initial test images every time before take-off. This was to ensure image capturing was started with the right exposure setting, which was calibrated before every flight missions. Images were captured with a flight path setting of 80% front overlap and 75% side lap. The ground sampling distance (GSD) varies with the flight altitude. The time required for the processing of aerial images at different flight altitudes with the same workstation was also recorded and summarized in Section 4.1.

3.2. Data Processing

The flowchart of the proposed method is summarized in Figure 4. The overall aim is to investigate the influence of UAV altitude using MicaSense RedEdge Multispectral sensor on the extracted oil palm's biophysical parameters. Therefore, a suitable spatial scale of data collection could be determined, which will be useful for precision agriculture applications. The collected multispectral images at altitude 20 m, 60 m, and 80 m were processed using the structure from motion (SfM) technique in PiX4D mapper software running in a workstation with an Intel Core i7–(9700) processor and with 16 GB random access memory (RAM). Standard image processing steps in the Pix4Dmapper software were followed. In the initial processing, the individual bands (B, G, R, R-Edge, and NIR) of the aerial images were radiometrically corrected using reference images of the calibration panel, which were also collected in the field before every flight. Then, followed by key point extraction, matching, camera optimization, and geolocations of GCPs occurred. The coordinate system used in this study is the local coordinate system of a Borneo rectified skew orthomorphic (BRSO) Timbalai 1948 in meters measurement unit. The processing was then followed by steps of creating point clouds with scale constraint defined setting and meshing in Pix4Dmapper with settings selected at high resolution. The final steps of the processing were the generation of outputs of multispectral orthomosaic (B, G, R, R-Edge, and NIR) in GeoTIFF format, point clouds in LAS format, and digital surface model (DSM) in GeoTIFF format.

The point clouds were used to produce DEM and to subtract the DSM for canopy height models (CHM) production and subsequently to delineate individual oil palm tree crowns for generating the CPA (details in Sections 3.2.1–3.2.3). The orthomosaics were used for the transformations to vegetation indices of NDVI and normalized difference red edge index (NDRE). Figure 5 shows the images in true color composite with the orthomosaics combination of RGB bands at a flight altitude of 20 m, 60 m, and 80 m. Figures A1 and A2 illustrate the output of 3D point clouds and 3D DSM for the three flights altitude, respectively. Statistical information of the biophysical parameters of CPA, height, including vegetation indices (NDVI and NDRE) of individual oil palm trees, were extracted and analyzed. The statistical analysis was performed with the help of central tendencies and histogram of difference to see the deviation of biophysical parameters at different flight altitudes.

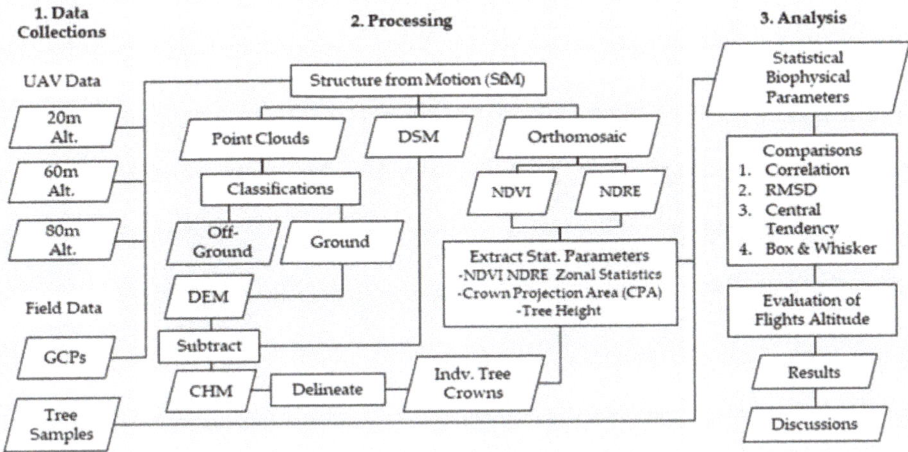

Figure 4. Flowchart of the methodology.

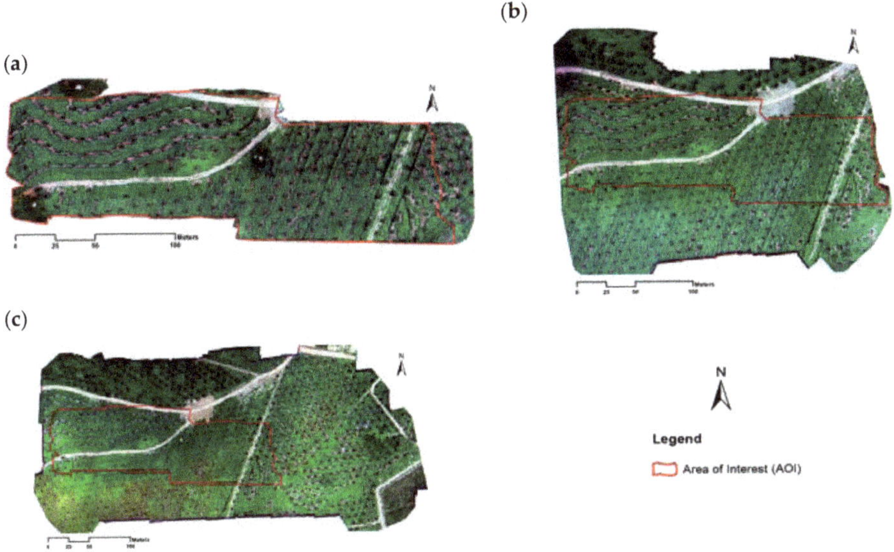

Figure 5. Multispectral orthomosaic aerial images in (RGB) true color composites at a flight altitude of (**a**) 20 m, (**b**) 60 m, and (**c**) 80 m with AOI overlay.

3.2.1. Classifications of Point Clouds and Production of DSM and DEM

The point clouds were classified to separate oil palm trees (off-ground points) and ground elevations (ground points) using automatic cloth simulation filter (CSF) in CloudCompare software. However, automatic CSF classification was not sufficient; some remaining off-ground points in the ground class were cleaned manually. It was observed that CSF was unable to totally clean off ground point clouds because of the inability to detect independent non-grouped off ground points. Cleaned ground point clouds were used to produce the digital elevation model (DEM) of the study area while the original was for productions of DSM directly. The term DEM was employed because the ground points did not indicate real bare ground. The ground was covered by the legume crop

Mucuna Bracteata at a height of approximately less than 25 cm. Figure 6a,b illustrate the original point clouds and classification operations to extract ground elevation, respectively.

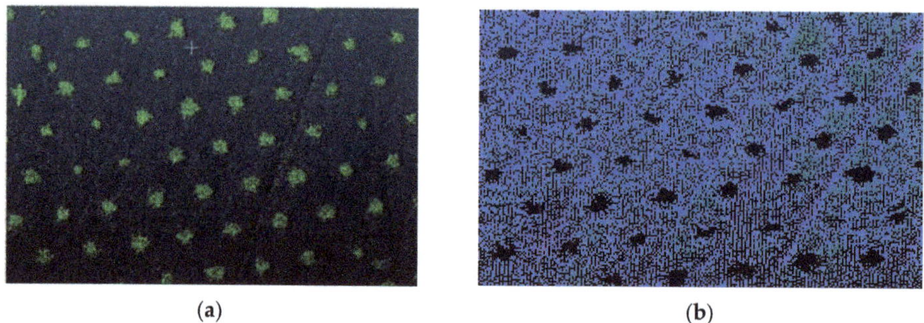

(a)   (b)

Figure 6. Point clouds: (**a**) original and (**b**) ground elevation after classification.

3.2.2. Production of Canopy Height Model (CHM)

The canopy height model (CHM) was derived by simple subtraction of the DEM from the DSM (i.e., CHM = DSM − DEM) computed in ArcMap using the Raster Calculator tools. The CHM process is illustrated in Figure 7, with all DEM, DSM, and CHM showed from 3D perspectives. It can be observed in the CHM that individual oil palm tree canopies with crown and height were depicted in black to white color height gradient.

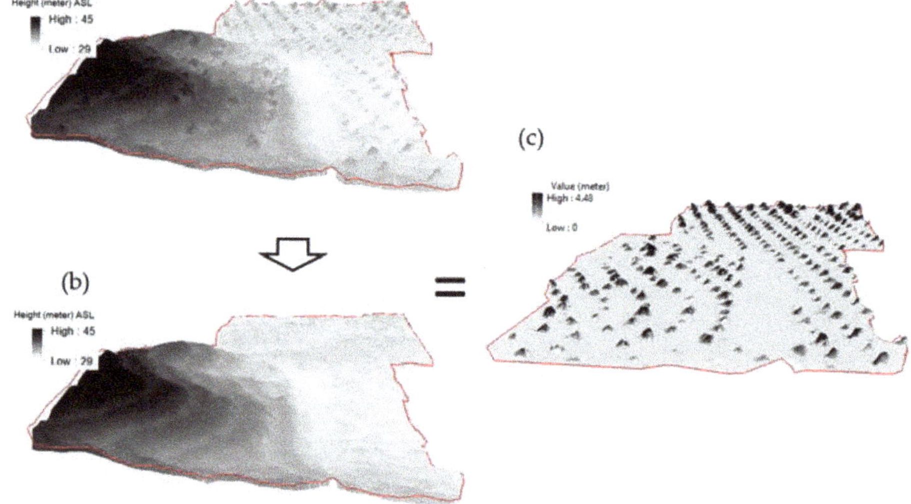

Figure 7. 3D view of (**a**) DSM subtracted using (**b**) DEM to produce (**c**) CHM.

3.2.3. Height, Crown Area, and Crown Projection Area (CPA)

The crown area of individual trees were extracted into vector format by generating contours at a specified height of crown edges of CHM for each altitude in Global Mapper software. In CHM, the pixels with high values represent the presence of the oil palm trees, as compared to the surrounding CHM, which has zero value after the terrain was completely removed (Figure 7c). The extracted individual tree crown area vectors were filtered to remove small polygons that did not represent oil palm trees. Subsequently, information of individual oil palm tree height was derived from the CHM-based on crown area maximum height using zonal statistics in ArcMap software. As a result, individual oil palm tree height was stored in a tabular format. Additionally, the CPA was generated using "minimum bounding geometry" in a circle that represents the generalized size of the area covered by the crown. Figure 8 represents the biophysical parameters of individual oil palm trees. Height, crown area, and CPA information were combined into the attribute table of individual trees.

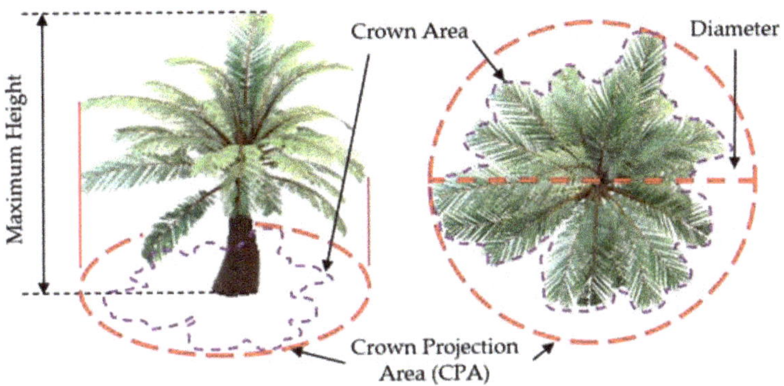

Figure 8. Illustration of height, crown area, and CPA.

3.2.4. Vegetation Indices (NDVI and NDRE) Transformations

The land use and land cover in the study site are mainly oil palm trees, *Mucuna Bracteata*, and exposed soil. Previous studies have used various vegetation indices based on UAV acquired data to delineate information about various vegetation parameters [31–33]. These vegetation indices have been widely used to identify vegetation and soil properties. Basically, plants interact with incident solar radiation by absorbing, transmitting, and/or reflecting electromagnetic radiation. The reflected radiation contained information about the plants' biophysical composition and physiological status and is measured with multispectral sensors [34]. In this study, NDVI and NDRE were computed using ArcMap software. NDVI is a standard spectral transformation technique used for monitoring vegetation health [35,36]. NDVI shows a strong sensitivity to the vegetation as compared to the background soil. The equation for calculation of NDVI is given below:

$$NDVI = \frac{NIR - Red}{NIR + Red} \tag{1}$$

where red represents reflectance in red band and NIR is the reflectance in the near-infrared band of the acquired data.

Meanwhile, NDRE is used to measure the stress and chlorophyll content in leaves [37]. It is more suitable to detect early stress as compared to NDVI [38]. NDRE is the ratio measurement between the near-infrared band with the red edge band. The equation to calculate NDRE is given below:

$$NDRE = \frac{NIR - Red\ Edge}{NIR + Red\ Edge} \tag{2}$$

Both NDVI and NDRE were computed using the multispectral orthomosaic aerial images in ArcMap software for the 20 m, 60 m, and 80 m flight altitudes. Figures A3 and A4 show the NDVI and NDRE images of the study area at UAV altitude 20 m, 60 m, and 80 m, respectively. NDVI and NDRE show a variation in the vegetative and soil areas. The NDVI and NDRE values of individual oil palm trees were extracted using the vector file of the crown area in ArcMap.

3.3. Data Analysis

The biophysical parameters (CPA diameter, tree height, including vegetation indices of NDVI, and NDRE) were statistically analyzed to evaluate the influence of flight altitudes. Firstly, we compared biophysical parameters extracted from different flight altitudes using central tendencies from histogram analysis. Histograms were used to see the tendency of value differences resulting from the deviation of each parameter between different flight altitudes. The central tendency of subtraction values was assessed by calculating mean and median values. Ground measured CPA diameter and tree height were used to validate UAV derived parameters. Standard error of estimation (SE) with a 95% confidence level was used to assess the accuracy of UAV extracted CPA diameter and tree height with the ground information [39,40]. For NDVI and NDRE, root mean squared deviation (RMSD) was calculated to determine the absolute difference between each UAV altitude [41]. The formula of RMSD is given as follows:

$$RMSD = \sqrt{\frac{\sum_{i=1}^{N}(y1i - y2i)^2}{N}} \tag{3}$$

where $y1$ denotes the value of parameter obtained from "1" UAV altitude, while $y2$ from "2" UAV altitude. N represents the total number of samples and i represents a specific tree sample.

4. Results

UAV collected data were processed to extract various biophysical parameters and transformations of vegetation indices from oil palm trees. Table 3 summarized the areal coverage, duration, ground sampling distance (GSD), and processing time (total and per acres) of images for different UAV flight altitude acquisition. Higher altitude flight of 60 m and 80 m yielded much larger areal coverage (12 and 22 acres) in just one flight as compared to the 20 m altitude, which required 4 flight missions to cover just 5.7 Acres. This is mainly because it was not possible to perform the whole flight mission with only one battery.

Table 3. Summary of flight parameters and details of collected images. GSD: ground sampling distance.

Flight Altitude	Area Covered (Acres)	Number of Flights	Planned GSD	Processed GSD	Number of Images	Total Processing Time	Processing Time per Acres
20 m	5.7	4	1.39 cm	1.37 cm	8800	4h and 22 m	46 m
60 m	12.2	1	4.17 cm	5.16 cm	2350	56 m	5 m
80 m	22.0	1	5.56 cm	5.68 cm	2195	1h and 6 m	3 m

On the other hand, a lower altitude (20 m) gathered more than three times the number of aerial images (8800) compared to 60 m (2350) and 80 m (2195). The processing time required for 20 m, 60 m, and 80 m altitude flight missions was about 4 hours 22 minutes, 56 minutes, and 1 hour 6 minutes, respectively (Table 3). The 60 m altitude required less processing time compared to the 80 m because the areal coverage was only about half of that obtained during the 80 m altitude flights. Nevertheless, the image acquired at 20 m flight altitude produced GSD of 1.37 cm, whereas 60 m and 80 m flight resulted in 5.16 cm and 5.68 cm GSD, respectively. A lower GSD (higher resolutions) implies that more ground details and, therefore, dense point clouds are available for the subsequent SfM analysis and DSM generation. There is a positive relationship between dense point cloud reconstruction and processing time, which agrees with the studies that reported the relationship between flight altitude and point cloud density [42]. In general, a higher flight altitude causes a decrease in processing time [27]. The results of the biophysical parameters extracted from different flight altitude missions are discussed in the following sections.

4.1. Crown Projection Area (CPA) Diameter

The crown area of individual oil palm trees was generated from the canopy height model (CHM); subsequently, the crown projection area (CPA) was generated using the "minimum bounding geometry" of the crown area (see Figure 8). The CPA is indeed a generalization representing the crown rather than the crown area itself, which has complex-shaped morphometry. It is worth noting that there are certain difficulties in obtaining uniform diameter measurements from the complex shaped crown. Estimating crown diameter based on CPA may result in under- or overestimation. Yet, CPA size variations within a uniform age oil palm plantation may provide insight into underlying health issues when combined with vegetation indices [43]. Therefore, deriving tree diameter from the CPA is suitable for this study.

The derived CPA diameters were statistically analyzed for the three different flight altitudes. The CPA diameter values were plotted on the x and y-axis (Figure 9a–c). From the results, it can be observed that the scatterplots exhibit normal distributions with all CPA values are clustered around the 1:1 line. Despite the difference in the altitudes, the derived CPA values have a strong linear relationship indicated by the high correlations with the value of the coefficient of determination (R^2) more than 0.61. The CPA derived from 20 m and 60 m altitudes show a weaker relationship ($R^2 = 0.616$), while CPA derived from 60 m and 80 m show a high correlation ($R^2 = 0.649$).

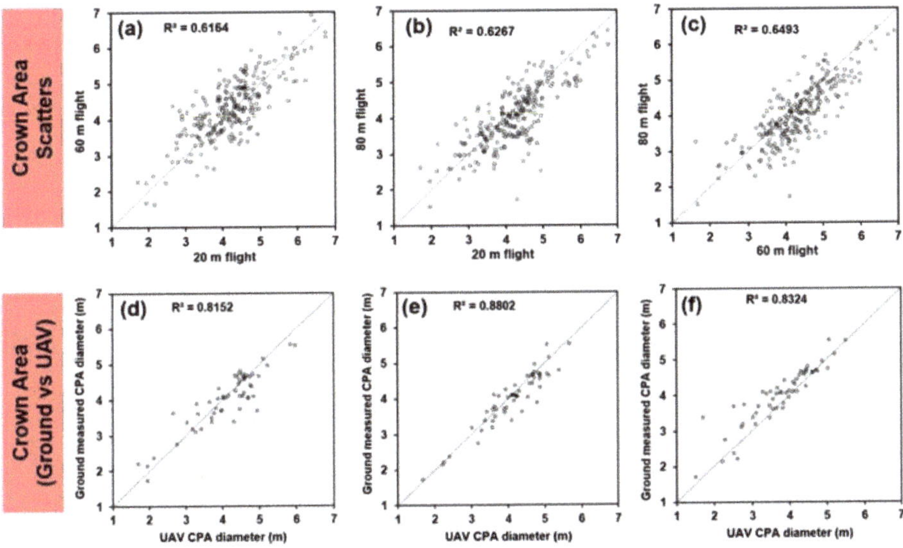

Figure 9. Scatterplot to compare crown projected area (CPA) diameter measured from different altitudes: (**a**) 20 m and 60 m, (**b**) 20 m and 80 m, and (**c**) 60 m and 80 m. A plot between ground measured CPA and UAV estimated CPA at flight altitude of (**d**) 20 m, (**e**) 60 m, and (**f**) 80 m.

Figure 10 shows the histogram of the difference for CPA to see the tendency at different altitudes. Difference values are the result of subtraction between the compared flight altitudes. It can be observed in the graphs that the CPA values derived from 80 m flight are generally less than 20 m and 60 m flight altitudes (see Figure 10b,c). CPA values at 60 m UAV altitude are higher among the others as the central tendency of histograms (Figure 10a,b) skewing towards 60 m altitudes.

Validation of CPA Diameter

The UAV derived CPA diameters were validated with the ground measured CPA diameters collected randomly in the field. The comparison of UAV derived CPA diameters with field data is given in Figure 9d–f, as well as Table 4. As shown in Table 4, all three flights show high correlations to the ground data, but the strongest correlation is observed from the 60 m altitude data (0.938). A high correlation indicates that the UAV derived CPA diameter has a good agreement with the ground data.

In addition, to assess the accuracy, SE for three UAV altitude were also calculated. The 60 m flight altitude produced the highest accuracy of 92.47%, among other altitudes (Table 4).

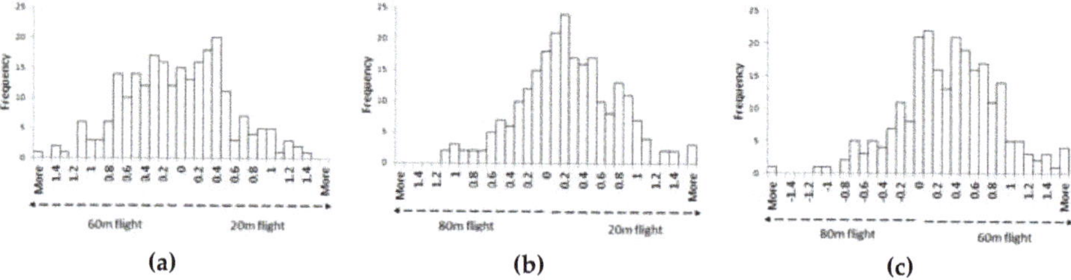

Figure 10. Histogram of difference for CPA diameter between different flight altitudes: (**a**) 60 m and 20 m, (**b**) 80 m and 20 m, and (**c**) 80 m and 60 m.

Table 4. Correlation and accuracy value of CPA diameter.

Flight Altitude	Correlation Coefficient (r)	Accuracy (%)
20 m	0.903	90.35
60 m	0.938	92.47
80 m	0.912	89.69

4.2. Tree Height Model

Unlike the tree crown relationships, the tree height scatterplots yielded a low correlation coefficient between different altitudes. As shown in Figure 11a–c, only the scatterplots of 60 m: 80 m had a good correlation (R^2 = 0.568). A lower coefficient obtained for 20 m: 60 m and 20 m:80 m suggest that the tree height measured from 20 m flight produced a large difference in tree height values compared to 60 m and 80 m flight. Histograms of difference also show that the tree heights at 20 m flight are generally higher than 60 m and 80 m, and that of 60 m is higher than 80 m flight (Figure 12).

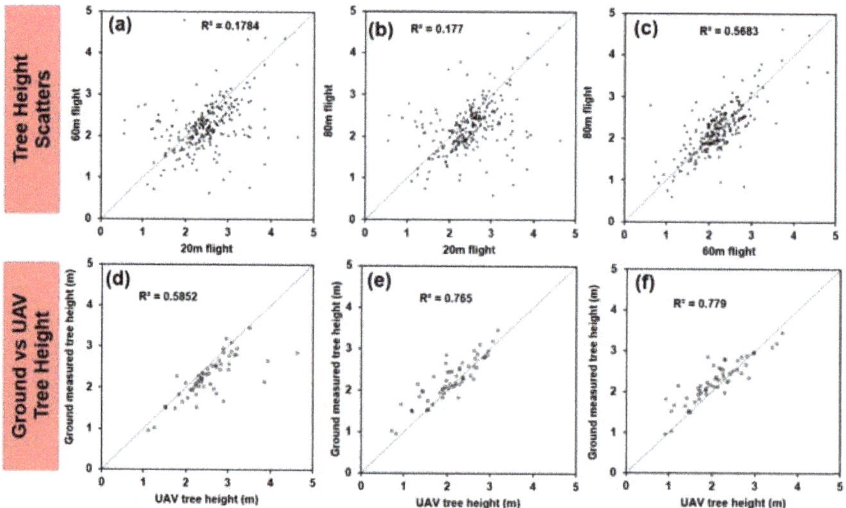

Figure 11. Scatterplot to compare tree height measured from different altitudes. Comparison between (**a**) 20 m and 60 m, (**b**) 20 and 80 m, and (**c**) 60 m and 80 m. Plots between ground measured tree height and UAV estimated tree height at flight altitude of (**d**) 20 m, (**e**) 60 m, and (**f**) 80 m.

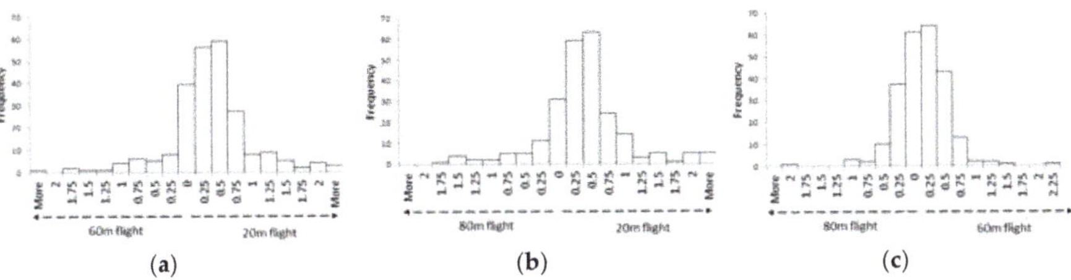

Figure 12. Histogram of difference for tree height between different flight altitudes: (**a**) 60 m and 20 m, (**b**) 80 m and 20 m, and (**c**) 80 m and 60 m.

Validation of Tree Height Model

Similar to CPA diameter, the tree heights derived from UAV were validated with ground measured tree heights as shown in Figure 11d–f. Based on the correlation and accuracy value (Table 5), our results showed that 20 m flight data has the lowest accuracy, reaching 78.10%. The 60 m UAV altitude had the highest accuracy (86.52%) among all three flight altitude missions (Table 5).

Table 5. Correlation and accuracy value of tree height.

Flight Altitude	Correlation Coefficient (r)	Accuracy (%)
20 m	0.765	78.10
60 m	0.875	86.52
80 m	0.883	85.70

4.3. Vegetation Indices (NDVI and NDRE) Comparison

Figure 13 illustrates the scatterplots to compare extracted NDVI and NDRE values from different flight altitudes. Only for the scatterplot of 60 m: 80 m were close to the 1:1 line for NDVI and NDRE (Figure 13c,f). The 60 m: 80 m also had the highest correlations for NDVI (r = 0.774) and NDRE (r = 0.696). The second highest correlation is observed for 20 m: 60 m with (r = 0.507) and (r = 0.402), respectively. The flight comparison of NDVI value at 20 m: 80 m shows the least correlation with a value of 0.435. The correlation value of NDRE at 20 m: 80 m is very low, reaching only 0.273 (p-value = 0.000018). Although having the same pattern, the relationship of each flight comparison on NDVI is stronger than that of NDRE. Similarly, the RMSD follows the same results as the scatterplots (Figure 14). Observation of scatterplots and RMSD reveals that numerous tree plots extracted from 20 m have a big difference in vegetation indices value when compared to 60 m and 80 m flight. Least RMSD value was observed between 60 m and 80 m flight altitudes.

Since the scatterplots are similar between NDVI and NDRE, the histograms of difference value provide the same pattern (Figure 15). The histograms in Figure 15c,f show that the small deviations were observed between measurements made at 60 m and 80 m, suggesting a better match between them.

4.4. Evaluation of Flight Altitude

We analyzed the optimal flight height by assessing the accuracy of CPA diameter and tree height, as well as RMSD value of NDVI and NDRE, as discussed in Sections 4.1–4.3. Furthermore, we also assessed the consistency of biophysical parameter extraction for each flight. With the growth of oil palm trees, there is growth in biophysical parameters such as crown diameter, tree height, etc. [43–45]. Hence, crown diameter and tree height should grow linearly. Since both parameters were generated from point clouds, we examined the relationship and observed whether flight altitude affects consistency. Table 6 shows a strong relationship between CPA diameter and tree height with a value of correlation coefficient is 0.568 and 0.583 at 60 m and 80 m flight altitude, respectively. The flight altitude at 20 m shows a low correlation of 0.277. Similarly, the relationship between vegetation indices

(NDVI and NDRE) was also assessed. All flight altitudes showed a strong relationship between NDVI and NDRE, with correlations of more than 0.85. The 20 m flight maintained produces the weakest relationship for both comparisons, which means having the lowest consistency. On the contrary, 60 m and 80 m were highly consistent for extracting the biophysical parameter.

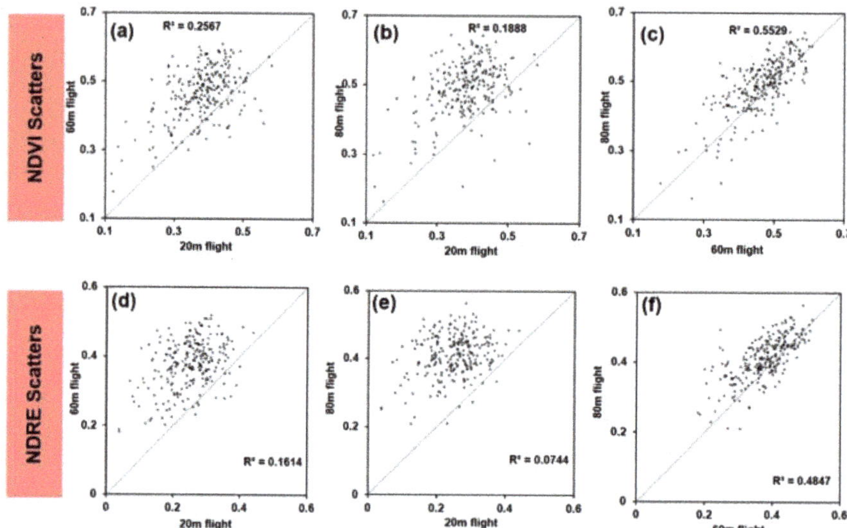

Figure 13. Scatterplot to compare NDVI and NDRE measured from different altitudes. Comparison between 20 m and 60 m (**a,d**), 20 m and 80 m (**b,e**), and 60 m and 80 m (**c,f**).

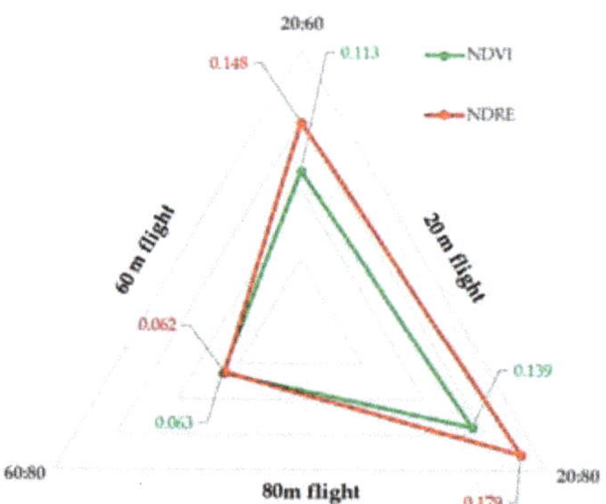

Figure 14. RMSD value of NDVI (blue) and NDRE (green) for each compared flight altitude.

Table 6. Correlation analysis between crown diameter and height, and NDVI and NDRE at different flight heights.

Variables of Comparison		Flight Altitude	Correlation Coefficient (r)
CPA Diameter	Height	20 m	0.277
		60 m	0.568
		80 m	0.583
NDVI	NDRE	20 m	0.863
		60 m	0.910
		80 m	0.924

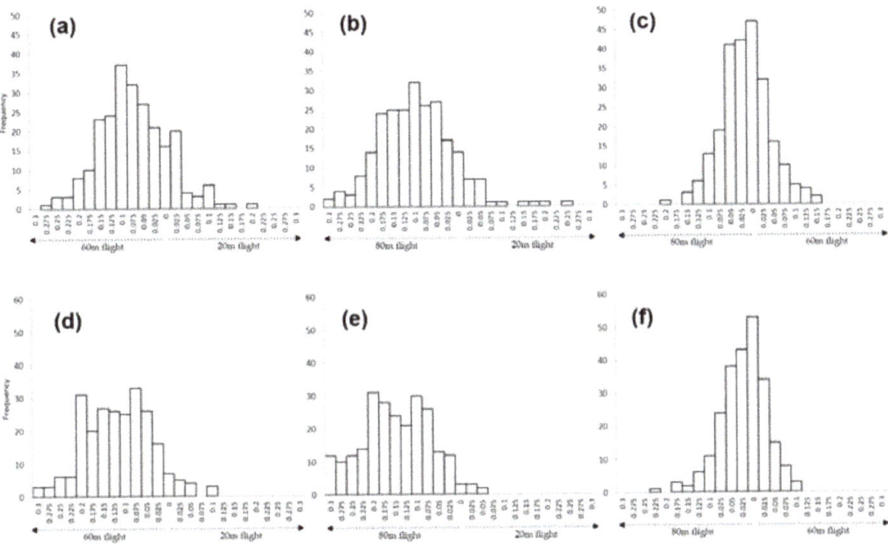

Figure 15. Histogram of difference for NDVI and NDRE between different flight altitudes: (**a,d**) between 60 m and 20 m, (**b,e**) between 80 m and 20 m, and (**c,f**) between 80 m and 60 m.

5. Discussion

This study demonstrates a systematic analysis of the influence of different UAV altitudes to extract biophysical parameters of the young oil palm plantation area in Malaysia. It provides a methodological approach to extract various biophysical parameters from UAV data. These parameters can be useful indicators to monitor plant growth and health. The objective of this study was to evaluate the suitable flying height to extract crown diameter and height using UAV-based aerial images. Obtaining the tree heights and crown diameter from satellite-based surface models have limitations due to their low spatial resolution. Further, obtaining tree height values with a GNSS device is difficult in denser forest areas [19,43]. Terrestrial light detection and ranging (LiDARs) and UAV-LiDARs, on the other hand, produce accurate results, but long processing time and heavier payload limit the gathering of base data [44]. Therefore, a low-weight DJI Phantom–4 UAV device mounted with the MicaSense RedEdge Multispectral sensor were used for obtaining the tree height and crown diameter at 20 m, 60 m, and 80 m flight altitudes. Since the RedEdge camera has the capability to obtain information in the NIR and RedEdge spectrum, we also calculated NDVI and NDRE in this study.

The important finding of this was that flight altitude at 60 m can provide more accurate results as compared to 20 m and 80 m. The highest accuracy to extract CPA diameter and height was produced at 60 m altitudes. UAV data at low altitude (20 m) with an increased number of point clouds can provide better height estimation, but it was not true in this study. The findings of this study is in contrast to Whitehead et al. [23] and Seifert et al. [22]. Whitehead et al. reported that the probability of detection of objects is better with higher point clouds data [23]. However, a higher spatial resolution is not necessary to obtain the desired accuracy, as noticed in this study. Seifert et al. [22] reported that low flight altitude could produce more details in forest areas. The reason might be that the authors also used high forward overlaps, which is not explored in this study. Even though 20 m flight altitude produced more point clouds than 60 m and 80 m, the systematic error propagation may also be higher while employing denser point clouds. However, this condition is not correct to extract CPA diameter. In this study, flights at all altitudes produced high accuracy for CPA measurements. This is mainly because the extraction of CPA diameter is not as sensitive as tree height, which relies on the maximum height value of the point clouds. According to Section 4.2, many errors (overestimations) are produced at 20 m flight altitudes, which makes the accuracy lower than tree height estimated at 60 m and 80 m flight altitudes.

To produce accurate point clouds, capturing UAV aerial images closer to objects is not always necessary. At a low flight altitude, the UAV captures many images because of the smaller field of view of the sensor. However, even though producing more point clouds, the images-alignment from many images at lower altitudes can result in an additional error. Tree structures (twigs and leaf) may change due to wind-induced movement. This shifts the relative position of objects between images, which causes mismatching in image alignment [22]. Therefore, more images captured may lead to more errors in image alignment. An additional consideration when capturing aerial images is the height of objects in the area of interest. For example, the maximum oil palm height in our study area was around 5 m and therefore only 15 m difference from the sensor. Moreover, the topography was not flat, which meant some trees would be captured less than 15 m from the UAV. These conditions also make the less consistent for biophysical parameters extraction was observed at 20 m height.

Based on our results, we determined that the NDVI and NDRE are best extracted from 60 m followed by 80 m flight altitudes. Mesas-Carrascosa et al. reported that the NDVI value is not significantly affected by the different flight altitudes [24]. Contrary to their observation, we found that the low flight altitude (20 m) produced larger RMSD values than 60 m and 80 m flight. This is probably because, at the highest pixel size (1.37 cm), more noise may also be captured when compared to the coarser pixel sizes. The higher altitude can maintain the spectral accuracy as it is observed that NDVI and NDRE values of 60 m and 80 m flight are well correlated. During field measurement, the UAV speed for all altitudes were at 5 m/s. Therefore, the closer sensor to the objects, the faster its relative speed to the object even the drone speed remains the same. As a result, some images were not clear and had to be eliminated. To overcome the limitations of the distance between objects and sensors, the use of a GNSS onboard system like Phantom4 RTK is useful. It can provide real-time, centimeter-level positioning data for improved absolute accuracy on image metadata. The use of real-time detection of young oil palm biophysical parameters using UAV is advantageous because at the young stage, there is a rapid growth of biophysical parameters and it can be helpful to monitor the health of oil palm in case of pest infestation [13].

This study examined different flight altitudes, but all flights were lower than 100 m above the ground. To cover a larger area, flights at higher altitudes are needed for higher efficiency, and, in this case, 60 m altitude and 80 m altitude flights will be more efficient than flights at 20 m altitudes. It is challenging to provide an optimum value of UAV and sensors parameters since each combination of sensors and drone parameters produce different results. Nonetheless, we attempted to consider only a few parameters in this study. Thus, we need to optimize these combinations based on our requirement by considering various trade-offs such as: altitudes, sensors resolution, point clouds, processing time, side and forward overlaps, etc. A further investigation of the effects of sensors and overlaps would be desirable to better understand their impact.

This study is focused on young oil palm plantation areas with limited coverage, while the old oil palm plantation area was not considered. Fawcett et al. reported that 100 m UAV altitude is the best for estimating the height of seven-year old oil palm trees [25]. Therefore, we can suggest that flight altitude should be increased in tall trees to minimize the high relative speed, as mentioned earlier. Moreover, the crowns of young oil palm plantations are still in the growing stage, which makes the gaps between individual trees are apparent in this study. Therefore, the estimation of crown size at the young stage is more accurate as compared to the old stage because at the old stage, there is a possibility of overlaps between the crowns of old trees. This challenge can be further explored to determine optimal flight altitudes for different growth stages of oil palm in the future.

6. Conclusions

This research was undertaken to evaluate the influence of UAV flight altitude on the extraction of biophysical parameters of oil palm plantations. Multispectral UAV aerial images over oil palm plantations were processed to produce multispectral orthomosaic, DSM, and point clouds. The study involved (i) detection of individual oil palm trees and extraction of biophysical parameters; (ii) comparison of biophysical parameters with ground data; and (iii) evaluation of UAV altitude for obtaining the most accurate biophysical parameters. CHM was generated by subtraction of DSM with DEM. Individual oil palm trees were segmented from the CHM, which was used for extractions of biophysical parameters such as tree height, crown diameter also vegetation indices of NDVI, and NDRE. Statistical methods were used for comparison of biophysical parameters at different UAV altitudes. The results of the statistical analysis show that 60 m altitude is best for measuring CPA diameter and extracting oil palm height. Moreover, NDVI and NDRE show good vigor at 60 m and followed by 80 m UAV altitudes. Based on the results presented in this study, flying at 60 m altitudes is suitable for extracting biophysical parameters of oil palm. However, 20 m UAV altitude tends to overestimate the biophysical parameters even though visually, it shows the best visual detail. The 60 m followed by 80 m altitudes are suitable for UAV aerial images collection, since biophysical parameters can be accurately measured at these altitudes and larger areas can be covered more efficiently. This is also important for commercial applications. The findings of this study can be useful for future research because the generation of DSMs from UAV is rapidly increasing. This study can contribute to finding optimal flight altitudes to extract biophysical parameters accurately and efficiently. In the future, real-time processing of UAV data can help in plant disease detection and fast response to support timely remediation.

Author Contributions: Conceptualization, R.A., S.A.S., and A.K.; methodology, R.A., S.A.S., and D.A.U.; software, R.A., S.A.S., and M.S.S.; validation, S.A.S. and M.S.S., and A.K.; formal analysis, R.A., S.A.S., and D.A.U.; investigation, R.A., M.S.S., and A.K.; resources, M.S.S., A.K., and R.A.; writing—original draft preparation, R.A., S.A.S., A.K., M.S.S., D.A.U., and A.P.Y.; writing—review and editing, S.A.S., R.A., M.S.S., D.A.U., and A.P.Y.; funding acquisition, R.A. and A.K. All authors have read and agreed to the published version of the manuscript.

Funding: This research received no external funding.

Acknowledgments: The authors would like to thank the management and staff, especially the managers, Asrif Mahmud and Henry Jubair of Ladang Sabahmas, Lahad Datu, Sabah, Malaysia, for the cooperation and logistics support which enabled this research to be done. The authors would also like to thank Hokkaido University L-station and SOUSEI support for Young Researchers.

Conflicts of Interest: The authors declare no conflict of interest.

Appendix A

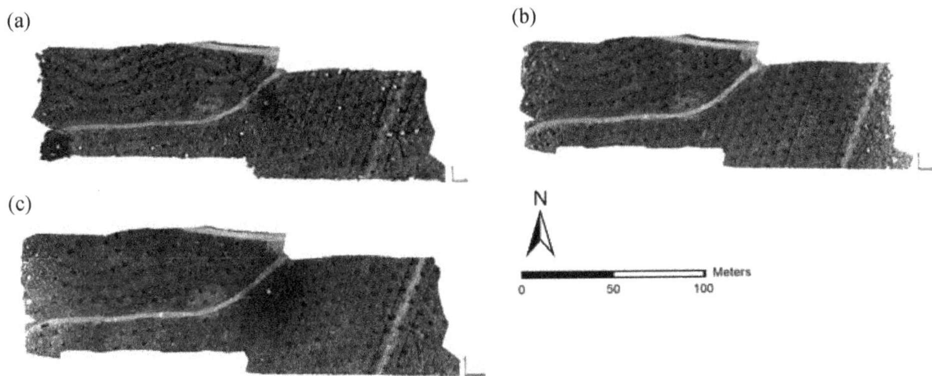

Figure A1. Point Clouds of the AOI at UAV altitude of (**a**) 20 m, (**b**) 60 m and (**c**) 80 m.

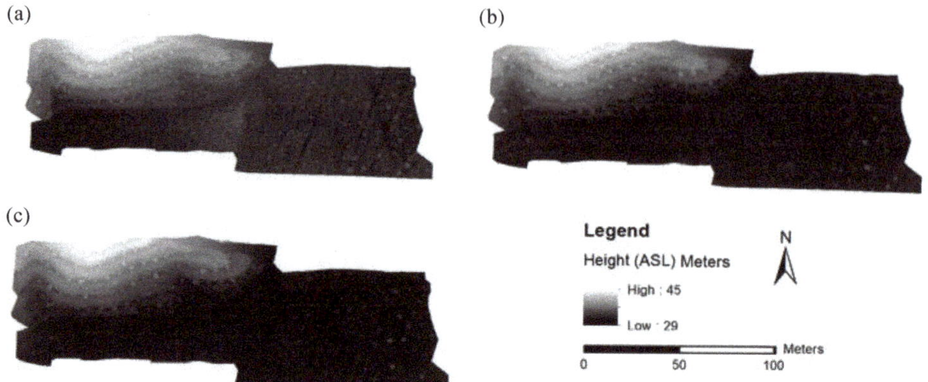

Figure A2. Digital Surface Model (DSM) of the AOI at UAV altitude of (**a**) 20 m, (**b**) 60 m and (**c**) 80 m.

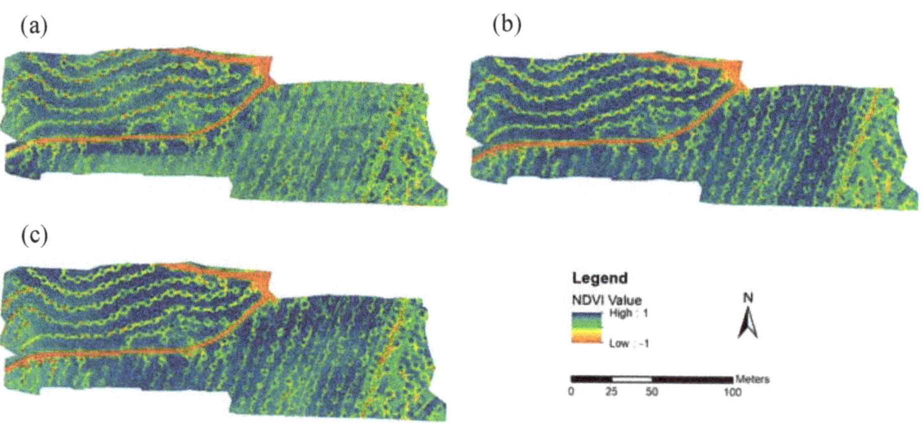

Figure A3. Normalized Difference Vegetation Index (NDVI) for (**a**) 20 m, (**b**) 60 m and (**c**) 80 m altitude.

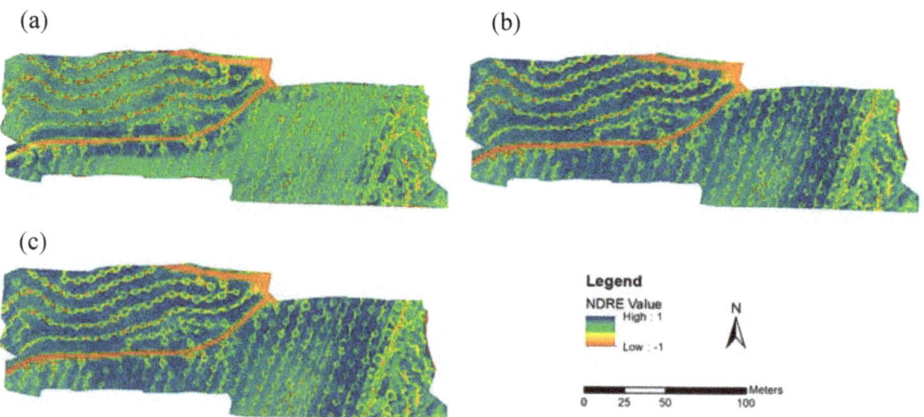

Figure A4. Normalized Difference Red Edge (NDRE) for (**a**) 20 m, (**b**) 60 m and (**c**) 80 m altitude.

References

1. See, B.D.; Hashim, S.J.; Shafri, H.Z.M.; Azrad, S.; Hassan, M.R. A new rapid, low-cost and GPS-centric unmanned aerial vehicle incorporating in-situ multispectral oil palm trees health detection. *J. Agric. Sci. Bot.* **2018**, *2*, 12–16.

2. The Oil Palm. Economic Contribution. Available online: https://theoilpalm.org/economic-contribution/ (accessed on 10 January 2019).

3. Department of Statistics, Malaysia. Available online: https://bit.ly/3hubR5h (accessed on 30 May 2019).

4. Woittiez, L.S.; van Wijk, M.T.; Slingerland, M.; van Noordwijk, M.; Giller, K.E. Yield gaps in oil palm: A quantitative review of contributing factors. *Eur. J. Agron.* **2017**, *83*, 57–77. [CrossRef]

5. Malaysia Palm Oil Council (MPOC). The Oil Palm Tree. Available online: http://mpoc.org.my/the-oil-palm-tree/ (accessed on 21 November 2019).

6. Toh, C.M.; Ewe, H.T.; Tey, S.H.; Tay, Y.H. A study on the influence of oil palm biophysical parameters on backscattering returns with ALOS-PALSAR2 image. In Proceedings of the IGARSS 2018–2018 IEEE International Geoscience and Remote Sensing Symposium, Valencia, Spain, 22–27 July 2018; pp. 8236–8239.

7. Rival, A. Breeding the oil palm (Elaeis guineensis Jacq.) for climate change. *OCL* **2017**, *24*, 1–7. [CrossRef]

8. Robert, P.C. Precision agriculture: A challenge for crop nutrition management. *Plant Soil* **2002**, *247*, 143–149. [CrossRef]

9. Friedl, M.A. *Remote Sensing of Croplands*; Elsevier: Amsterdam, The Netherlands, 2018.

10. Priwiratama, H.; Susanto, A. Utilization of Fungi for the Biological Control of Insect Pests and Ganoderma Disease in the Indonesian Oil Palm Industry. *J. Agric. Sci. Technol.* **2014**, *4*, 103–111.

11. Avtar, R.; Yunus, A.P.; Saito, O.; Kharrazi, A.; Kumar, P.; Takeuchi, K. Multi-temporal remote sensing data to monitor terrestrial ecosystem responses to climate variations in Ghana. *Geocarto Int.* **2020**, *0*, 1–17. [CrossRef]

12. Avtar, R.; Ishii, R.; Kobayashi, H.; Fadaei, H.; Suzuki, R.; Herath, S. Efficiency of multi-frequency, multi-polarized SAR data to monitor growth stages of oilpalm plants in Sarawak, Malaysia. In Proceedings of the 2013 IEEE International Geoscience and Remote Sensing Symposium-IGARSS, Melbourne, Australia, 21–26 July 2013; pp. 2137–2140.

13. Meng, L.; Peng, Z.; Zhou, J.; Zhang, J.; Lu, Z.; Baumann, A.; Du, Y. Real-time detection of ground objects based on unmanned aerial vehicle remote sensing with deep learning: Application in excavator detection for pipeline safety. *Remote Sens.* **2020**, *12*, 182. [CrossRef]

14. Maes, W.H.; Steppe, K. Perspectives for Remote Sensing with Unmanned Aerial Vehicles in Precision Agriculture. *Trends Plant Sci.* **2019**, *24*, 152–164. [CrossRef]

15. Bansod, B.; Singh, R.; Thakur, R.; Singhal, G. A comparision between satellite based and drone based remote sensing technology to achieve sustainable development: A review. *J. Agric. Environ. Int. Dev.* **2017**, *111*, 383–407.

16. Suab, S.A.; Avtar, R. *Unmanned Aerial Vehicle System (UAVS) Applications in Forestry and Plantation Operations: Experiences in Sabah and Sarawak, Malaysian Borneo*; Springer: Cham, Germany, 2020.

17. Dong, X.; Zhang, Z.; Yu, R.; Tian, Q.; Zhu, X. Extraction of Information about Individual Trees from High-Spatial-Resolution UAV-Acquired Images of an Orchard. *Remote Sens.* **2020**, *12*, 133. [CrossRef]

18. Hunt, E.R.; Daughtry, C.S.T. What good are unmanned aircraft systems for agricultural remote sensing and precision agriculture? *Int. J. Remote Sens.* **2018**, *39*, 5345–5376. [CrossRef]

19. Díaz-Varela, R.A.; de la Rosa, R.; León, L.; Zarco-Tejada, P.J. High-resolution airborne UAV imagery to assess olive tree crown parameters using 3D photo reconstruction: Application in breeding trials. *Remote Sens.* **2015**, *7*, 4213–4232. [CrossRef]

20. Zarco-Tejada, P.J.; Diaz-Varela, R.; Angileri, V.; Loudjani, P. Tree height quantification using very high resolution imagery acquired from an unmanned aerial vehicle (UAV) and automatic 3D photo-reconstruction methods. *Eur. J. Agron.* **2014**, *55*, 89–99. [CrossRef]

21. Khokthong, W.; Zemp, D.C.; Irawan, B.; Sundawati, L.; Kreft, H.; Hölscher, D. Drone-Based Assessment of Canopy Cover for Analyzing Tree Mortality in an Oil Palm Agroforest. *Front. For. Glob. Chang.* **2019**, *2*, 1–10. [CrossRef]

22. Seifert, E.; Seifert, S.; Vogt, H.; Drew, D.; van Aardt, J.; Kunneke, A.; Seifert, T. Influence of drone altitude, image overlap, and optical sensor resolution on multi-view reconstruction of forest images. *Remote Sens.* **2019**, *11*, 1252. [CrossRef]

23. Whitehead, K.; Hugenholtz, C.H.; Myshak, S.; Brown, O.; LeClair, A.; Tamminga, A.; Barchyn, T.E.; Moorman, B.; Eaton, B. Remote sensing of the environment with small unmanned aircraft systems (UASs), part 2: Scientific and commercial applications. *J. Unmanned Veh. Syst.* **2014**, *2*, 86–102. [CrossRef]

24. Mesas-Carrascosa, F.J.; Torres-Sánchez, J.; Clavero-Rumbao, I.; García-Ferrer, A.; Peña, J.M.; Borra-Serrano, I.; López-Granados, F. Assessing optimal flight parameters for generating accurate multispectral orthomosaicks by uav to support site-specific crop management. *Remote Sens.* **2015**, *7*, 12793–12814. [CrossRef]

25. Fawcett, D.; Azlan, B.; Hill, T.C.; Kho, L.K.; Bennie, J.; Anderson, K. Unmanned aerial vehicle (UAV) derived structure-from-motion photogrammetry point clouds for oil palm (Elaeis guineensis) canopy segmentation and height estimation. *Int. J. Remote Sens.* **2019**, *40*, 7538–7560. [CrossRef]

26. Nex, F.; Remondino, F. UAV for 3D Mapping Applications: A Review. *Appl. Geomatics* **2013**, *6*, 1–15. [CrossRef]

27. Torres-Sánchez, J.; López-Granados, F.; Borra-Serrano, I.; Peña, J.M. Assessing UAV-collected image overlap influence on computation time and digital surface model accuracy in olive orchards. *Precis. Agric.* **2018**, *19*, 115–133. [CrossRef]

28. Climate data org. Lahad Datu Climate. Available online: https://bit.ly/32ozhoi (accessed on 31 May 2019).

29. Wawan, W.; Dini, I.R.; Hapsoh, H. The effect of legume cover crop Mucuna bracteata on soil physical properties, runoff and erosion in three slopes of immature oil palm plantation. *IOP Conf. Ser. Earth Environ. Sci.* **2019**, *250*. [CrossRef]

30. MicaSense. What Is the Accuracy of the Included GPS for RedEdge? Available online: https://support.micasense.com/hc/en-us/articles/115005911207-What-is-the-accuracy-of-the-included-GPS-for-RedEdge (accessed on 30 June 2020).

31. Tucker, J.C. Red and photographic infrared linear combinations for monitoring vegetation. *Remote Sens. Environ.* **1979**, *8*, 127–150. [CrossRef]

32. Bendig, J.; Yu, K.; Aasen, H.; Bolten, A.; Bennertz, S.; Broscheit, J.; Gnyp, M.L.; Bareth, G. Combining UAV-based plant height from crop surface models, visible, and near infrared vegetation indices for biomass monitoring in barley. *Int. J. Appl. Earth Obs. Geoinf.* **2015**, *39*, 79–87. [CrossRef]

33. Avtar, R.; Suab, S.A.; Yunus, A.P.; Kumar, P. *Applications of UAVs in Plantation Health and Area Management in Malaysia Chapter 7 Applications of UAVs in Plantation Health and Area Management in Malaysia*; Springer: Cham, Germany, 2020.

34. Segarra, J.; Buchaillot, M.L.; Araus, J.L.; Kefauver, S.C. Remote sensing for precision agriculture: Sentinel-2 improved features and applications. *Agronomy* **2020**, *10*, 641. [CrossRef]

35. Stroppiana, D.; Pepe, M.; Boschetti, M.; Crema, A.; Candiani, G.; Giordan, D.; Baldo, M.; Allasia, P.; Monopoli, L. Estimating Crop Density From Multi-spectral UAV Imagery in Maize Crop. In Proceedings of the ISPRS Geospatial Week 2019, Enschede, The Netherlands, 10–14 June 2019; pp. 619–624.

36. Wijitdechakul, J.; Sasaki, S.; Kiyoki, Y.; Koopipat, C. UAV-based multispectral image analysis system with semantic computing for agricultural health conditions monitoring and real-time management. In Proceedings of the 2016 International Electronics Symposium (IES), Denpasar, Indonesia, 29–30 September 2016; pp. 459–464.

37. Barnes, E.M.; Clarke, T.R.; Richards, S.E.; Colaizzi, P.D.; Haberland, J.; Kostrzewski, M.; Waller, P.; Choi, C.; Riley, E.; Thompson, T.; et al. Coincident Detection of Crop Water Stress, Nitrogen Status and Canopy Density Using Ground Based Multispectral Data. In Proceedings of the Fifth International Conference on Precision Agriculture, Madison, WI, USA, 16–19 July 2000.

38. Eitel, J.U.H.; Vierling, L.A.; Litvak, M.E.; Long, D.S.; Schulthess, U.; Ager, A.A.; Krofcheck, D.J.; Stoscheck, L. Broadband, red-edge information from satellites improves early stress detection in a New Mexico conifer woodland. *Remote Sens. Environ.* **2011**, *115*, 3640–3646. [CrossRef]

39. Wicaksono, P.; Danoedoro, P.; Hartono, H.; Nehren, U.; Ribbe, L. Preliminary work of mangrove ecosystem carbon stock mapping in small island using remote sensing: Above and below ground carbon stock mapping on medium resolution satellite image. In Proceedings of the Remote Sensing for Agriculture, Ecosystems, and Hydrology XIII, Prague, Czech Republic, 7 October 2011.

40. Umarhadi, D.A.; Danoedoro, P. The effect of topographic correction on canopy density mapping using satellite imagery in mountainous area. *Int. J. Adv. Sci. Eng. Inf. Technol.* **2020**, *10*, 1317–1325. [CrossRef]

41. Zhang, X.; Zhang, T.; Zhou, P.; Shao, Y.; Gao, S. Validation analysis of SMAP and AMSR2 soil moisture products over the United States using ground-based measurements. *Remote Sens.* **2017**, *9*, 104. [CrossRef]

42. Dandois, J.P.; Olano, M.; Ellis, E.C. Optimal altitude, overlap, and weather conditions for computer vision uav estimates of forest structure. *Remote Sens.* **2015**, *7*, 13895–13920. [CrossRef]
43. Suab, S.A.; Syukur, M.S.; Avtar, R.; Korom, A. Unmanned Aerial Vehicle (UAV) Derived Normalised Difference Vegetation Index (NDVI) and Crown Projection Area (CPA) to Detect Health Conditions of Young Oil Palm Trees for Precision Agriculture. In Proceedings of the 2019 6th International Conference on Geomatics and Geospatial Technology, Kuala Lumpur, Malaysia, 1–3 October 2019; pp. 611–614.
44. Tan, K.P.; Kanniah, K.D.; Cracknell, A.P. Use of UK-DMC 2 and ALOS PALSAR for studying the age of oil palm trees in southern peninsular Malaysia. *Int. J. Remote Sens.* **2013**, *34*, 7424–7446. [CrossRef]
45. Chemura, A.; van Duren, I.; van Leeuwen, L.M. Determination of the age of oil palm from crown projection area detected from WorldView-2 multispectral remote sensing data: The case of Ejisu-Juaben district, Ghana. *ISPRS* **2015**, *100*, 118–127. [CrossRef]

Article

Assessing the Performance of ICESat-2/ATLAS Multi-Channel Photon Data for Estimating Ground Topography in Forested Terrain

Yanqiu Xing [1], Jiapeng Huang [1,*], Armin Gruen [2] and Lei Qin [1]

[1]   Centre for Forest Operations and Environment, Northeast Forestry University, Harbin 150040, China; yanqiuxing@nefu.edu.cn (Y.X.); qinlei@nefu.edu.cn (L.Q.)

[2]   Institute of Theoretical Physics, Federal Institute of Technology (ETH), 8092 Zurich, Switzerland; armin.gruen@geod.baug.ethz.ch

*   Correspondence: huangjp@nefu.edu.cn; Tel.: +86-13946053718

Received: 26 May 2020; Accepted: 26 June 2020; Published: 29 June 2020

Abstract: As a continuation of Ice, Cloud, and Land Elevation Satellite-1 (ICESat-1), the ICESat-2/Advanced Topographic Laser Altimeter System (ATLAS) employs a micro-pulse multi-beam photon counting approach to produce photon data for measuring global terrain. Few studies have assessed the accuracy of different ATLAS channels in retrieving ground topography in forested terrain. This study aims to assess the accuracy of measuring ground topography in forested terrain using different ATLAS channels and the correlation between laser intensity parameters, laser pointing angle parameters, and elevation error. The accuracy of ground topography measured by the ATLAS footprints is evaluated by comparing the derived Digital Terrain Model (DTM) from the ATL03 (Global Geolocated Photon Data) and ATL08 (Land and Vegetation Height) products with that from the airborne Light Detection And Ranging (LiDAR). Results show that the ATLAS product performed well in the study area at all laser intensities and laser pointing angles, and correlations were found between the ATLAS DTM and airborne LiDAR DTM (coefficient of determination—-R^2 = 1.00, root mean squared error—-RMSE = 0.75 m). Considering different laser intensities, there is a significant correlation between the tx_pulse_energy parameter and elevation error. With different laser pointing angles, there is no significant correlation between the tx_pulse_skew_est, tx_pulse_width_lower, tx_pulse_width_upper parameters and the elevation error.

Keywords: ATLAS; ground topography in forested terrain; laser intensity; laser pointing angle

---

1. Introduction

The spatial structure of forested terrain is listed as an important indicator for monitoring carbon stocks by the International Union of Forest Research Organizations (IUFRO) [1,2]. Assessing ground topography in forested terrain is a prerequisite to accurately determine the forest spatial structure; therefore, high spatial resolution modeling is necessary to characterize forest ecosystems [3,4]. Most optical remote sensing systems can measure ground topography in forested terrain; however, they have poor measurement accuracy (elevation difference = 2.9 m to 4.9 m) [5,6]. Spaceborne [7], airborne [8,9], and terrestrial [10,11] Light Detection And Ranging (LiDAR) systems have shown great potential for acquiring accurate topographic information in this field. Although airborne and terrestrial LiDAR can accurately quantify ground topography in forested terrain, these methods remain largely impractical at large spatial scales due to high data acquisition costs [8,12–14]. Spaceborne LiDAR is unique since it comes with low acquisition costs and provides a synoptic perspective of certain plot-level details from orbit [15].

The Ice, Cloud, and land Elevation satellite-1 (ICESat-1) [16], the Global Ecosystem Dynamics Investigation (GEDI) [17], and the Ice, Cloud, and land Elevation satellite-2 (ICESat-2) [18] are typical spaceborne LiDAR systems. ICESat-1 and GEDI carry large-footprint and waveform LiDAR systems. The Geoscience Laser Altimeter System (GLAS) instrument aboard ICESat-1 was launched in 2003 and decommissioned in 2009 [19]. GLAS is the first spaceborne LiDAR instrument designed to make global observations. GLAS operates a single laser beam from a ~600 km orbit at 40 Hz and has a 70 m diameter footprint and a ~170 m sampling rate along track.

GLAS waveform data has been successfully used to estimate the vertical structure of forest terrain, including ground topography and canopy heights [19–21]. Harding et al. [20] found that the waveform was an accurate representation of the canopy height distribution within a GLAS footprint. Lefsky et al. [21] observed that the models combing GLAS waveforms and Shuttle Radar Topography Mission (SRTM) could explain ~59%–68% of the variance in the field-measured forest canopy height (root mean squared error—RMSE = 4.85–12.66 m); however, sloped ground in forested terrain reduced the canopy height accuracy by using waveform data. Chen [22] found that the ground topography in forested terrain was the critical factor affecting the accurate measurement of canopy height using waveform data, and with increasing forest terrain complexity, the accuracy of estimating forest canopy height decreased. Fang et al. [23] found that in forested terrain with complex ground topography, the GLAS waveform was characterized by multiple energy peaks, in which the ground topography might be broadened and mixed, making the extraction of canopy height difficult. In order to quantify the influence of ground topography on canopy height estimation using GLAS waveform data, Lee et al. [24] found that without slope correction, the canopy height could be overestimated by 3 m over a 15 degree slope. Removing the ground topography in forested terrain from large LiDAR footprint could improve the accuracy of canopy height estimates. Claudia et al. [25] revealed that GLAS height estimates were accurate for areas with a slope up to 10 degrees, whereas the waveform results for areas above 15 degrees were problematic. Ten-to-fifteen degree slopes have been found to be a critical crossover point. The aforementioned studies demonstrated that it was feasible to extract ground topography in forested terrain and canopy height from spaceborne waveform data at stand level; however, the accuracy of canopy height estimation was largely determined by the ground topography, and extracting canopy height across a large LiDAR footprint using waveform data over hilly or mountainous regions is a great challenge. The GEDI was launched on 5 December 2018; however, the GEDI spaceborne data has just recently been released, and no related study was found [26].

The Advanced Topographic Laser Altimeter System (ATLAS) instrument aboard ICESat-2 was launched on 15 September 2018, and data was released on 30 May 2019. ATLAS is the first spaceborne photon-counting LiDAR instrument designed for continuous global observation of Earth [27–29]. Different from the GLAS waveform-digitizing LiDAR system, ATLAS only responds to the presence of return signals and records the time tags with an output of 0 or 1; however, it does not record the return waveform [30–32]. ATLAS operates six laser beams from a ~600 km orbit at 10 kHz and has a footprint (17 m in diameter) sampling rate of ~0.7 m along-track [33,34]. The center-to-center spacing along a track for ATLAS is narrower than that of GLAS (170 m). The high repetition rate enables ATLAS to obtain nearly continuous tracking information, which is necessary to measure the ground topography in forested terrain. While the GLAS LiDAR system uses a laser beam, the ATLAS configuration uses a diffractive optical element to split the laser into six beams arranged as three beam pairs, each of which consists of a strong and weak energy beam at a 4:1 ratio, allowing for local slope determination between each beam pair as well as compensation for varying surface reflectance [27,33,34]. The travel time of each detected photon is used to determine a unique XYZ location on the Earth's surface [35,36]. After ATLAS data was released, Neuenschwander found good correlations between matching Digital Terrain Model (DTM) from airborne LiDAR data and ATLAS data (R^2 = 0.99, RMSE = 0.85 m) [37]. Wang et al. found that the overall mean difference and RMSE values between the ground elevations retrieved from the ICESat-2 data and the airborne LiDAR-derived ground elevations are −0.61 m and 1.96 m, respectively [38]. However, he primarily examined the retrieved canopy height accuracy from the ICESat-2 strong beam and did not analyze the accuracy of the ICESat-2 weak beam. Under the same orbital conditions, ATLAS can acquire more continuous photon cloud data using the six-beam instrument with different laser pointing angles and laser intensities. The measurement accuracy of the different ATLAS channels remains to be quantified [39]. To the authors' knowledge, only a few studies have been carried out to analyze the multi-beam geometrical features for measuring ground topography in forested terrain from photon-counting data onboard ICESat-2. Therefore, the effective quantifying of the ground topography in forested terrain using the six-beam photon-counting data is essential to quantify the performance of the unique photon-counting instrument onboard ICESat-2.

The objective of this study is to assess the performance of the ICESat-2/ATLAS multi-channel photon data for estimating ground topography in forested terrain by comparing the derived ground topography from different ATLAS beam photon-counting data with that from Goddard's LiDAR, Hyperspectral and Thermal imager (G-LiHT) data. The paper also analyzes the correlation between laser intensity parameters, laser pointing angle parameters, and estimated ground topography error in forested terrain.

2. Materials and Methods

2.1. Study Area

The study area (33.564°N, 81°684′W) is a forested area within the City of Aiken, South Carolina, USA (Figure 1). Vegetation footprint types in the study area include cultivated land (0.04%), forest (88.52%), shrubland (0.51%), wetland (6.97%), and artificial surfaces (3.95%) [39]. The upland forest has many tree species, including sand post oak (*Quercus margaretta*), loblolly pine, water oak (*Quercus nigra*), hickory (*Carya*), and turkey oak (*Quercus laevis*) [40]. The elevation of the study area ranges from 91 m to 164 m. Vegetation coverage in the study area ranges from 25% to 66%.

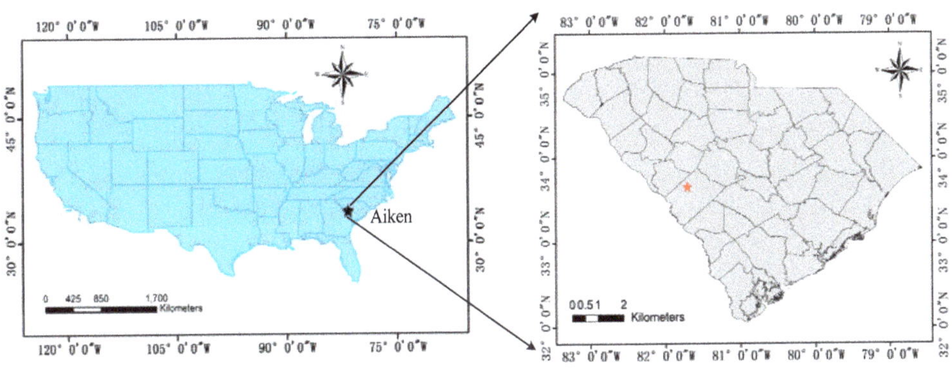

Figure 1. Location of the study area at the national and state levels. Figure 1 shows a part of the study site using an image of a USA map. The left part of the image shows the location of the study area in the USA, the right part shows it relative to South Carolina.

2.2. Data

The ICESat-2 mission produces along-track ground topography in forested terrain that includes telemetry data (ATL00), reformatted telemetry (ATL01), science unit converted telemetry (ATL02), global geolocated photon data (ATL03), land vegetation elevation (ATL08), and a land/canopy grid (ATL18) [37–41]. ATL00, ATL01, and ATL02 are original photon data sets without scientific algorithms. ATL03 is the geolocated photon cloud and serves as the input data for each of the higher-level data products such as ATL08 and ATL18. The ATL08 algorithm was developed specifically for the extraction of terrain and canopy heights from the ATL03 photon cloud data, and the ATL08 geophysical data product has a 100 m step size in the along-track direction [33]. The ATL03 product not only includes latitude, longitude, height, and signal photon confidence level of each received photon, but also includes tx_pulse_energy, tx_pulse_skew_est, tx_pulse_width_lower, and tx_pulse_width_upper parameters, which may be related to laser intensity and laser pointing angle [37]. All ICESat-2 data products were acquired from https://search.earthdata.nasa.gov.

Here the ATL03 product parameters were used, including lat_ph, lon_ph, h_ph, geoid, delta_time, signal_conf_ph, sc_orient, tx_pulse_energy, tx_pulse_skew_est, tx_pulse_width_lower, and tx_pulse_width_upper. The names and corresponding descriptions of product parameters are listed in Table 1 [37–39]. In order to reduce the influence of noise photons in forested terrain ground topography measurement, a signal_conf_ph of 4 was used as the signal photon parameter evaluation standard.

Table 1. The statistical indicators of global geolocated photon (ATL03) data [39].

ATL03 Product Parameter Name	Description
lat_ph	Latitude of each received photon. Computed from the ECF Cartesian coordinates of the bounce point.
lon_ph	Longitude of each received photon. Computed from the ECF Cartesian coordinates of the bounce point.
h_ph	Height of each received photon, relative to the WGS-84 ellipsoid.
geoid	Geoid height above WGS-84 reference ellipsoid (range −107 to 86 m).
delta_time	Elapsed seconds from the ATLAS SDP GPS Epoch, corresponding to the transmit time of the reference photon.
signal_conf_ph	Confidence level associated with each photon event selected as signal. 0 = noise. 1 = added to allow for buffer but algorithm classifies as background; 2 = low; 3 = med; 4 = high).
sc_orient	This parameter tracks the spacecraft orientation between forward, backward and transitional flight modes.
tx_pulse_energy	The average transmit pulse energy, measured by the internal laser energy monitor, split into per-beam measurements.
tx_pulse_skew_est	The difference between the averages of the lower and upper threshold crossing times. This is an estimate of the transmit pulse skew.
tx_pulse_width_lower	The average distance between the lower threshold crossing times measured by the Start Pulse Detector.
tx_pulse_width_upper	The average distance between the upper threshold crossing times measured by the Start Pulse Detector.

A diagram of ICESat-2 for estimating ground topography in forested terrain is illustrated in Figure 2. The forward orientation (sc_orient=1) corresponds to ATLAS traveling along the +x direction in the ATLAS instrument reference frame [41–43]. The ATLAS signal photon shown in the yellow square represents the photons detected from the gt3r laser channel. The ATLAS footprint shown in the red square represents the photons detected from the gt3l laser channel. The number of photons in the gt3l channel is less than in the gt3r channel, which is due to the backward orientation of ATLAS [43]. The photon level in the along-track direction was selected to calculate the ground topography in forested terrain. In the right figures, the two laser beams are 90 m apart. The ground topography measured by the two laser beams is similar, and the terrain elevation ranges from 130 m–145 m.

Due to the influence of sunlight as well as atmospheric and system noise, a large number of noise photons are present in the ATLAS data, which seriously reduce the ground elevation measurement accuracy. In order to improve the estimation accuracy of ATLAS photon data, NASA proposed a Differential, Regressive, and Gaussian Adaptive Nearest Neighbor (DRAGANN) method and ATL08 data classified algorithm to filter out noise photon data and classify ground photons [41–43]. In order to explore the estimation accuracy of forested terrain from ATLAS data, this contribution chose to associate the ATL08 classified label with the ATL03 photon data and used the ground signal photons flag mentioned in ATL08 as ground photons to establish an ATLAS-based DTM (Table 2) [37].

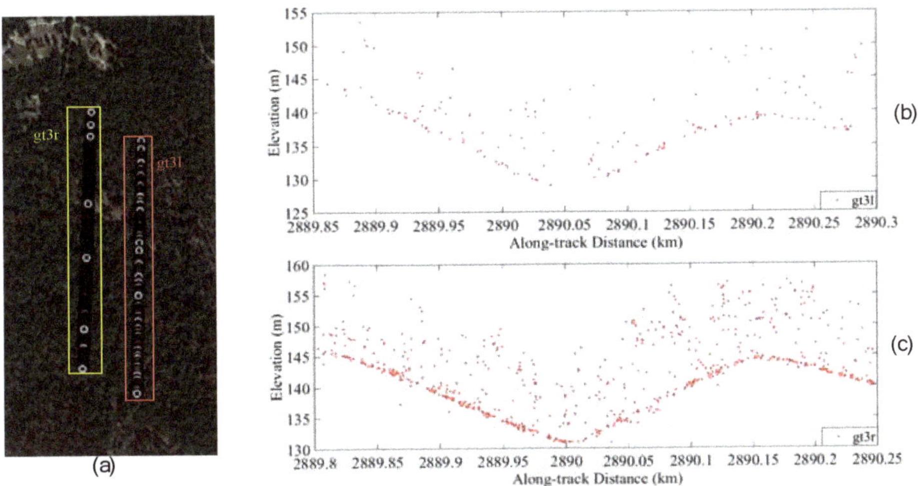

Figure 2. (**a**) The location of gt3r and gt3l data in site 1, City of Aiken, USA. gt3r photons (yellow) and gt3l photons (red) show the location of the Ice, Cloud, and Land Elevation Satellite-2 (ICESat-2) track in Google Earth for context [43]. This illustration only includes signal photons (signal_conf_ph=4) and is located in City of Aiken, USA. (**b**): Profile of ATL03 photons from the weak beam (gt3l). This data was collected on 26 December 2018 at 05:31. (**c**): Profile of ATL03 photons from the strong beam (gt3r), which has a greater number of signal photons above the surface than in gt3l.

Table 2. The statistical indicators of land vegetation elevation (ATL08) product [42].

ATL08 Product Parameter Name	Description
classed_pc_flag	Land Vegetation ATBD classification flag for each photon as either noise, ground, canopy, and top of canopy. 0 = noise, 1 = ground, 2 = canopy, or 3 = top of canopy.
classed_pc_indx	The unique identifier for tracing each ATL08 signal photon to the corresponding photon record on ATL03 is the segment_id, orbit, cycle, and classed_pc_indx.
ph_segment_id	Segment ID of photons tracing back to specific 20 m segment_id on ATL03. The unique identifier for tracing each ATL08 signal photon to the photon on ATL03 is the segment_id, orbit, and classed_pc_indx. The unique identifier for tracing each ATL08 signal photon to the corresponding photon record on ATL03 is the segment_id, orbit, cycle, and classed_pc_indx.

To assess the accuracy of the six beam-ATLAS DTM, the DTM obtained from the ATLAS data was compared with airborne discrete-return LiDAR data, collected for the same longitude and latitude using the multi-sensor instrument G-LiHT [44]. G-LiHT provides distributed laser pulses for measuring ground topography and canopy heights (Table 3) [45].

Table 3. The Goddard's LiDAR, Hyperspectral and Thermal imager (G-LiHT) product levels [44].

G-LiHT Product	Product Level
Trajectory data	Level 1
Classified return data	Level 2
Above Ground Level (AGL) height	Level 2
LiDAR returns	Level 3
DTM	Level 3
Canopy Height Model (CHM)	Level 3

Both Level 2 and Level 3 products along the flight transects were generated from airborne LiDAR data from the G-LiHT science team. The DTM has a 1 m-resolution and was released as a Tag Image File Format (TIFF) profile. The DTM was assessed to validate the ground topography accuracy [44,45]. The trajectory of the G-LiHT KML (Keyhole Markup Language) data (blue line) and ATLAS data (green line) illustrates the location of the study area of the NASA EARTHDATA (Figure 3). This illustration also includes two G-LiHT DTM profiles used in the study, AMIGACarb_Augusta_FIA_Sep2011_l16s597_DTM.tif and AMIGACarb_Augusta_FIA_ Sep2011_ l40s557_DTM.tif, respectively.

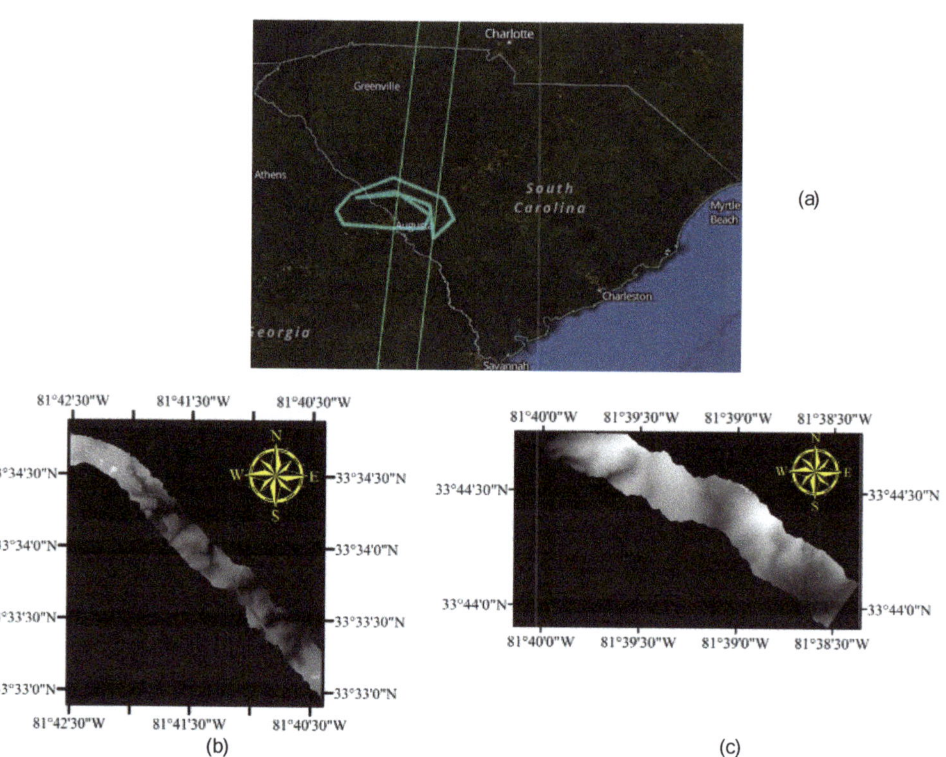

Figure 3. (**a**): Trajectories of Advanced Topographic Laser Altimeter (ATLAS) data (green) and G-LiHT data (blue) depict the location of the study area NASA EARTHDATA. (**b**): G-LiHT data named AMIGACarb_Augusta_FIA_Sep2011_l16s597_DTM.tif as reference airborne data. (**c**): G-LiHT data named AMIGACarb_Augusta_FIA_Sep2011_l40s557_DTM.tif as reference airborne data.

2.3. Methodology

The primary challenge of this contribution centers on reducing the influence of noise photons on the ground elevation data derived from ICESat-2 data, distinguishing canopy signal photons and ground signal photons, and matching the ATL03, ATL08, and G-LIHT data. Although ATLAS data has more noise photons, the NASA official team used the DRAGANN and an algorithm for determining Land Vegetation along-track to provide classification labels (classed_pc_flag) for the photon data [43]. In this contribution, the ground signal photon classification label ATL08 is used for ground photons, and the DTM$_{ATLAS}$ will be established based on ATL03 data and ATL08 label. This contribution presents a quantitative assessment of the ground topography in forested terrain using ATL03 and ATL08 data compared to airborne G-LiHT LiDAR data.

The geolocation between the ATLAS and G-LiHT data is not completely along orbit; therefore, this paper proposes an approach based on the ATL03 profile to match these two datasets (Figure 3).

To clearly illustrate the proposed methodology, an overview of the major steps is exhibited in Figure 4 and described as follows:

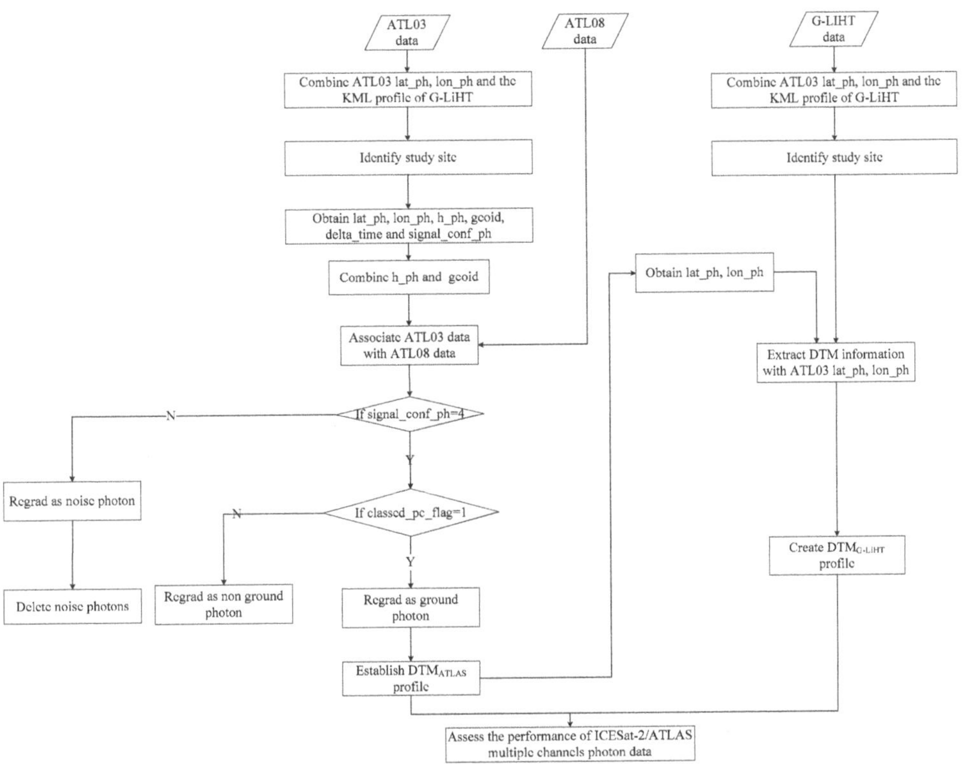

Figure 4. The flowchart of the analysis process.

(1) Identifying study site. In this step, we combine ATL03 lat_ph, lon_ph, and G-LiHT KML profile and identify the study site.

(2) Obtain parameters from the ATL03 and ATL08 data by matching different channels under the same orbit conditions using time tags (delta_time in ATL03 HDF5 profile). To extract the photon's height (h_ph which is relative to the WGS-84 ellipsoid), latitude (lat_ph) and longitude (lon_ph), signal_conf_ph, sc_orient, tx_pulse_energy, tx_pulse_skew_est, tx_pulse_width_lower, and tx_pulse_width_upper parameters from the ATL03 HDF5 profile, combine the geoid and h_ph by interpolating. Extract the photon classification parameters (classed_pc_flag), classed_pc_indx, and ph_segment_id parameters from the ATL08 HDF5 profile.

(3) Establishing the relationship between ATL03 and ATL08 data photon classification parameters by classed_pc_indx, ph_segment_id and applying each photon classification label from ATL08 to each photon data from ATL03.

(4) Establishing the DTM_{ATLAS}. The photons with a signal confidence flag from high confidence (signal_confidence = 4 in ATL03 HDF5 profile) and photon classification parameter (classed_pc_flag=1 in ATL08 HDF5 profile) were used to establish the DTM_{ATL03}.

(5) Obtaining the DTM_{G-LiHT}. In this step, we extract the latitude-longitude information from DTM_{ATL03} to match the DTM_{G-LiHT} profile corresponding position generated from G-LiHT and generated DTM_{G-LiHT} with the corresponding ATLAS footprint latitude-longitude. If the absolute

difference between the elevation of ATLAS ground photons and the corresponding elevation of the $DTM_{G\text{-}LiHT}$ is more than 20 m, this photon was classified as an invalid ground photon.

(6) Assessing the performance of ICESat-2/ATLAS multiple channels photon data. In this final step, we compare the DTM_{ATL03} profile with the corresponding $DTM_{G\text{-}LiHT}$ profile, compute and analyze the evaluating indicator from different channels. In order to quantify the influence of different laser intensity parameters and laser pointing angle parameters on the estimation accuracy of ground elevation, corresponding four parameters as follow: tx_pulse_energy, tx_pulse_skew_est, tx_pulse_width_lower and tx_pulse_width_upper are extracted and analyzed the relationship between the four parameters and elevation errors.

The Figure 5 shows the DTM files of G-LiHT and ATLAS photon data corresponding to the two tracks in the study area. The green dots are ATLAS footprints.

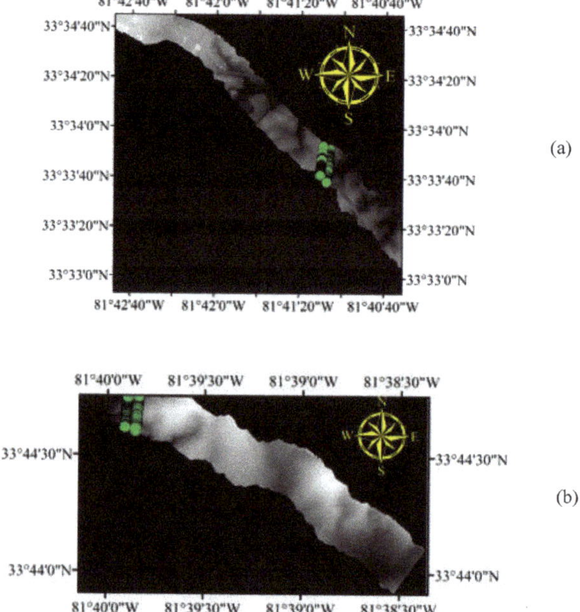

Figure 5. Two experimental data (**a**,**b**) sets in study area. The green dots on this figure are ATLAS footprints. The gray block is the G-LiHT Digital Terrain Model (DTM).

2.4. Accuracy Evaluation

DTM data derived from airborne G-LiHT LiDAR data were utilized to assess the accuracy of the ATLAS-derived ground elevations. The ground elevation errors in the ATLAS data mentioned in this contribution are calculated by subtracting the G-LiHT elevations from the ATLAS elevations [38]. Seven statistical variables, including root mean squared error (RMSE), mean absolute error (MAE), coefficient of determination (R^2), mean error (ME), Pearson correlation coefficient, Spearman correlation coefficient, and Kendall correlation coefficient between the ground elevations and the corresponding G-LiHT's DTM values were calculated to quantitatively evaluate the accuracy of the ATLAS-derived ground elevations. Three statistical variables, Pearson correlation coefficient, Spearman correlation coefficient, and Kendall correlation coefficient between the elevation errors and the corresponding laser intensity parameter and laser pointing angle parameters were calculated to quantitatively evaluate the correlation of laser intensities and laser pointing angles with the elevation error in estimating ground topography in forested terrain.

3. Results

Result Comparisons

The G-LiHT data in the study was acquired in 2011 and are not temporally coincident with ICESat-2/ATLAS. However, the availability of aerial-based LiDAR data for the study area makes it feasible to validate the ground topography from ATL03. The ATL03 gt3l channel photon data and gt3r channel photon data over the study are plotted with the G-LiHT data shown in Figures 6 and 7, respectively. The scatterplots of the ATL03 (gt3l and gt3r) ground elevations versus the G-LiHT ground elevations are shown in Figures 8 and 9. The statistical indicators, namely the RMSE, MAE, R^2, ME, Pearson correlation coefficient, Spearman correlation coefficient, and Kendall correlation coefficient are listed in Table 4.

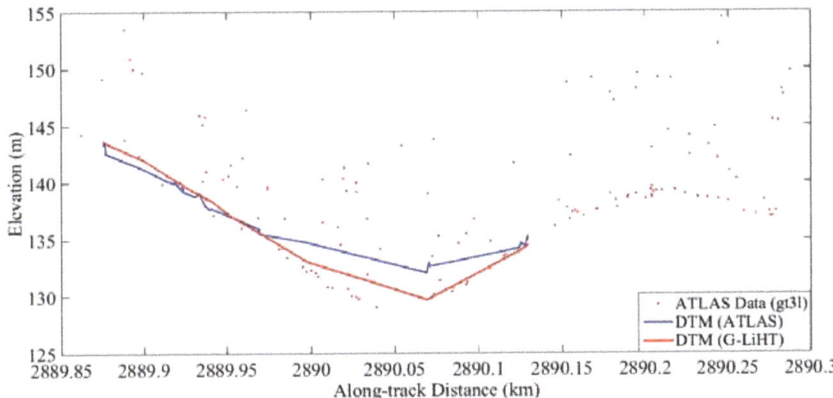

Figure 6. A subset of the forested terrain ATL03 (gt3l) ground topography validation results from track ATL03_20181226053112_13530106_001_01. ATLAS photon data are shown as red dots. The ATLAS DTM (blue line) and the G-LiHT DTM (red line) are also shown.

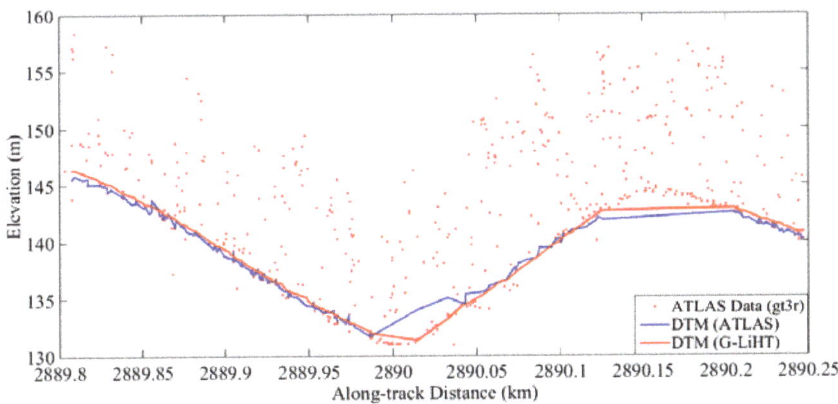

Figure 7. A subset of the forested terrain ATL03 (gt3r) ground topography validation results from track ATL03_20181226053112_13530106_001_01. The ATLAS photon data are shown as red dots. The ATLAS DTM (blue line) and the G-LiHT DTM (red line) are also shown.

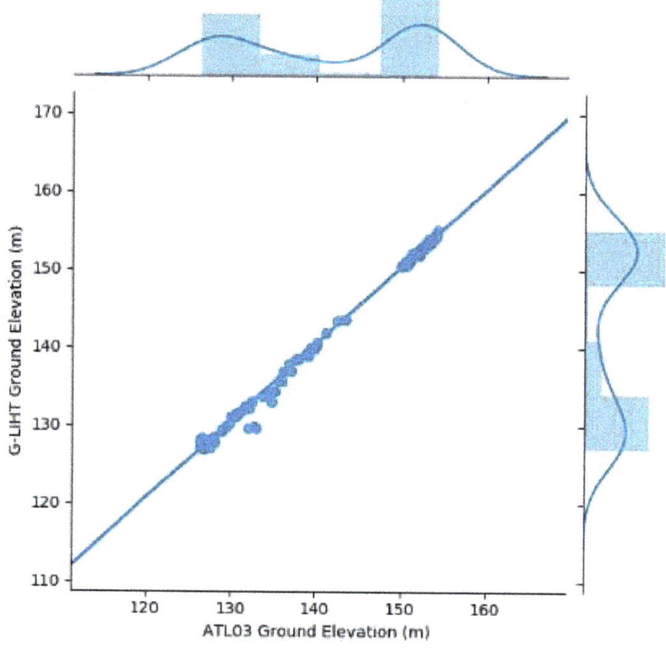

Figure 8. Scatterplot of ATL03 (gt3l) ground topography compared to G-LiHT DTM.

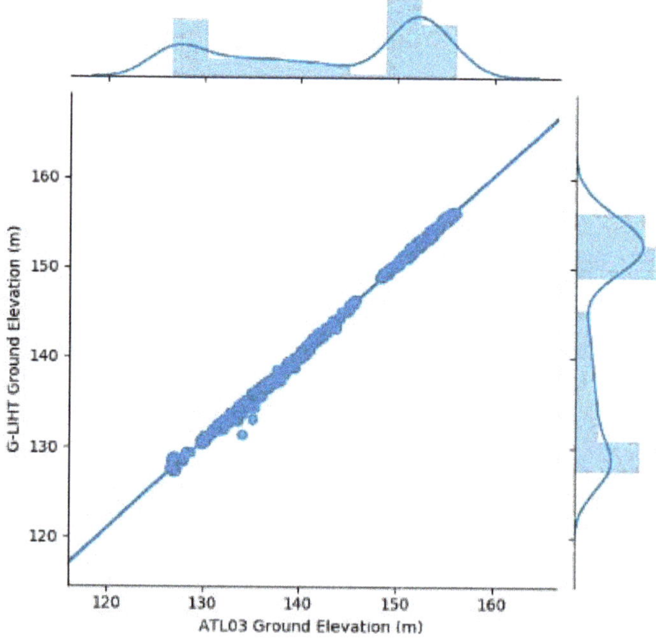

Figure 9. Scatterplot of ATL03 (gt3r) ground topography compared to G-LiHT DTM.

Table 4. The statistical indicators of the various laser channels.

Different Laser Channels	RMSE (m)	MAE (m)	R^2	ME (m)	Pearson	Spearman	Kendall
gt1l	0.69	0.35	1.00	−0.05	1.00	0.99	0.92
gt1r	0.55	0.45	1.00	0.27	1.00	0.99	0.92
gt2l	0.80	0.67	1.00	0.59	1.00	0.96	0.84
gt2r	1.03	0.58	0.99	0.04	1.00	0.99	0.93
gt3l	0.78	0.60	1.00	0.45	1.00	0.99	0.93
gt3r	0.64	0.52	1.00	0.49	1.00	0.99	0.94
Mean of channel	0.75	0.53	1.00	0.30	1.00	0.98	0.91

Note: RMSE—root mean squared error; MAE—mean absolute error; R^2—coefficient of determination; ME—mean error.

Figures 6 and 7 show the comparisons of elevations from the ATLAS DTM and G-LiHT DTM for multiple channels, and both are referenced to the WGS84 geoid. The ATLAS DTM (blue line) and G-LiHT's DTM (red line) had a similar performance (Figures 6 and 7). The ATLAS gt3l channel (weak beam) photon cloud contains fewer signal photons than the gt3r channel (strong beam). The photons near the blue line and red line denote the ground topography photons. A t-test showed that no significant difference exists between these two regression lines in Figures 8 and 9 at a 95% confidence level. The ATLAS gt3r channel (Figures 7 and 9) signal photons show a clearer depiction of the ground topography than the ATLAS gt3l channel (Figures 6 and 8).

The R^2 of the all experiments are higher than 0.99 (Table 4). The mean ME and RMSE values of all ground photons are 0.3 m and 0.75 m, respectively. For gt1l, gt1r, gt2l, gt2r, gt3l, and gt3r, the ME values are −0.05 m, 0.27 m, 0.59 m, 0.04 m, 0.45 m, and 0.49 m, and the RMSE values are 0.69 m, 0.55 m, 0.80 m, 1.03 m, 0.78 m, and 0.64 m, respectively. In general, the statistical indicators for pointing right (namely strong beam) perform better than those for pointing left (namely weak beam) for all data sets. A possible reason is that the signal photon density from the strong beam channels is significantly higher than that of the weak beam channels.

All laser intensities and laser pointing angles from the ATL03 product performed well in the study site (Table 4, Figures 8 and 9). The ATL03 data have good ground elevation estimation accuracy, and the ATL08 algorithm can effectively filter out noise photon and classify ground photons in the forested terrain ground topography estimation process.

4. Discussion

In order to study the influence of laser intensity and laser pointing angle on ground elevation estimation accuracy, the ATLAS data errors under different laser intensities and laser pointing angles are analyzed respectively, and the correlation between elevation errors and corresponding laser parameters is examined.

4.1. Retrieved Ground Topography in Forested Terrain for Different Laser Intensities

The mean statistical indicators of the different laser intensities are listed in Table 5 and the correlation coefficient statistics between tx_pulse_energy parameters and the ATLAS data elevation errors are listed in Table 6.

Table 5. Statistical indicators for different laser intensities.

Different Laser Intensities Types	RMSE (m)	MAE(m)	R^2	ME (m)	Pearson	Spearman	Kendall
Weak beam	0.76	0.54	1.00	0.33	1.00	0.98	0.89
Strong beam	0.74	0.51	1.00	0.27	1.00	0.99	0.93

Table 6. Correlation coefficient statistics between tx_pulse_energy parameters and ATLAS data. error.

Different Laser Intensities Types	Pearson	Spearman	Kendall
Weak beam	0.58	0.76	0.54
Strong beam	0.59	0.74	0.51
Mean of different laser intensities Types	0.59	0.75	0.53

The estimated ground topography for different intensity beams shows significant agreement with the reference DTM elevations. For the weak beam and strong beam, the mean R^2 values are 1, the ME values are 0.33 m and 0.27 m, and the RMSE values are 0.76 m and 0.74 m, respectively. The correlation coefficient of all types is greater than 0.89.

For the varying laser intensities in this data set, the statistical indicators for the strong beam performed better than that of weak beam with the lower RMSE ($RMSE_{strong\ beam}$ = 0.74 m and $RMSE_{weak\ beam}$ = 0.76 m), lower MAE ($MAE_{strong\ beam}$ = 0.51 m and $MAE_{weak\ beam}$ = 0.54 m), lower ME ($ME_{strong\ beam}$ = 0.33 m and $ME_{weak\ beam}$ = 0.27 m), and higher correlation coefficient. A possible reason is that the weak beam channel has fewer signal photons compared to the strong channel, making measuring ground topography in forested terrain using the weak beam more difficult. Using a strong beam, ATLAS could produce more signal photons than under the weak beam. The laser intensity ratio of strong beam to weak beam is 4:1. Depending upon the surface reflectance and atmospheric conditions, up to 16 photons per outgoing shot could be detected for the strong beam, while the weak beam could detect only 4 photons. However, the strong and weak beams can both provide useable data for measuring ground topography in forested terrain.

To further explore the influence of laser intensities on elevation errors, we calculated three correlation coefficients between tx_pulse_energy parameters and elevation errors (Table 6). For all the data, the correlation coefficients for the elevation errors and tx_pulse_energy parameters are greater than 0.5. In addition, the Spearman correlation coefficient values for the various laser intensities are greater than 0.74, indicating there is a significant correlation between the tx_pulse_energy parameters and elevation error. However, this contribution only explores the correlation between tx_pulse_energy and elevation error for a laser intensity ranging from 0.02 mJ to 0.09 mJ. Future studies with the tx_pulse_energy parameters will need to perform a more detailed analysis on the effect of tx_pulse_energy on elevation error.

Compared to the forested terrain ground topography estimation method in proposed by Neuenschwander et al. [37], higher R^2 values and lower RMSE values were observed in the strong beam mode and weak beam mode. However, this study proposes a photon level which is different to the Neuenschwander et al. method [37], which notes that the result of a photon subset using a strong beam and a weak beam can reasonably explain the ground topography in the forest study area.

4.2. Retrieved Ground Topography in Forested Terrain Elevation with Different Laser Pointing Angles

The mean statistical indicators for the different laser pointing angles are listed in Table 7 and the correlation coefficient statistics between tx_pulse_skew_est, tx_pulse_width_lower, tx_pulse_width_upper parameters and the elevation errors of ATLAS data are listed in Table 8.

Table 7. The statistical indicators for different laser pointing angles.

Different Laser Channels	RMSE (m)	MAE(m)	R^2	ME (m)	Pearson	Spearman	Kendall
gt1	0.62	0.40	1.00	0.11	1.00	0.99	0.92
gt2	0.92	0.62	1.00	0.31	1.00	0.97	0.88
gt3	0.71	0.56	1.00	0.47	1.00	0.99	0.93

Table 8. The correlation coefficient statistics between tx_pulse_skew_est, tx_pulse_width_lower, tx_pulse_width_upper parameters and the elevation errors of ATLAS data.

Statistical Indicators	Different Laser Pointing Angles	Pearson	Spearman	Kendall
tx_pulse_skew_est	gt1	0.12	0.05	0.04
	gt2	0.11	0.21	0.12
	gt3	0.15	0.15	0.11
	Mean of different pointings	0.13	0.14	0.09
tx_pulse_width_lower	gt1	0.20	0.10	0.07
	gt2	0.10	0.07	0.05
	gt3	0.20	0.21	0.14
	Mean of different pointings	0.17	0.13	0.09
tx_pulse_width_upper	gt1	0.24	0.31	0.23
	gt2	0.09	0.12	0.09
	gt3	0.11	0.13	0.09
	Mean of different pointings	0.15	0.19	0.14

The estimated ground topography in forested terrain using different laser pointing angles shows strong agreement with the reference $DTM_{G\text{-LiHT}}$ elevations, as demonstrated by the R^2 values equaling 1.00 and the RMSE values less than 0.92 m.

Results indicated that the gt1 channel pointing ($R^2_{gt1\ channel}$ = 1.00, $RMSE_{gt1\ channel}$=0.62 m, $MAE_{gt1\ channel}$ = 0.4m) performed better than gt2 channel pointing ($R^2_{gt2\ channel}$ = 1.00, $RMSE_{gt2\ channel}$ = 0.92 m, $MAE_{gt2\ channel}$ = 0.62m) and gt3 channel pointing ($R^2_{gt3\ channel}$ = 1.00, $RMSE_{gt3\ channel}$ = 0.71 m, $MAE_{gt3\ channel}$ = 0.56m), which was due to several hardware reasons. On one hand, the gt1 channel could achieve more effective forested terrain signal photons than other channels in the study area. More signal photons can give a clearer depiction of ground topography in forested terrain. On the other hand, the photon rates of the gt1 and gt3 channels are higher than the gt2 channel, which is consistent with the description of the different laser pointing angles [46].

To further explore the influence of laser pointing angles on elevation error, we calculated three correlation coefficients between tx_pulse_skew_est, tx_pulse_width_lower, tx_pulse_width_upper parameters and elevation error. In the ATL03 Algorithm Theoretical Basis Document (ATBD), these parameters may be related to the laser pointing angles. The quantitative results of the correlation coefficients are summarized in Table 8. For all the data, the mean correlation coefficients for the elevation errors are less than 0.20. There is no significant correlation between the tx_pulse_skew_est, tx_pulse_width_lower, tx_pulse_width_upper parameters and the elevation error. However, this contribution only explores the correlation between tx_pulse_skew_est, tx_pulse_width_lower, tx_pulse_width_upper and the elevation error. Future studies needed to analyze other laser pointing angles parameters' relative elevation errors.

The results of this contribution performed better than that proposed by Neuenschwander et al. [37] (R^2 = 0.99, RMSE = 0.85), which notes that the result of a photon subset using different laser pointing angles can reasonably explain the ground topography in the study area.

4.3. Retrieved Ground Topography in Forested Terrain Elevation with ATLAS

Most optical remote sensing systems could provide images of the horizontal distribution of ground topography, and the product generally follows the uppermost surface elevation (i.e., representing a digital surface model, DSM). However, the optical remote sensing images do not provide detailed information on the vertical distribution of ground topography in forested terrain, without regard to whether the surface is comprised of forest or not [5,6,47]. In contrast, the LiDAR photon counting signature from ICESat-2/ATLAS could provide a direct depiction for ground topography in forested terrain. In this contribution, the close correspondence between the ATLAS and G-LiHT (R^2 = 1.00, RMSE = 0.75 m) confirms that the received photon data are an accurate representation of the ground topography forested terrain elevation within the ATLAS footprints. As a comparison, in areas of low relief (slope $\leq 5°$) and middle dense tree cover (tree cover = 20%–40%), the mean and standard deviation of elevation differences between the ICESat/GLAS centroid and SRTM are -2.48 ± 4.04 m [48]. Thus, the estimation of ground topography for forest-covered areas is able to be accomplished with ICESat-2/ATLAS.

The results from this contribution indicate that the ground topography in forested terrain elevation can be estimated using photon data from ICESat-2/ATLAS multi-channel. We were able to retrieve terrain elevation successfully in forest-covered areas. Prior work showing the correlation between spaceborne LiDAR-measured canopy height and ground topography in forested terrain [21–23], which provide confidence that ICESat-2/ATLAS photon data in combination with GEDI data can substantially contribute to a global inventory of forest biomass. The work also provides insights for future work to improve the accuracy of the canopy height estimations.

5. Conclusions

In this contribution, ICESat-2 data is used to measure ground topography in forested terrain using different channels. The retrieved ground topography was validated by experiments with G-LiHT airborne data at different laser pointing angles and laser intensity types on the same route. Based on the results, the following conclusions can be drawn:

(1) Both qualitative and quantitative results indicate that at all laser intensities and laser pointing types resulted in a mean $R^2 = 1.00$ and mean RMSE = 0.75 m, highlighting the ability of the ATL03 and ATL08 data to retrieve ground elevations.

(2) A significant correlation exists between the tx_pulse_energy parameters and elevation error. There is no significant correlation between tx_pulse_skew_est, tx_pulse_width_lower, tx_pulse_width_upper parameters and elevation error.

These conclusions give valuable insight into the ground topography in forested terrain using different ATLAS channels. Nevertheless, there are still many issues to be addressed in the future. Since ATLAS data is still in the research stage, we only considered the effects of laser pointing angles and laser intensity on retrieving ground topography. Other factors (e.g., canopy height, canopy cover, etc.) influencing the results were not considered. Therefore, the effects of other factors on retrieving ground topography over forested terrain using ATLAS data should be thoroughly examined in the future.

Author Contributions: J.H., Y.X. and L.Q. conceived and designed the experiments; J.H. performed the experiments; J.H. and Y.X. analyzed the data; L.Q. contributed materials; J.H., Y.X. and A.G. wrote the paper; Y.X. secured funding for the project; A.G. polished the language of the manuscript. All authors have read and agreed to the published version of the manuscript.

Funding: This work is supported by the National Key R&D Program of China (Grant Number: 2017YFD0600904), the Fundamental Research Funds for the Central Universities (Grant Number:2572019AB18), and the Key Laboratory of Satellite Mapping Technology and Application, National Administration of Surveying, Mapping and Geoinformation (KLSMTA-201706).

Acknowledgments: We thank the editor and anonymous reviewers for reviewing our paper.

Conflicts of Interest: The authors declare no conflict of interest.

References

1. Thom, D.; Keeton, W.S. Stand structure drives disparities in carbon storage in northern hardwood-conifer forests. *For. Ecol. Manag.* **2019**, *442*, 10–20. [CrossRef]
2. Ferreira, A.C.; Bezerra, L.E.A.; Cascon, H.M. Aboveground carbon stock in a restored neotropical mangrove: Influence of management and brachyuran crab assemblage. *Wetl. Ecol. Manag.* **2019**, *27*, 223–242. [CrossRef]
3. Quegan, S.; le Toan, T.; Chave, J.; Dall, J.; Exbrayat, J.F.; Minh, D.H.T.; Lomas, M.; d'Alessandro, M.M.; Paillou, P.; Papathanassiouet, K.; et al. The European Space Agency BIOMASS mission: Measuring forest above-ground biomass from space. *Remote Sens. Environ.* **2019**, *227*, 44–60. [CrossRef]

4. Ni, W.; Zhang, Z.; Sun, G.; Liu, Q. Modeling the Stereoscopic Features of Mountainous Forest Landscapes for the Extraction of Forest Heights from Stereo Imagery. *Remote Sens.* **2019**, *11*, 1222. [CrossRef]

5. Hu, T.; Su, Y.; Xue, B.; Liu, J.; Zhao, X.; Fang, J.; Guo, Q. Mapping Global Forest Aboveground Biomass with Spaceborne LiDAR, Optical Imagery, and Forest Inventory Data. *Remote Sens.* **2016**, *8*, 565. [CrossRef]

6. Wang, S.; Ren, Z.; Wu, C.; Lei, Q.; Gong, W.; Ou, Q.; Zhang, H.; Ren, G.; Li, C. DEM generation from Worldview-2 stereo imagery and vertical accuracy assessment for its application in active tectonics. *Geomorphology* **2019**, *336*, 107–118. [CrossRef]

7. Yu, Y.; Yang, X.G.; Fan, W.Y. Estimates of forest structure parameters from GLAS data and multi-angle imaging spectrometer data. *Int. J. Appl. Earth Obs. Geoinf.* **2015**, *38*, 65–71. [CrossRef]

8. Zhao, F.; Yang, X.Y.; Strahler, A.H.; Schaaf, C.L.; Yao, T.; Wang, Z.S.; Román, M.O.; Woodcock, C.E.; Ni-Meister, W.; Jupp, D.L.B.; et al. A comparison of foliage profiles in the Sierra National Forest obtained with a full-waveform under-canopy EVI lidar system with the foliage profiles obtained with an airborne full-waveform LVIS lidar system. *Remote Sens. Environ.* **2013**, *136*, 330–341. [CrossRef]

9. Bouvier, M.; Durrieu, S.; Fournier, R.A.; Renaud, J.P. Generalizing predictive models of forest inventory attributes using an area-based approach with airborne LiDAR data. *Remote Sens. Environ.* **2015**, *156*, 322–334. [CrossRef]

10. Dandois, J.P.; Ellis, E.C. High spatial resolution three-dimensional mapping of vegetation spectral dynamics using computer vision. *Remote Sens. Environ.* **2013**, *136*, 259–276. [CrossRef]

11. Disney, M.; Burt, A.; Calders, K.; Schaaf, C.; Stovall, A. Innovations in Ground and Airborne Technologies as Reference and for Training and Validation: Terrestrial Laser Scanning (TLS). *Surv. Geophys.* **2019**, *40*, 937–958. [CrossRef]

12. Disney, M. Terrestrial LiDAR: A three-dimensional revolution in how we look at trees. *N. Phytol.* **2019**, *222*, 1736–1741. [CrossRef] [PubMed]

13. Wijesingha, J.; Moeckel, T.; Hensgen, F.; Wachendorf, M. Evaluation of 3D point cloud-based models for the prediction of grassland biomass. *Int. J. Appl. Earth Obs. Geoinf.* **2019**, *78*, 352–359. [CrossRef]

14. Schneider, F.D.; Kükenbrink, D.; Schaepman, M.E.; Schimel, D.S.; Morsdorf, F. Quantifying 3D structure and occlusion in dense tropical and temperate forests using close-range LiDAR. *Agric. For. Meteorol.* **2019**, *268*, 249–257. [CrossRef]

15. Bakx, T.R.; Koma, Z.; Seijmonsbergen, A.C.; Kissling, W.D. Use and categorization of Light Detection and Ranging vegetation metrics in avian diversity and species distribution research. *Divers. Distrib.* **2019**, *25*, 1045–1059. [CrossRef]

16. Zwally, H.J.; Schutz, B.; Abdalati, W.; Abshire, J.; Bentley, C.; Brenner, A.; Bufton, J.; Dezio, J.; Hancock, D.; Harding, D.; et al. ICESat's laser measurements of polar ice, atmosphere, ocean, and land. *J. Geodyn.* **2002**, *34*, 405–445. [CrossRef]

17. Kellner, J.R.; Armston, J.; Birrer, M.; Cushman, K.C.; Duncanson, L.; Eck, C.; Falleger, C.; Imbach, B.; Král, K.; Krůček, M.; et al. New Opportunities for Forest Remote Sensing Through Ultra-High-Density Drone Lidar. *Surv. Geophys.* **2019**, *40*, 959–977. [CrossRef]

18. Popescu, S.C.; Zhou, T.; Nelson, R.; Neuenschwander, A.; Sheridan, R.; Narine, L.; Walsh, K.M. Photon counting LiDAR: An adaptive ground and canopy height retrieval algorithm for ICESat-2 data. *Remote Sens. Environ.* **2018**, *208*, 154–170. [CrossRef]

19. Wang, Y.; Ni, W.; Sun, G.; Chi, H.; Zhang, Z.; Guo, Z. Slope-adaptive waveform metrics of large footprint lidar for estimation of forest aboveground biomass. *Remote Sens. Environ.* **2019**, *224*, 386–400. [CrossRef]

20. Harding, D.J.; Carabajal, C.C. ICESat waveform measurements of within footprint topographic relief and vegetation vertical structure. *Geophys. Res. Lett.* **2005**, *32*, L21S10. [CrossRef]

21. Lefsky, M.A.; Harding, D.J.; Keller, M.; Cohen, W.B.; Carabajal, C.C.; Espirito-Santo, F.D.B.; Hunter, M.O.; de Oliveira, R., Jr. Estimates of forest canopy height and above ground biomass using ICESat. *Geophys. Res. Lett.* **2005**, *32*, L22S02. [CrossRef]

22. Chen, Q. Assessment of terrain elevation derived from satellite laser altimetry over mountainous forest areas using airborne lidar data. *ISPRS J. Photogramm. Remote Sens.* **2010**, *65*, 111–122. [CrossRef]

23. Fang, Z.; Cao, C.X. Estimation of Forest Canopy Height Over Mountainous Areas Using Satellite Lidar. *IEEE J. Sel. Top. Appl. Earth Obs. Remote Sens.* **2014**, *7*, 3157–3166. [CrossRef]

24. Lee, S.; Ni-Meister, W.; Yang, W.; Chen, Q. Physically based vertical vegetation structure retrieval from ICESat data: Validation using LVIS in White Mountain National Forest, New Hampshire, USA. *Remote Sens. Environ.* **2011**, *115*, 2776–2785. [CrossRef]

25. Claudia, H.; Christiane, S. Influence of Surface Topography on ICESat/GLAS Forest Height Estimation and Waveform Shape. *Remote Sens.* **2012**, *4*, 2210–2235. [CrossRef]

26. Saarela, S.; Holm, S.; Healey, S.P.; Andersen, H.-E.; Petersson, H.; Prentius, W.; Patterson, P.L.; Næsset, E.; Gregoire, T.G.; Ståhl, G. Generalized Hierarchical Model-Based Estimation for Aboveground Biomass Assessment Using GEDI and Landsat Data. *Remote Sens.* **2018**, *10*, 1832. [CrossRef]

27. Markus, T.; Neumann, T.; Martino, A.; Abdalati, W.; Brunt, K.; Csatho, B.; Farrell, S.; Fricker, H.; Gardner, A.; Harding, D.; et al. The Ice, Cloud, and land Elevation Satellite-2 (ICESat-2): Science requirements, concept, and implementation. *Remote Sens. Environ.* **2017**, *190*, 260–273. [CrossRef]

28. Brunt, K.M.; Neumann, T.A.; Walsh, K.M.; Markus, T. Determination of Local Slope on the Greenland Ice Sheet Using a Multibeam Photon-Counting Lidar in Preparation for the ICESat-2 Mission. *IEEE Geosci. Remote Sens. Lett.* **2014**, *11*, 935–939. [CrossRef]

29. Magruder, L.A.; Brunt, K.M. Performance Analysis of Airborne Photon Counting Lidar Data in Preparation for the ICESat-2 Mission. *IEEE Trans. Geosci. Remote Sens.* **2018**, *56*, 2911–2918. [CrossRef]

30. Ma, Y.; Li, S.; Zhang, W.; Zhang, Z.; Liu, R.; Wang, X.H. Theoretical ranging performance model and range walk error correction for photon counting lidars with multiple detectors. *Opt. Express* **2018**, *26*, 15924–15934. [CrossRef]

31. Li, S.; Zhang, Z.; Ma, Y.; Zeng, H.; Zhao, P.; Zhang, W. Ranging performance models based on negative-binomial (NB) distribution for photon-counting lidars. *Opt. Express* **2019**, *27*, A861–A877. [CrossRef] [PubMed]

32. Ma, Y.; Liu, R.; Li, S.; Zhang, W.; Yang, F.; Su, D. Detecting the ocean surface from the raw data of the MABEL photon-counting lidar. *Opt. Express* **2018**, *26*, 24752–24762. [CrossRef] [PubMed]

33. Neuenschwander, A.L.; Magruder, L.A. The Potential Impact of Vertical Sampling Uncertainty on ICESat-2/ATLAS Terrain and Canopy Height Retrievals for Multiple Ecosystems. *Remote Sens.* **2016**, *8*, 1039. [CrossRef]

34. Abdalati, W.; Zwally, H.J.; Bindschadler, R.; Csatho, B.; Farrell, S.L.; Fricker, H.A.; Harding, D.; Kwok, R.; Lefsky, M.; Markus, T.; et al. ICESat-2 Laser Altimetry Mission. *Proc. IEEE* **2010**, *98*, 735–751. [CrossRef]

35. Nie, S.; Wang, C.; Xi, X.; Luo, S.; Li, G.; Tian, J.; Wang, H. Estimating the vegetation canopy height using micro-pulse photon-counting LiDAR data. *Opt. Express* **2018**, *26*, A520–A540. [CrossRef]

36. Narine, L.L.; Popescu, S.C.; Malambo, L. Synergy of ICESat-2 and Landsat for Mapping Forest Aboveground Biomass with Deep Learning. *Remote Sens.* **2019**, *11*, 1503. [CrossRef]

37. Neuenschwander, A.L.; Magruder, L.A. Canopy and Terrain Height Retrievals with ICESat-2: A First Look. *Remote Sens.* **2019**, *11*, 1721. [CrossRef]

38. Wang, C.; Zhu, X.; Nie, S.; Xi, X.; Li, D.; Zheng, W.; Chen, S. Ground elevation accuracy verification of ICESat-2 data: A case study in Alaska, USA. *Opt. Express.* **2019**, *27*, 38168–38179. [CrossRef]

39. Neumann, T.; Brenner, A.; Hancock, D.; Robbins, J.; Saba, J.; Harbeck, K.; Gibbons, A. *ICE, CLOUD, and Land Elevation Satellite-2(ICESat-2) Project Algorithm Theoretical Basis Document (ATBD) for Global Geolocated Photons ATL03*; National Aeronautics and Space Administration: Washington, DC, USA; Goddard Space Flight Centre: Greenbelt, MD, USA, 2018.

40. Harrington, T.B.; Edwards, M.B. Understory vegetation, resource availability, and litterfall responses to pine thinning and woody vegetation control in longleaf pine plantations. *Can. J. For. Res.* **1999**, *29*, 1055–1064. [CrossRef]

41. Huang, J.P.; Xing, Y.Q.; You, H.T.; Qin, L.; Tian, J.; Ma, J.M. Particle Swarm Optimization-Based Noise Filtering Algorithm for Photon Cloud Data in Forest Area. *Remote Sens.* **2019**, *11*, 980. [CrossRef]

42. Neuenschwander, A.; Popescu, S.; Nelson, R.; Harding, D.; Pitts, K.; Robbins, J.; Pederson, D.; Sheridan, R. *ICE, CLOUD, and Land Elevation Satellite (ICESat-2) Algorithm Theoretical Basis Document (ATBD) for Land-Vegetation Along-Track Products*; National Aeronautics and Space Administration: Washington, DC, USA; Goddard Space Flight Centre: Greenbelt, MD, USA, 2019.

43. Neumann, T.; Brenner, A.; Hancock, D.; Luthcke, S.; Lee, J.; Robbins, J.; Harbeck, K.; Saba, J.; Brunt, K. *ATLAS/ICESat-2 L2A Global Geolocated Photon Data, Version 2*; NSIDC National Snow and Ice Data Center: Boulder, CO, USA, 2019.

44. Cook, B.D.; Corp, L.A.; Nelson, R.F.; Middleton, E.M.; Morton, D.C.; McCorkel, J.T.; Masek, J.G.; Ranson, K.J.; Ly, V.; Montesano, P.M. NASA Goddard's LiDAR, Hyperspectral and Thermal (G-LiHT) Airborne Imager. *Remote Sens.* **2013**, *5*, 4045–4066. [CrossRef]
45. Chen, B.; Pang, Y.; Li, Z.; North, P.; Rosette, J.; Sun, G.; Suárez, J.; Bye, I.; Lu, H. Potential of Forest Parameter Estimation Using Metrics from Photon Counting LiDAR Data in Howland Research Forest. *Remote Sens.* **2019**, *11*, 856. [CrossRef]
46. Kwok, R.; Markus, T.; Kurtz, N.T.; Petty, A.A.; Farrell, T.A.N.S.L.; Cunningham, G.F.; Hancock, D.W.; Ivanoff, A.; Wimert, J.T. Surface Height and Sea Ice Freeboard of the Arctic Ocean From ICESat-2: Characteristics and Early Results. *J. Geophys. Res. Oceans* **2019**, *124*. [CrossRef]
47. Meddens, A.J.; Vierling, L.A.; Eitel, J.U.; Jennewein, J.S.; White, J.C.; Wulder, M.A. Developing 5 m resolution canopy height and digital terrain models from WorldView and ArcticDEM data. *Remote Sens. Environ.* **2018**, *218*, 174–188. [CrossRef]
48. Carabajal, C.C.; Harding, D.J. ICESat validation of SRTM C-band digital elevation models. *Geophys. Res. Lett.* **2005**, *32*, L22S01. [CrossRef]

 *remote sensing*

Article

Canopy Height Estimation Using Sentinel Series Images through Machine Learning Models in a Mangrove Forest

Sujit Madhab Ghosh, Mukunda Dev Behera * and Somnath Paramanik

Centre for Oceans, Rivers, Atmosphere and Land Sciences, Indian Institute of Technology Kharagpur, West Bengal 721302, India; sujitmghosh@iitkgp.ac.in (S.M.G.); somnathpu06@iitkgp.ac.in (S.P.)
* Correspondence: mukundbehera@gmail.com or mdbehera@coral.iitkgp.ac.in

Received: 11 March 2020; Accepted: 6 May 2020; Published: 9 May 2020

Abstract: Canopy height serves as a good indicator of forest carbon content. Remote sensing-based direct estimations of canopy height are usually based on Light Detection and Ranging (LiDAR) or Synthetic Aperture Radar (SAR) interferometric data. LiDAR data is scarcely available for the Indian tropics, while Interferometric SAR data from commercial satellites are costly. High temporal decorrelation makes freely available Sentinel-1 interferometric data mostly unsuitable for tropical forests. Alternatively, other remote sensing and biophysical parameters have shown good correlation with forest canopy height. The study objective was to establish and validate a methodology by which forest canopy height can be estimated from SAR and optical remote sensing data using machine learning models i.e., Random Forest (RF) and Symbolic Regression (SR). Here, we analysed the potential of Sentinel-1 interferometric coherence and Sentinel-2 biophysical parameters to propose a new method for estimating canopy height in the study site of the Bhitarkanika wildlife sanctuary, which has mangrove forests. The results showed that interferometric coherence, and biophysical variables (Leaf Area Index (LAI) and Fraction of Vegetation Cover (FVC)) have reasonable correlation with canopy height. The RF model showed a Root Mean Squared Error (RMSE) of 1.57 m and R^2 value of 0.60 between observed and predicted canopy heights; whereas, the SR model through genetic programming demonstrated better RMSE and R^2 values of 1.48 and 0.62 m, respectively. The SR also established an interpretable model, which is not possible via any other machine learning algorithms. The FVC was found to be an essential variable for predicting forest canopy height. The canopy height maps correlated with ICESat-2 estimated canopy height, albeit modestly. The study demonstrated the effectiveness of Sentinel series data and the machine learning models in predicting canopy height. Therefore, in the absence of commercial and rare data sources, the methodology demonstrated here offers a plausible alternative for forest canopy height estimation.

Keywords: symbolic regression; random forest; Sentinel-1; Sentinel-2; ICESat; Bhitarkanika

---

1. Introduction

Understanding the role of forest carbon emissions and sequestration is needed to build a robust framework for international agreements to limit the concentration of greenhouse gases in the atmosphere [1]. The function of tropical forests is critical in the global carbon cycle because they are carbon-dense and highly productive [2]. Above-Ground Biomass (AGB) is the best indicator of the carbon content of tropical forests [3]. AGB estimation models for tropical forests generally ignore canopy height as a factor [4]. However, studies have shown that the inclusion of canopy height in the allometric models tends to improve the estimation accuracy of AGB in tropical forests [4–6]. Therefore, the tree canopy height of tropical forests is an essential factor in estimating its biomass, and an inaccurate estimate of canopy height can result in over- or underestimation of AGB [7].

The ground-based canopy height measurement instruments exploit the planimetric distance and the angle between the device to the base and the top of the tree, to estimate canopy height using a trigonometric relationship [8]. The laser rangefinder is quite a standard instrument that uses this method in field-based canopy height measurements [9,10]. Apart from that, instruments like altimeter and clinometer are used in some studies [11–13]. However, in dense tropical forests, it is often difficult to identify the top and base of the tree due to lack of direct line of sight to the canopy top, limited accessibility in rough terrains, blockage of the crown top by adjacent trees, and presence of understorey vegetation [8,14]. Therefore, the field estimated height and actual tree height often show a wide variation in the tropics [15].

The digital photogrammetry method has been used to measure canopy height in earlier remote sensing-based studies. The data used in those studies varied from aerial photogrammetry in historical cases to multispectral satellite data [16–18]. The advancement of LiDAR technologies was found to be more useful in measuring the tree height [19,20]. However, most of the studies were based on airborne LiDAR data, which has limited area coverage and which is costly to acquire in tropical regions [21,22]. Till now, only NASA's Ice, Cloud, and land Elevation Satellite (ICESat-1) mission has provided world-wide spaceborne LiDAR data, obtained through Geoscience Laser Altimeter System (GLAS), which has been extensively used for vegetation mapping [23]. Many studies found that the GLAS data can be used efficiently for vegetation height monitoring of different types of forests [24–28]. Therefore, GLAS data alongside other remote sensing and ancillary data has been used broadly in AGB estimation [29–32]. In 2018, the Ice, Cloud, and land Elevation Satellite-2 (ICESat-2) became operational [33]. Data obtained through its Advanced Topographic Laser Altimeter System (ATLAS) sensor was made available recently. Preliminary studies with simulated ICESat-2 data have shown its effectiveness in canopy height estimation [34]. Despite having many advantages, LiDAR data, mainly spaceborne data, usually has limited spatial and temporal coverage [35]. LiDAR data is also not suitable for the wall to wall mapping as this data is usually acquired for specific footprints [36].

One of the ways by which spatially continuous height mapping can be done is by using the Synthetic Aperture Radar (SAR) interferometry (InSAR) method [37,38]. InSAR measures the surface topography and height of the surface features by using the phase information of the radar signal [39]. Coherence is measured as the magnitude of the complex cross-correlation between the constituent images of an interferometric pair [40]. Decorrelation or reduction in coherence values occurs with changes in the ground condition between the two acquisitions of the interferometric pair; thereby, reducing its ability to measure height correctly using interferrometric information. Shorter wavelength SARs have a greater tendency for greater temporal decorrelation, even for a time gap as short as one day [39]. The interferometric data with global spatial coverage became freely available with the launch of the European Space Agency's Sentinel-1 mission. However, the Sentinel-1 mission has a maximum temporal resolution of six days [41]. There can be significant decorrelation between images while mapping dense tropical forests using Sentinel-1 repeat-pass interferometry, which may result in loss of coherence, thereby severely affecting the interferometric height measurement results.

In recent years researchers have found newer ways to estimate mean canopy height (Table 1). It was suggested that rather than using phase information of SAR images, interferometric coherence can be used to model tree height [42]. Tree height inversion from coherence data also showed improvement in biomass estimation accuracy [43]. Apart from coherence values, field-measured tree height can be correlated with SAR backscatter of different polarization to establish a canopy height model [44]. In addition to SAR, multispectral band values were also used in establishing relationship with tree canopy height. Multispectral bands of Landsat-7 and 8 showed good promise while measuring tree heights in the range of 5–20 m [45].

Fraction of Vegetation Cover (FVC) has been used for vegetation height estimation. FVC is defined as the percentage of a given area that is covered by vegetation canopy [46]. MODIS derived tree cover showed a good correlation with tree height derived from GLAS data [47]. Recently, Sentinel-1 based FVC demonstrated its effectiveness in canopy height estimation [35]. LAI is measured as the total one-sided leaf area per unit ground surface area, demonstrating good correlation with tree height [48,49]. LiDAR based studies found that canopy height has a good correlation with LAI [50,51] and one could serve as proxy to other [52]. Canopy height estimation with/without a Digital Elevation Model (DEM) has been benefitted from the Shuttle Radar Topography Mission (SRTM) [53,54].

Table 1. Compilation of canopy height estimation studies using different instruments and sensors.

Method	Site	Instrument/Sensor	Main Predictor Variable	Reference
Field-based	Panama	Rangefinder	Not applicable	[14]
Field-based	Finland	Clinometer	Not applicable	[13]
Field and RS-based	USA	Altimeter, Airborne LiDAR, Airborne SAR, SRTM	Digital Terrain Model values	[12]
Field and RS-based	Finland	Terrestrial Laser Scanning, Airborne LiDAR	Digital Terrain Model values	[8]
Field-based	Global	Not available	Field height	[52]
RS-based	India	ICESat-1	LiDAR waveform estimated height	[25]
RS-based	Tanzania	Airborne LiDAR, TanDEM-X	InSAR height	[38]
RS-based	Estonia	Airborne LiDAR, TanDEM-X	InSAR coherence	[42]
RS-based	Brazil	RADARSAT-2	SAR backscatter	[44]
RS-based	China	Sentinel-1, Sentinel-2	SAR backscatter, FVC	[35]
RS-based	Global	MODIS, ICESat-1	LiDAR height, FVC	[47]
RS-based	Korea	Airborne LiDAR, PALSAR, DEM	Digital Terrain Model values	[9]
RS-based	Canada	Airborne LiDAR, WorldView-2 Multispectral	LAI, LiDAR height	[50]

Modeling the relationship of forest parameters such as canopy height from field measured values, and remote sensing derived variables can be done in several ways. The most common are parametric regression methods [55–57] or machine learning models like Random Forest (RF) [58,59]. Due to their simplicity, parametric regression-based methods are widely used for modeling biophysical parameters and remote sensing variables. One major problem in parametric regression is that it assumes a standard relationship before the analysis, which may not be true, in reality. RF is a decision tree-based machine learning model used for estimation of biophysical parameters [60]. The RF model can capture the actual non-linear relationship between the predictor and predicted variables. However, the interpretation of an RF model is complicated as it consists of many trees and numerous sets of rules.

Symbolic Regression (SR) using genetic programming is a relatively modern technique that estimates a straightforward best-fit model for a given dataset by minimizing error rates while searching through all possible regression models [61,62]. Conventional machine learning algorithms work as black-boxes; implying that the internal mechanisms of these algorithms are hard to comprehend and difficult to reproduce desired results. The SR model works with the principle to determine the input-output relationship and selection of variables, which are most effective in predicting outputs [63]. The SR model's inclination towards finding correct solutions makes it distinct from other types of regressions. SR is capable of constructing a robust and interpretable formula, which is not possible by other linear and nonlinear regressions or machine learning models.

The main objective of this study was to establish a methodology by which forest canopy height can be estimated using SAR and optical remote sensing data using machine learning models. The mangrove forest in the Bhitarkanika Wildlife Sanctuary (BWS), Odisha, India, was chosen as the study area. First, we attempted to establish canopy height models using SR and RF, and Sentinel-1 and Sentinel-2 derived parameters. Further, the canopy heights estimated using the two models were compared and validated with field measured and ICESat-2 derived data.

2. Materials and Methods

2.1. Study Area and Filed Measurement of Canopy Heights

The Bhitarkanika Wildlife Sanctuary (BWS), with 58 species, is considered as one of the vital mangrove ecosystems in the world due to its genetic diversity [64]. Dense and moderately dense mangroves cover an area of 165 km^2 [65] of BWS. Some of the dominant species of BWS are *Heritiera* sp., *Excoecaria* sp., *Avicennia* sp., and *Sonneratia* sp. [66]. This area receives a high average annual rainfall of about 1642 mm, and most of it is received during June to October. BWS experiences a typical warm and humid tropical climate with temperature maxima in May and minima in January [67]. Multiple field surveys were conducted during November–December 2018 for canopy height measurement. Height measurement of all the trees in an inventory plot is time consuming, and redundant, especially if it has to be correlated with satellite derived canopy height pixels. Sullivan et al. [68] have mentioned that although an increase in the number of sampled trees results in better accuracy, the inclusion of the ten largest trees is most important. Thus, we observed the sampling frequency of the 10 largest trees per plot and measured tree height using a laser range finder instrument. The center locations were obtained using a handheld GPS for all 185 plots (Figure 1).

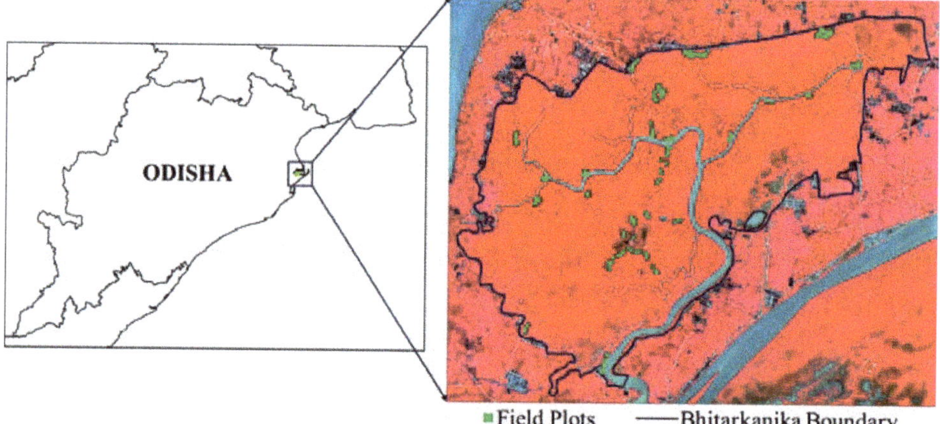

Figure 1. Location of the Bhitarkanika Wildlife Sanctuary (Odisha state) on the eastern coast of India. The false color composite image of BWS is shown with the boundary demarcated over which the canopy height measured plot positions are overlaid with green markers.

2.2. Satellite Data and Processing

Sentinel-1 and Sentinel-2 data were downloaded from the European Space Agency's (ESA) Copernicus Hub. The Sentinel-1 mission consists of two satellites, Sentinel-1A and Sentinel-1B, which image the earth in C-Band SAR with a six day revisit period between them. A total of six images, three each for Sentinel-1A and Sentinel-1B, were in the Single Look Complex (SLC) format. They were acquired in the interferometric wide-swath mode (IW), which contains both amplitude and phase information of the backscattered SAR signal (Tables 2 and 3). Sentinel-1 uses the advanced Terrain Observation by Progressive Scans (TOPS) SAR mode to capture images in three sub-swaths in a total swath width of 250 km [69]. The study area falls in the second sub-swath of all the SLC images. So, each image was split to subset only the second sub-swath. Each six day pair of Sentinel-1 SLC images was co-registered in the ESA's Sentinel Application Platform (SNAP) using orbital information and SRTM 1-sec DEM. Only the VV polarisation band of the Sentinel-1 data were selected for the study as the VH polarisation tends to lower the coherence value by introducing decorrelation due to cross-polarization noise [70]. The Sentinel-2 mission also consists of two satellites, Sentinel-2A and Sentinel-2B, with a five day revisit period in combination. Seven cloud free Sentinel-2 data were acquired in the L1C processing level, with the top of the atmosphere reflectance values. Sentinel-2 data were atmospherically corrected to the L2A processing level using the SEN2COR processor [71]. It adjusted the image to yield the bottom of the atmosphere-surface reflectance. The pre-processed Sentinel-1 and Sentinel-2 data were further used for specific parameter extraction and canopy height modeling (Figure 2).

Table 2. The acquisition details of the Sentinel dataset used in this study.

Data Type	Platform	Processing Level	Acquisition Date
SAR	Sentinel-1A	SLC	29 November; 11, 23 December 2018
	Sentinel-1B	SLC	5, 17, 29 December 2018
Optical	Sentinel-2A	L1C	16, 26 November; 26 December 2018
	Sentinel-2B	L1C	21 November; 11, 31 December 2018

Table 3. Perpendicular baselines of all interferometric pairs used in the study.

DOP of Interferometric Pairs	Perpendicular Baseline (in m)
29 November–5 December	122.4
5 December–11 December	76.26
11 December–17 December	27.92
17 December–23 December	79.34
23 December–29 December	28.49

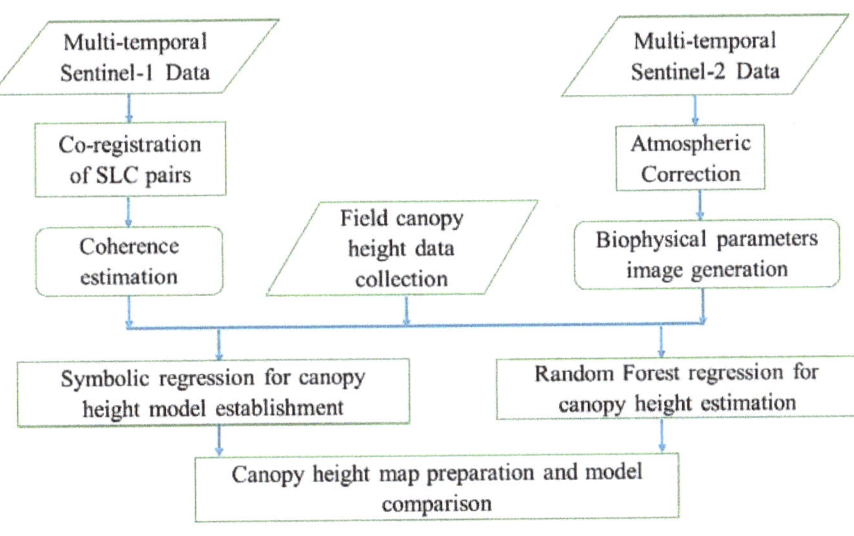

Figure 2. Methodology flow diagram depicting the steps adopted for generating canopy height maps and comparison of model outputs.

Launched in September 2018, ICESat-2 can measure the earth surface with a 17 m diameter footprint and with 91 days temporal resolution. ICESat-2 has three pairs of beams. Each pair of beams footprint is separated by about 3 km of cross-track with a pair spacing of 90 m [72]. The ICESat-2 data were downloaded in hdf5 format from the National Snow and Ice Data Centre (NSIDC) [73]. The data product selected was ALT08, which includes the height of the surface, including the canopy. An R package was used to extract the canopy height information from the ALT08 data.

2.3. Interferometric Coherence and Biophysical Parameters Extraction

Coherence for all co-registered SLC pairs was estimated using the SNAP tool. The coherence of an image pair depends on the baseline length. A massive baseline results in low coherence and vice versa. The longest baseline for which coherence becomes zero is known as the critical baseline [39]. It can be expressed as a function of band chirp width, orbital height, and sensor operating frequency [74], as shown in Equation (1).

$$B_{cr} = (b \times h \times \sin\theta)/(f \times cos^2\theta) \tag{1}$$

where, B_{cr} = critical baseline; b = chirp width; f = operating frequency; θ = incidence angle.

Sentinel-1 sensor characteristics vary according to observation mode [75]. The critical baseline, obtained using optimal values of parameters stands at 3.65 Km. As all the pairs have a baseline length significantly lower than the critical baseline, the coherence values fall within the acceptable range (Table 3). Backscatter images were also generated from the SLC images by converting them from the SLC image to Ground Range Detected (GRD) image.

Sentinel-2 L2A images were used to calculate LAI and FVC using the SNAP toolbox biophysical variable processor. All Sentinel-2 images were resampled to 20 m pixel size. The biophysical processor algorithm was implemented to generate biophysical products from a range of sensors [76]. An extensive database was prepared to include the top of canopy reflectance and associated vegetation characteristics in the biophysical processor algorithm. This database was used to train neural networks to estimate the canopy characteristics from the top of canopy reflectance along with the observational configuration. In SNAP, the prediction of the new variable was made based on the set of coefficients computed during the training phase.

2.4. Canopy Height Modeling

After generating the necessary variables, the next step was to establish the relationship between the variables and canopy height through regression. The values of coherence and other biophysical parameters were extracted at field plot locations to build the dataset for regression. Non-linear regression was implemented via two machine learning models, first, using the RF and then using SR. Primarily, the whole canopy height dataset was divided randomly into two parts using data partition function of CARET package. Seventy percent of it was used for model building, and the other 30 percent was used in model validation. The model building data was further divided into the model training and testing datasets in accordance with the model characteristics. Therefore, for both the models, we had separate model building, i.e., training and testing, and validation data.

2.4.1. Random Forest

The RF model was implemented via the CARET package in R [77]. In an RF model, hundreds of trees are built based on a bootstrap sample of the original data [78]. Variables were chosen randomly at each node for the split, and the final value was predicted by averaging the prediction of all the trees. The importance of the variables in the RF model was measured to find the most influential variables. As the first step of the importance measurement, the Mean Squared Error (MSE) was measured for each tree using an Out Of the Bag (OOB) sample. Thereafter a new error rate was calculated using the same procedure after permuting a variable. The difference between the two accuracies were averaged for all trees, and normalized by the standard error. This value was termed as the importance of the permuted variable to the model. The exclusion of a variable with positive importance increases the error rate of the model, while it is opposite for the variables with negative importance. After the initial run of the model, variables with negative importance were removed from the predictor list. The final model was built only on the variables with positive importance. Here one more fold cross-validation was done to reduce the chance of over-fitting. Initially, the field-measured canopy heights were correlated with the remote sensing derived variables, followed by model prediction of canopy heights.

2.4.2. Symbolic Regression

The use of SR, in retrieval of biophysical parameters has not been explored earlier. The success of machine learning algorithms in the biophysical parameters' retrieval problem showed the existence of the non-linear relationship between remotely sensed variables and biophysical parameters. However, the establishment of an interpretable model describing the relationship was not possible with machine learning regression. The working principle of SR makes it a viable option for the establishment of such a model.

SR was implemented through genetic programming, and it consisted of several steps. First was the selection of the terminals, i.e., independent variables. Coherence, SRTM DEM, FVC, and LAI were selected as terminals for prediction of canopy height. The second step was to identify a set of functions that would be used to build the models. In this study, constant, addition, multiplication, subtraction, division, exponents and natural logarithms were selected as the primitive functions. Each of these functions has an associated complexity. The first four have a complexity of 1, division has a complexity of 2, and exponents and natural logarithm have a complexity of 4 each. The total complexity of the solution is the sum of the complexity of the functions used in the solution [79]. Each symbolic expression proposed by the genetic programming was evaluated based on its fitness, which in this case, was measured by the mean squared error between observed and predicted values. Probability values were assigned to the initial models based on their fitness. After that, a new generation of models were created by reproduction, i.e., copying an existing model to the new population and genetic recombination, i.e., building a new model by recombining parts from existing models [61]. The trial version of Eureqa pro software was used for the SR model [80]. In the absence of specific termination criteria in the software, the final model was chosen when the 50th generation was reached. After getting the final canopy height model, the efficiency of the model was checked using the test dataset. Further, canopy heights were predicted using the model for the study area.

One of the goals of the SR model was to identify the variables that provided the most significant explanatory power for the dataset. Sensitivity analysis was used to identify the variables which have the greatest contribution to the regression equation [81]. The partial derivative of the dependent variable with respect to an independent variable was taken as the first step of sensitivity analysis. The final sensitivity was obtained by multiplying the partial derivative with the ratio of the standard deviation of the independent variable and the dependent variable [79]. The sensitivity of variable x for the function $y = f(x)$ is estimated as follows:

$$\left| \frac{\partial y}{\partial x} \right| \cdot \frac{\sigma_x}{\sigma_y} \tag{2}$$

where $\frac{\partial y}{\partial x}$ = partial derivative of y with respect to x; σ_x = standard deviation of the independent variable x, and σ_y = standard deviation of the dependent variable y.

The sensitivity indicates the direction, either positive or negative, and the magnitude of the correlation between input and output variables. A positive sensitivity suggests that an increase in the input variable will increase the value of the output variable and vice versa. The magnitude of the sensitivity determines the amount of increment or reduction of the output variable due to a unit increase in the input variable. A higher magnitude of positive sensitivity denotes a high amount of increase in the output variable value and vice versa for negative sensitivity.

3. Results

3.1. Field Measured Canopy Height

The distribution of the field-measured canopy height showed that the height range of 8 to 10 m has the largest number of plots (Figure 3). The tallest canopy heights observed during the field measurement were in the range of 14–16 m and occurred in three plots, whereas the lowest measured from 2 plots were in the range of 2–3 m.

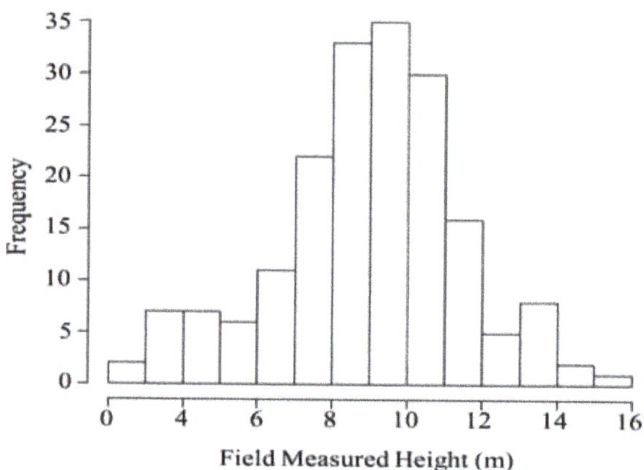

Figure 3. Distribution of field-measured canopy heights displayed against their frequency of occurrence.

3.2. Isnterferometric Coherence and Biophysical Psarameters

The average values of SAR coherence and biophysical variable images were used as inputs to reduce the effect of temporal variation. The spatial distribution of pixel values of all the variables showed different patterns (Figure 4). The coherence values found higher and close to '1' over fully correlated areas, and near '0', where there is no statistical relationship between the images [40]. As vegetation loss is a significant reason explaining loss of coherence, the coherence values remained low in dense mangrove areas. The FVC and LAI images provided idea about presence and absence of vegetation. Lower values of FVC and LAI indicated sparse or no vegetation. A comparison of the coherence images with LAI and FVC images showed that areas (within yellow ellipse in Figure 4), with lower FVC and LAI values, showed a relatively higher coherence corresponding to lower DEM values. However, for some areas (under the red ellipse), FVC, and LAI showed higher values, but the coherence of the region remained high, and DEM values remained on the lower range.

The frequency of pixel values in the coherence image almost shaped like a gaussian distribution (Figure 5). Most of the pixel values fall between 0.3 and 0.5. The rest of the values were distributed quite evenly on each side. Distribution of both LAI and FVC values followed a similar pattern. Due to the dense nature of mangroves, FVC also showed higher values for most of the pixels, with a much smaller number of pixels with values <0.4 and >0.6. However, LAI values remained low, between 1.5 and 2. The DEM image demonstrated flat topography with maxima of 16 m.

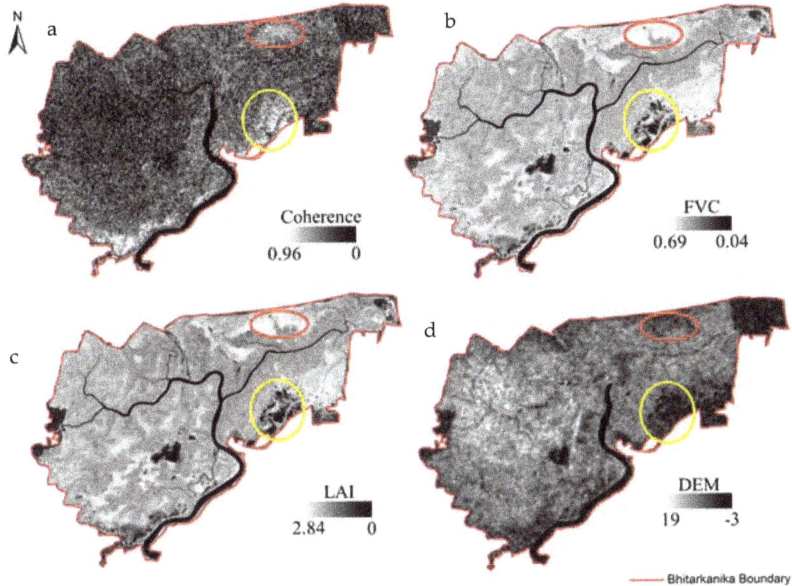

Figure 4. (**a**) Coherence, (**b**) Fraction of Vegetation Cover (FVC), (**c**) Leaf Area Index (LAI) images and (**d**) Digital Elevation Model (DEM), with a vegetated and non-vegetated patch shown under red and yellow ellipses respectively.

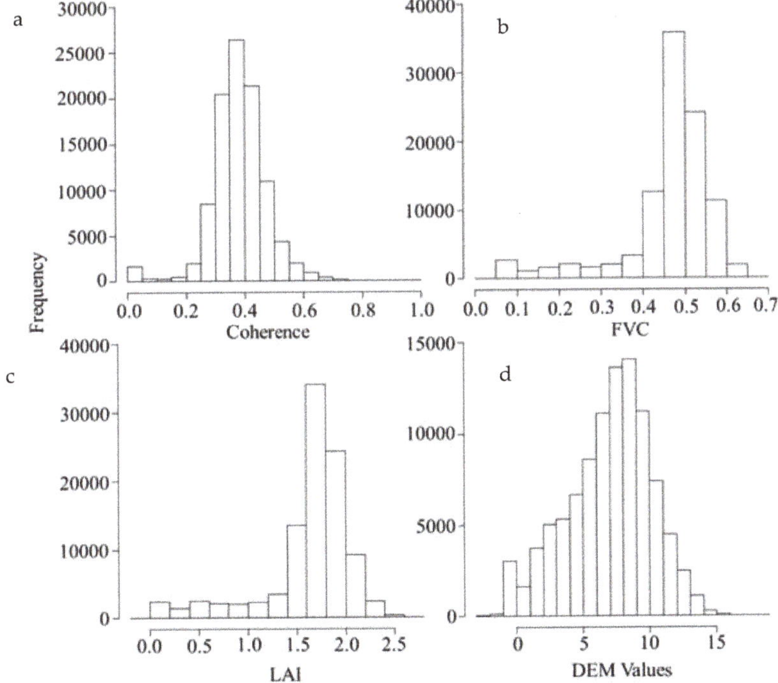

Figure 5. Frequency distribution of the input images for the canopy height model: (**a**) coherence; (**b**) FVC; (**c**) LAI; (**d**) DEM.

3.3. Canopy Height Model Establishment Using SR

The progression of the SR showed how the mean squared error decreased by selecting different relationships between the variables (Figure 6). The final regression model (Equation (3)) is a genetic combination of seven primary relationships formed with the input data. The final model by the SR used multiplication, addition, subtraction and division to build the relationships between the dependent and independent variables.

$$
\begin{aligned}
Height = 147.7 \quad &\times Coh + 0.000924 \times DEM^3 + 29.27 \times FVC \times Coh \times VH \\
&+ 15.87 \times Coh \times LAI^2 - 10.82 \times FVC \times VH - 21.98 \times FVC \times LAI \\
&- 45.05
\end{aligned}
\tag{3}
$$

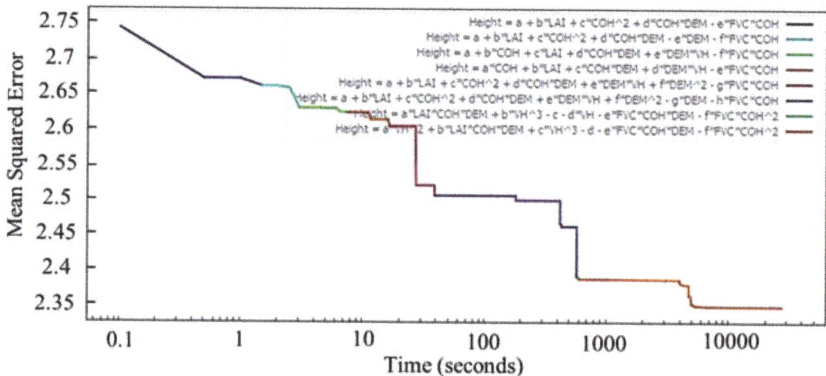

Figure 6. Progress of Symbolic Regression (SR) over time, for canopy height model establishment, including the relationship assumed for the regression.

Considering the operators and constants used in the model, the total complexity of the model was 32. Equation (3) was used to predict canopy height for the test dataset.

The 'variable sensitivity' analysis gave an idea about the critical variables and their impacts on the regression model (Table 4). FVC had the highest sensitivity that means among all the variables, a unit increase in FVC caused the most significant change in estimated height. Additionally, the direction of FVC sensitivity was negative for 100% of cases, which suggested negative correlation with canopy height. A unit increase in FVC value caused a decrease of 1.122 m in estimated canopy height. LAI was also highly sensitive, but it was positively correlated to the estimated canopy height for all the cases. Thus, a unit increase in LAI increased the estimated canopy height by an amount of 1.108 m. Coherence and DEM had relatively lower sensitivity. Coherence was negatively correlated with height with a sensitivity of 0.57056. For DEM, the correlation was positive, with a sensitivity of 0.34082. VH backscatter had very low sensitivity with the estimated canopy height.

Table 4. Variable sensitivity of the SR model that explained the relative impact a variable has on the target variable within the model.

Variable	Sensitivity	% Positive	Positive Magnitude	% Negative	Negative Magnitude
FVC	1.122	0%	0	100%	1.122
LAI	1.108	100%	1.108	0%	0
COH	0.57056	0%	0	100%	0.57056
DEM	0.34082	100%	0.34082	0%	0
VH	0.02177	100%	0.02177	0%	0

The observed and predicted values of canopy height closely followed the identity line (Figure 7). The R^2 value between observed and predicted canopy height was 0.62, and RMSE value was 1.48 m.

A trend of overestimation for lower height and underestimation for higher height can be observed in the correlation plot. However, the magnitude of deviation from the identity line was pretty low, and the number of overestimated and underestimated points was almost evenly distributed. The normalized RMSE was 13.7% concerning the range of field measured canopy heights.

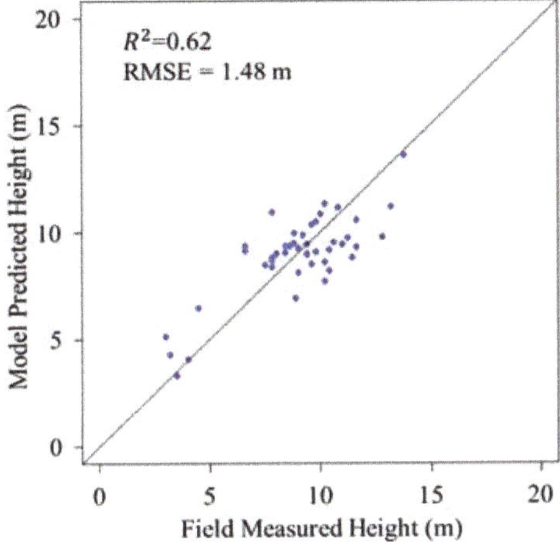

Figure 7. Correlation plot of canopy heights between field measurements and SR model based predictions.

3.4. Canopy Height Model Establishment Using RF

RF regression was run for canopy height model establishment. The same set of training and the test dataset were used. Coherence was the most critical variable in the RF model, followed by LAI and FVC. DEM and VH backscatter acted as the variables with the lowest importance (Figure 8a). The importance of VH backscatter found much less than other variables, similar to SR model (Table 4). The correlation plot between field-measured canopy heights and model predicted canopy heights, showed that the magnitude of over- and underestimation of canopy height was more or less similar, like the SR model (Figure 8b). However, the result showed a lower R^2 of 0.6, between field measured and model predicted canopy height values, while the RMSE value of RF model was higher (1.57 m) and the normalized RMSE was 14.54%.

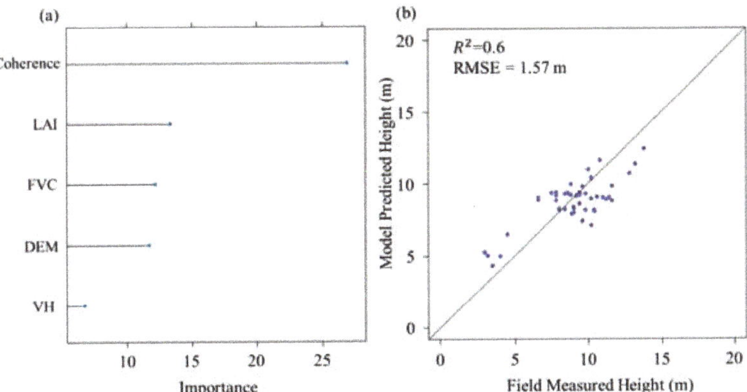

Figure 8. (a) Variable importance of the RF model for canopy height estimation; (b) correlation plot of canopy height between field measured and RF-based model prediction.

3.5. Comparison of Canopy Height Maps Derived Using SR and RF Models

The predicted canopy height map using SR and RF models demonstrated a range between 0 to 18 m and 3 to 15 m respectively (Figure 9a,c). Areas showing the upper canopy height range distribution was less in the RF model based predicted map, while for SR prediction, a larger area was found with upper canopy height ranges. Overall, both the maps showed similar trends with the medium range of canopy height values being the same for both. The difference in the canopy height as per the two models were found largely in lower range values (Figure 9e,f).

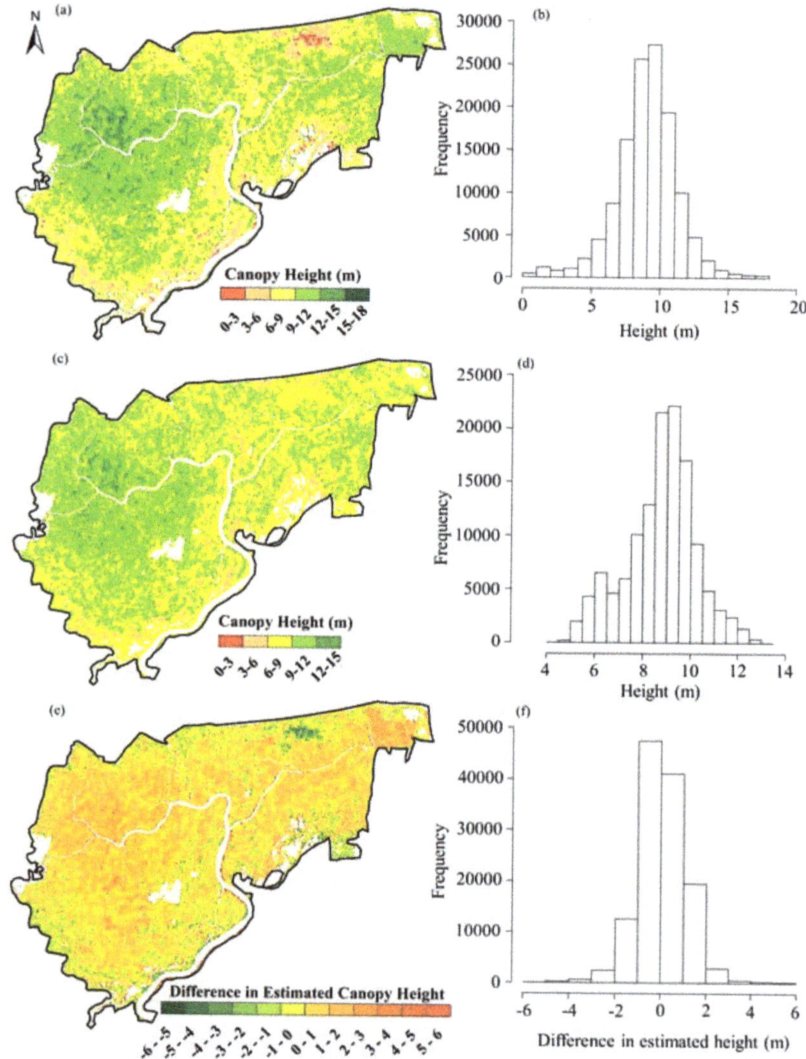

Figure 9. (**a,b**) Canopy height map using the SR model and its histogram distribution; (**c,d**) canopy height map using RF model and its histogram distribution; (**e,f**) canopy height difference map derived using estimations from two models (SR–RF) and its histogram distribution.

3.6. Comparison of Canopy Heights Derived from Model Predictions with ICESat-2 Estimates

The distribution of canopy heights from ICESat-2 showed a similar pattern with SR model predictions (Figure 10a,b). However, canopy heights from ICESat-2, data demonstrated a peak at 10–12 m, whereas it was at 9–10 m for both model predictions. It can be observed that the ICESat-2 footprints lie mostly in the areas with higher canopy heights. Further, the canopy height distribution was more similar to the SR-based predictions than RF.

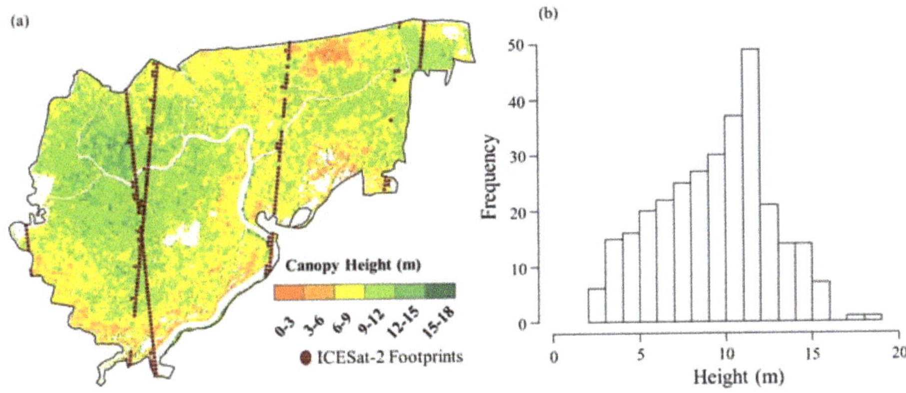

Figure 10. (**a**) ICESat-2 footprints shown over the BWS on the canopy height map using SR prediction, and (**b**) Frequency distribution of canopy height pixels from ICESat-2 data.

The canopy height values from SR prediction showed a better correlation with ICESat-2 estimated canopy heights with an R^2 value of 0.45 and RMSE of 2.24 m. The relationship with ICESat-2 based canopy height was much weaker for the RF model predicted canopy height values, where the R^2 value between observed and predicted height was 0.34 with an RMSE of 2.69 m (Figure 11). There were quite a high number of footprints with extreme values, for which over- and under-estimation can be observed.

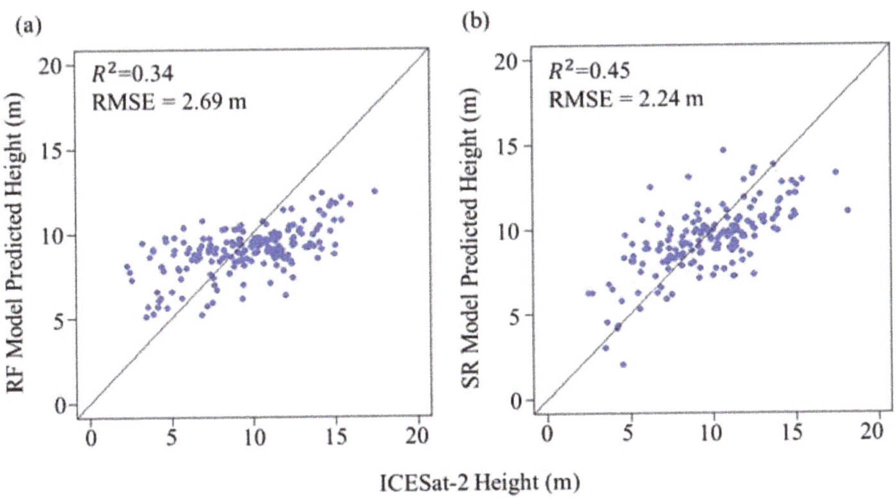

Figure 11. Correlation plots of canopy height values between ICESat-2 footprints and (**a**) RF model, and (**b**) SR model predictions.

4. Discussion

4.1. Comparison of SR and RF Model Based Canopy Height Estimates

Although, both SR and RF can be termed as a machine learning models, the fundamental working principle is entirely different from one another. The variable importance of RF and variable sensitivity of SR varied significantly for the canopy height models, as both methods had a different perception regarding the importance of variables. The RF model determines variable importance by the change in the regression error through variable permutation [82], i.e., the change in prediction accuracy due to the presence and absence of a variable was used as a measure of importance [83]. The size of the increase or decrease in regression error due to the absence of a predictor variable measured the magnitude of importance in the RF model. A more significant increase means that the variable was more important compared to other variables. There was no separate variable importance measurement procedure in the SR model. In SR, the model undergoes through a continuous evolutionary process. Models incorporating unimportant variables will perform worse than individuals using only relevant variables. Those unimportant variables will in turn, have a lower chance of being chosen to produce highly accurate symbolic expressions [84]. Therefore, the presence of irrelevant variables was discouraged throughout the process. Hence, the presence of a variable in a sufficiently evolved population will indicate the necessity of that variable in the model.

Stijven et al. [84] had argued that the variable selection method of SR is more reliable than in RF regression. With the detailed analysis of four different datasets, they have listed several reasons for which variable selection by RF may not always be reliable. As the first reason, they concluded that when multiple variables had almost equal importance, the RF model struggles to differentiate between them and assigns random variable importance. RF can also assign considerably less significance to a variable than expected due to its correlation with other irrelevant variables present in the model. Data distribution often can influence the variable importance. SR was found to be free from all these obstacles, which were held back in the RF model. Thus, in the SR model, there was a lower chance of omitting the vital variables. Chen et al. [85] also confirmed that SR was more efficient in variable selection than RF.

4.2. Importance of Variables in the Canopy Height Models

While building the canopy height model using SR, it was found that FVC, LAI, and coherence worked as the most sensitive variable for canopy height estimation. Although, FVC was an indicator of the tree crown property, it had shown a good correlation with canopy height in some earlier studies. Simrad et al. [86] concluded that while canopy cover shows a correlation with tree height, it might not hold for tall mature tropical forests due to the saturation of canopy cover with tree height. Wang et al. [47] also produced a global canopy height map using GLAS data and RF regression. In their study, they confirmed that tree cover was closely related to canopy height. Liu et al. [35] also showed the relationship between tree height and its predictors would be a non-linear one. Korhonen et al. [87] in their study, found that on reaching a certain height, canopy height has a strong non-linear relationship with canopy cover. One more reason for this kind of result can be the species distribution in the study area. It was also observed from the results that FVC had a negative sensitivity with canopy height, which means an increase in FVC will result in a decrease of canopy height. In BWS, it was observed that trees with relatively lower height such as *Excoecaria agallocha* were densely packed than other taller trees, like *Avicennia officinalis*, and *Heritiera fomes*. Thus, densely packed patches could have lower canopy heights, leading to negative sensitivity of FVC to canopy height.

Several studies found that coherence showed reasonable correlation with canopy height. Olesk et al. [42] used four different models to illustrate the relationship between coherence and canopy height and found that all the models performed well in describing the relationships. In general, the coherence of an interferometric pair decreases due to the volumetric effect, among other reasons [74]. As tree height is a significant indicator of aboveground volume in forested areas, and the canopy height is highly correlated with aboveground volume, coherence must have a correlation with canopy height. Therefore, coherence can be used to estimate canopy height. Schlund et al. [88] found that volume decorrelation can be traced back to the canopy height. As a result, we found it as an essential variable in both models. In the SR model, it is negatively correlated with the canopy height. It happened so because, with the increase in height, the volume decorrelation increases leading to decrease in coherence. Therefore, decrease in coherence could indicate increase in canopy height and vice versa.

The estimation of canopy height with the help of LAI was not largely explored. Some studies have shown a possible relationship between them [52,89]. Pope and Treitz [50] showed that LiDAR predicted height could be used for LAI estimation to an acceptable extent for boreal forests. Qu et al. [51] found that even for a tropical forest site, height metrics derived from LiDAR data can estimate LAI quite efficiently. They also showed that it even performs better than MODIS LAI. Thus, these studies showed that there is a relationship that exists between canopy height and LAI. However, remote sensing based studies have not tried to use the relationship to estimate canopy height from LAI. Here, LAI was found as the second most important variable for both models, which showed good correlation of LAI with canopy height.

4.3. Canopy Height Models and Maps

Both SR and RF models demonstrated capability in predicting canopy heights with the former having better efficiency. The SR model established here is complicated with more functions and higher complexity. A complex model is more likely to map the inherent non-linearity among the predictors, while a simpler model is preferred for its easy interpretability. The relationship observed between remotely sensed parameters, and biophysical variables were generally non-linear and complex [90]. Although different machine learning algorithms, including RF, can estimate biophysical parameters efficiently, the lack of interpretability restricts the replication of their results; which can be overcome with the use of SR based models.

Though the canopy height maps differ in values, they showed similar trends for most of the areas, i.e., higher values in one map correspond to higher values in other maps, and vice versa. The difference in canopy height map showed that the values were generally higher for SR model prediction than the RF model prediction values. However, there could be exceptions (marked by a red ellipse in Figure 3), due to mainly two reasons. First, the different variable importance in different models as the RF and SR models used different set of essential variables leading to slightly different results. Second, some differences occurred due to the extrapolation problem of the RF model [91]. RF regressions cannot predict values outside the range of the training data as it is based on averaging the values of multiple outputs. In RF regression, final predictions were derived by averaging the results of many tree canopies. Additionally, each canopy output was derived as the mean values in each terminal node of a tree. The average for a set of values must be well within the value range. Therefore, the highest canopy height value remained below 14 m in the RF model predicted map (Figure 9c), whereas field-measured data reported canopy height values beyond 14 m. The RF model training sample tend to underestimate the higher canopy height values and overestimate the lower range values. The SR model can extrapolate values beyond the range of training data, therefore, a large area observed with a canopy height between 14 and 18 m for the SR model prediction (Figure 9a). Therefore, for the upper and lower canopy height values that lie beyond the training data range, the SR model prediction could be erroneous. Castillo et al. [92] also reported discouraging extrapolation with SR model.

The canopy height maps showed limited correlation with the ICESat-2 estimated canopy height. The first release of the ALT08 data product has some known issues which may affect the estimated canopy height [93]. ICESat-2 data is also found to have a vertical RMSE of 3.2 m for canopy height retrieval [94]. So, these reasons may affect the correlation between model estimated canopy height and ICESat-2 estimated height. In future, with the availability of an increased amount of more accurate footprints, an improved relationship between the two can be expected. Additionally, the canopy height models were built on field measured inputs with a laser rangefinder, which may have some instrumental error [8].

5. Conclusions

The mangroves are one of the critical storages of aboveground carbon, and they are experiencing considerable alternations due to climate change. Accurate information about the carbon storage proxies, such as canopy height, will help estimate AGB and carbon sequestration. Forest canopy height models, especially for the mangroves, were generally prepared by using airborne LiDAR, high spatial resolution stereo imagery, or SAR interferometry [20,56,95]. Nonetheless, most of these methods were not always applicable to all the areas due to a lack of proper data availability. The current study proposed a method that can be applied to anywhere else in the world as it is based on Sentinel series data having global coverage.

In this study, we have analysed the potential of Sentinel-1 interferometric coherence, Sentinel-2 biophysical parameters in predicting the canopy height for mangroves. The interferometric coherence and biophysical parameters act as good predictors due to their relationship with the canopy height. Machine learning models were found to be an excellent method for canopy height modeling. Although the RF model demonstrated its efficiency in canopy height estimation, the SR model through genetic programming was found to be the most effective. The SR also established an interpretable model, which is not possible via any other machine learning algorithms. The SR-based model outperforms commonly used machine learning models like the RF. The fraction of vegetation cover (FVC) was found to be an essential variable for predicting canopy height. It was also found that the canopy height map correlates with ICESat-2 estimated canopy height, albeit modest. Overall, this study demonstrated the effectiveness of Sentinel series data and the SR in predicting canopy height.

Author Contributions: Conceptualization, S.M.G.; and M.D.B.; methodology, S.M.G.; and S.P.; formal analysis, S.M.G.; writing—original draft preparation, S.M.G.; writing—review and editing, M.D.B.; visualization, S.P.; supervision, M.D.B. All authors have read and agreed to the published version of the manuscript.

Funding: S.M.G. and S.P. thanks Ministry of Human resources Development (MHRD), India for Fellowships for PhD research. Canopy height measurements were taken during execution of a research project funded by Space Applications Centre (ISRO), India.

Acknowledgments: We acknowledge the support received from IIT Kharagpur authorities and State Forest and Wildlife Department of Odisha, India for the study.

Conflicts of Interest: The authors declare no conflict of interest.

References

1. Pan, Y.; Birdsey, R.A.; Fang, J.; Houghton, R.; Kauppi, P.E.; Kurz, W.A.; Phillips, O.L.; Shvidenko, A.; Lewis, S.L.; Canadell, J.G.; et al. A Large and Persistent Carbon Sink in the World's Forests. *Science* **2011**, *333*, 988–993. [CrossRef] [PubMed]

2. Lewis, S.L. Tropical forests and the changing earth system. *Philos. Trans. R. Soc. B Biol. Sci.* **2006**, *361*, 195–210. [CrossRef] [PubMed]

3. Behera, S.K.; Sahu, N.; Mishra, A.K.; Bargali, S.S.; Behera, M.D.; Tuli, R. Aboveground biomass and carbon stock assessment in Indian tropical deciduous forest and relationship with stand structural attributes. *Ecol. Eng.* **2017**, *99*, 513–524. [CrossRef]

4. Feldpausch, T.R.; Lloyd, J.; Lewis, S.L.; Brienen, R.J.W.; Gloor, M.; Monteagudo Mendoza, A.; Lopez-Gonzalez, G.; Banin, L.; Abu Salim, K.; Affum-Baffoe, K.; et al. Tree height integrated into pantropical forest biomass estimates. *Biogeosciences* **2012**, *9*, 3381–3403. [CrossRef]

5. Valbuena, R.; Heiskanen, J.; Aynekulu, E.; Pitkänen, S.; Packalen, P. Sensitivity of Above-Ground Biomass Estimates to Height-Diameter Modelling in Mixed-Species West African Woodlands. *PLoS ONE* **2016**, *11*, e0158198. [CrossRef] [PubMed]

6.   Mutwiri, F.K.; Odera, P.A.; Kinyanjui, M.J. Estimation of Tree Height and Forest Biomass Using Airborne LiDAR Data: A Case Study of Londiani Forest Block in the Mau Complex, Kenya. *Open J. For.* **2017**, *7*, 255–269. [CrossRef]

7.   Kearsley, E.; De Haulleville, T.; Hufkens, K.; Kidimbu, A.; Toirambe, B.; Baert, G.; Huygens, D.; Kebede, Y.; Defourny, P.; Bogaert, J.; et al. Conventional tree height-diameter relationships significantly overestimate aboveground carbon stocks in the Central Congo Basin. *Nat. Commun.* **2013**, *4*, 1–8. [CrossRef]

8.   Wang, Y.; Lehtomäki, M.; Liang, X.; Pyörälä, J.; Kukko, A.; Jaakkola, A.; Liu, J.; Feng, Z.; Chen, R.; Hyyppä, J. Is field-measured tree height as reliable as believed – A comparison study of tree height estimates from field measurement, airborne laser scanning and terrestrial laser scanning in a boreal forest. *ISPRS J. Photogramm. Remote Sens.* **2019**, *147*, 132–145. [CrossRef]

9.   Lee, W.-J.; Lee, C.-W. Forest Canopy Height Estimation Using Multiplatform Remote Sensing Dataset. *J. Sens.* **2018**, *2018*, 1–9. [CrossRef]

10.   Verma, N.K.; Lamb, D.W.; Reid, N.; Wilson, B. Comparison of canopy volume measurements of scattered eucalypt farm trees derived from high spatial resolution imagery and LiDAR. *Remote Sens.* **2016**, *8*, 388. [CrossRef]

11.   St-Onge, B.A.; Achaichia, N. Measuring Forest Canopy Height Using a Combination. *Int. Arch. Photogramm. Remote Sens. Spat. Inf. Sci.* **2001**, *XXXIV*, 131–137.

12.   Sexton, J.O.; Bax, T.; Siqueira, P.; Swenson, J.J.; Hensley, S. A comparison of lidar, radar, and field measurements of canopy height in pine and hardwood forests of southeastern North America. *For. Ecol. Manag.* **2009**, *257*, 1136–1147. [CrossRef]

13.   Luoma, V.; Saarinen, N.; Wulder, M.A.; White, J.C.; Vastaranta, M.; Holopainen, M.; Hyyppä, J. Assessing precision in conventional field measurements of individual tree attributes. *Forests* **2017**, *8*, 38. [CrossRef]

14.   Larjavaara, M.; Muller-Landau, H.C. Measuring tree height: A quantitative comparison of two common field methods in a moist tropical forest. *Methods Ecol. Evol.* **2013**, *4*, 793–801. [CrossRef]

15.   Hunter, M.O.; Keller, M.; Victoria, D.; Morton, D.C. Tree height and tropical forest biomass estimation. *Biogeosciences* **2013**, *10*, 8385–8399. [CrossRef]

16.   Véga, C.; St-Onge, B. Height growth reconstruction of a boreal forest canopy over a period of 58 years using a combination of photogrammetric and lidar models. *Remote Sens. Environ.* **2008**, *112*, 1784–1794. [CrossRef]

17.   Jensen, J.R.; Lin, H.; Yang, X.; Iii, E.R.; Davis, B.A.; Ramsey, E. The measurement of mangrove characteristics in southwest Florida using spot multispectral data The Measurement of Mangrove Characteristics in Southwest Florida Using SPOT Multispectral Data. *Geocart. Int.* **1991**, *2*, 13–21. [CrossRef]

18.   Miller, D.R.; Quine, C.P.; Hadley, W. An investigation of the potential of digital photogrammetry to provide measurements of forest characteristics and abiotic damage. *For. Ecol. Manag.* **2000**, *135*, 279–288. [CrossRef]

19.   Lee, S.; Ni-meister, W.; Yang, W.; Chen, Q. Remote Sensing of Environment Physically based vertical vegetation structure retrieval from ICESat data: Validation using LVIS in White Mountain National Forest, New Hampshire, USA. *Remote Sens. Environ.* **2011**, *115*, 2776–2785. [CrossRef]

20.   Lagomasino, D.; Fatoyinbo, T.; Lee, S.K.; Feliciano, E.; Trettin, C.; Simard, M. A comparison of mangrove Canopy height using multiple independent measurements from land, air, and space. *Remote Sens.* **2016**, *8*, 327. [CrossRef]

21.   Ballhorn, U.; Jubanski, J.; Kronseder, K.; Siegert, F. Airborne LiDAR measurements to estimate tropical peat swamp forest above Ground Biomass. In Proceedings of the 2012 IEEE International Geoscience and Remote Sensing Symposium, Munich, German, 22–27 July 2012.

22.   Csillik, O.; Kumar, P.; Mascaro, J.; O'Shea, T.; Asner, G.P. Monitoring tropical forest carbon stocks and emissions using Planet satellite data. *Sci. Rep.* **2019**, *9*, 1–12. [CrossRef] [PubMed]

23.   Neuenschwander, A.L.; Urban, T.J.; Gutierrez, R.; Schutz, B.E. Characterization of ICESat/GLAS waveforms over terrestial ecosystems: Implications for vegetation mapping. *J. Geophys. Res. Biogeosci.* **2008**, *113*, 1–18. [CrossRef]

24.   Xing, Y.; de Gier, A.; Zhang, J.; Wang, L. An improved method for estimating forest canopy height using ICESat-GLAS full waveform data over sloping terrain: A case study in changbai mountains, China. *Int. J. Appl. Earth Obs. Geoinf.* **2010**, *12*, 385–392. [CrossRef]

25.   Ghosh, S.M.; Behera, M.D. Forest canopy height estimation using satellite laser altimetry: A case study in the Western Ghats, India. *Appl. Geomatics* **2017**, *9*, 159–166. [CrossRef]

26. Lefsky, M.A.; Harding, D.J.; Keller, M.; Cohen, W.B.; Carabajal, C.C.; Del Bom Espirito-Santo, F.; Hunter, M.O.; de Oliveira, R. Estimates of forest canopy height and aboveground biomass using ICESat. *Geophys. Res. Lett.* **2005**, *32*, 1–4. [CrossRef]

27. Lefsky, M.A.; Keller, M.; Pang, Y.; De Camargo, P.B.; Hunter, M.O. Revised method for forest canopy height estimation from Geoscience Laser Altimeter System waveforms. *J. Appl. Remote Sens.* **2007**, *1*, 013537.

28. Tripathi, P.; Behera, M.D. Plant height profiling in western India using LiDAR data. *Curr. Sci.* **2013**, *7*, 970–977.

29. Zhang, Y.; Liang, S.; Sun, G. Forest biomass mapping of northeastern china using GLAS and MODIS data. *IEEE J. Sel. Top. Appl. Earth Obs. Remote Sens.* **2014**, *7*, 140–152. [CrossRef]

30. Boudreau, J.; Nelson, R.F.; Margolis, H.A.; Beaudoin, A.; Guindon, L.; Kimes, D.S. Regional aboveground forest biomass using airborne and spaceborne LiDAR in Québec. *Remote Sens. Environ.* **2008**, *112*, 3876–3890. [CrossRef]

31. Nelson, R.; Ranson, K.J.; Sun, G.; Kimes, D.S.; Kharuk, V.; Montesano, P. Estimating Siberian timber volume using MODIS and ICESat/GLAS. *Remote Sens. Environ.* **2009**, *113*, 691–701. [CrossRef]

32. Mitchard, E.T.A.; Saatchi, S.S.; White, L.J.T.; Abernethy, K.A.; Jeffery, K.J.; Lewis, S.L.; Collins, M.; Lefsky, M.A.; Leal, M.E.; Woodhouse, I.H.; et al. Mapping tropical forest biomass with radar and spaceborne LiDAR in Lopé National Park, Gabon: Overcoming problems of high biomass and persistent cloud. *Biogeosciences* **2012**, *9*, 179–191. [CrossRef]

33. Abdalati, W.; Zwally, H.J.; Bindschadler, R.; Csatho, B.; Farrell, S.L.; Fricker, H.A.; Harding, D.; Kwok, R.; Lefsky, M.; Markus, T.; et al. The ICESat-2 laser altimetry mission. *Proc. IEEE* **2010**, *98*, 735–751. [CrossRef]

34. Narine, L.L.; Popescu, S.; Neuenschwander, A.; Zhou, T.; Srinivasan, S.; Harbeck, K. Estimating aboveground biomass and forest canopy cover with simulated ICESat-2 data. *Remote Sens. Environ.* **2019**, *224*, 1–11. [CrossRef]

35. Liu, Y.; Gong, W.; Xing, Y.; Hu, X.; Gong, J. Estimation of the forest stand mean height and aboveground biomass in Northeast China using SAR Sentinel-1B, multispectral Sentinel-2A, and DEM imagery. *ISPRS J. Photogramm. Remote Sens.* **2019**, *151*, 277–289. [CrossRef]

36. Su, Y.; Guo, Q.; Xue, B.; Hu, T.; Alvarez, O.; Tao, S.; Fang, J. Spatial distribution of forest aboveground biomass in China: Estimation through combination of spaceborne lidar, optical imagery, and forest inventory data. *Remote Sens. Environ.* **2016**, *173*, 187–199. [CrossRef]

37. Treuhaft, R.; Lei, Y.; Gonçalves, F.; Keller, M.; dos Santos, J.R.; Neumann, M.; Almeida, A. Tropical-forest structure and biomass dynamics from TanDEM-X radar interferometry. *Forests* **2017**, *8*, 277. [CrossRef]

38. Solberg, S.; Hansen, E.H.; Gobakken, T.; Næssset, E.; Zahabu, E. Biomass and InSAR height relationship in a dense tropical forest. *Remote Sens. Environ.* **2017**, *192*, 166–175. [CrossRef]

39. Rosen, P.A.; Hensley, S.; Joughin, I.R.; Li, F.; Madsen, S.N.; Rodriguez, E.; Goldstein, R.M. Synthetic Aperture Radar Interferometry. *Proc. IEEE* **1999**, *14*, R1–R54. [CrossRef]

40. Richards, J.A. *Remote Sensing with Imaging Radar*; Springer: Berlin/Heidelberg, Germany, 2009.

41. Torres, R.; Snoeij, P.; Geudtner, D.; Bibby, D.; Davidson, M.; Attema, E.; Potin, P.; Rommen, B.; Floury, N.; Brown, M.; et al. GMES Sentinel-1 mission. *Remote Sens. Environ.* **2012**, *120*, 9–24. [CrossRef]

42. Olesk, A.; Praks, J.; Antropov, O.; Zalite, K.; Arumäe, T.; Voormansik, K. Interferometric SAR coherence models for Characterization of hemiboreal forests using TanDEM-X dssata. *Remote Sens.* **2016**, *8*, 700. [CrossRef]

43. Torano Caicoya, A.; Kugler, F.; Hajnsek, I.; Papathanassiou, K. Boreal forest biomass classification with TanDEM-X. In Proceedings of the 2012 IEEE International Geoscience and Remote Sensing Symposium, Munich, German, 22–27 July 2012.

44. Cougo, M.; Souza-Filho, P.; Silva, A.; Fernandes, M.; Santos, J.; Abreu, M.; Nascimento, W.; Simard, M. Radarsat-2 Backscattering for the Modeling of Biophysical Parameters of Regenerating Mangrove Forests. *Remote Sens.* **2015**, *7*, 17097–17112. [CrossRef]

45. Hansen, M.C.; Potapov, P.V.; Goetz, S.J.; Turubanova, S.; Tyukavina, A.; Krylov, A.; Kommareddy, A.; Egorov, A. Mapping tree height distributions in Sub-Saharan Africa using Landsat 7 and 8 data. *Remote Sens. Environ.* **2016**, *185*, 221–232. [CrossRef]

46. Zhang, S.; Chen, H.; Fu, Y.; Niu, H.; Yang, Y.; Zhang, B. Fractional vegetation cover estimation of different vegetation types in the Qaidam Basin. *Sustainability* **2019**, *11*, 864. [CrossRef]

47. Wang, Y.; Li, G.; Ding, J.; Guo, Z.; Tang, S.; Wang, C.; Huang, Q.; Liu, R.; Chen, J.M. A combined GLAS and MODIS estimation of the global distribution of mean forest canopy height. *Remote Sens. Environ.* **2016**, *174*, 24–43. [CrossRef]

48. Welles, J.M.; Norman, J.M. Instrument for Indirect Measurement of Canopy Architecture. *Agron. J.* **1991**, *83*, 818. [CrossRef]

49. Watson, D.J. Comparative Physiological Studies on the Growth of Field Crops. *Ann. Bot.* **1947**, *11*, 41–76. [CrossRef]

50. Pope, G.; Treitz, P. Leaf Area Index (LAI) estimation in boreal mixedwood forest of Ontario, Canada using Light detection and ranging (LiDAR) and worldview-2 imagery. *Remote Sens.* **2013**, *5*, 5040–5063. [CrossRef]

51. Qu, Y.; Shaker, A.; Silva, C.A.; Klauberg, C.; Pinagé, E.R. Remote sensing of leaf area index from LiDAR height percentile metrics and comparison with MODIS product in a selectively logged tropical forest area in Eastern Amazonia. *Remote Sens.* **2018**, *10*, 970. [CrossRef]

52. Yuan, Y.; Wang, X.; Yin, F.; Zhan, J. Examination of the Quantitative Relationship between Vegetation Canopy Height and LAI. *Adv. Meteorol.* **2013**, *2013*, 1–6. [CrossRef]

53. Kenyi, L.W.; Dubayah, R.; Hofton, M.; Schardt, M. Comparative analysis of SRTM-NED vegetation canopy height to LIDAR-derived vegetation canopy metrics. *Int. J. Remote Sens.* **2009**, *30*, 2797–2811. [CrossRef]

54. Sadeghi, Y.; St-Onge, B.; Leblon, B.; Prieur, J.F.; Simard, M. Mapping boreal forest biomass from a SRTM and TanDEM-X based on canopy height model and Landsat spectral indices. *Int. J. Appl. Earth Obs. Geoinf.* **2018**, *68*, 202–213. [CrossRef]

55. Wicaksono, P.; Danoedoro, P.; Hartono; Nehren, U. Mangrove biomass carbon stock mapping of the Karimunjawa Islands using multispectral remote sensing. *Int. J. Remote Sens.* **2016**, *37*, 26–52. [CrossRef]

56. Feliciano, E.A.; Wdowinski, S.; Potts, M.D.; Lee, S.K.; Fatoyinbo, T.E. Estimating mangrove canopy height and above-ground biomass in the Everglades National Park with airborne LiDAR and TanDEM-X data. *Remote Sens.* **2017**, *9*, 702. [CrossRef]

57. Berninger, A.; Lohberger, S.; Stängel, M.; Siegert, F. SAR-based estimation of above-ground biomass and its changes in tropical forests of Kalimantan using L- and C-band. *Remote Sens.* **2018**, *10*, 831. [CrossRef]

58. Pham, L.T.H.; Brabyn, L. Monitoring mangrove biomass change in Vietnam using SPOT images and an object-based approach combined with machine learning algorithms. *ISPRS J. Photogramm. Remote Sens.* **2017**, *128*, 86–97. [CrossRef]

59. Ghosh, S.M.; Behera, M.D. Aboveground biomass estimation using multi-sensor data synergy and machine learning algorithms in a dense tropical forest. *Appl. Geogr.* **2018**, *96*, 29–40. [CrossRef]

60. Breiman, L. Random forests. *Mach. Learn.* **2001**, *45*, 5–32. [CrossRef]

61. Koza, J.R. Genetic programming as a means for programming computers by natural selection. *Stat. Comput.* **1994**, *4*, 87–112. [CrossRef]

62. Iba, H.; Feng, J.; Izadi Rad, H. GP-RVM: Genetic Programing-Based Symbolic Regression Using Relevance Vector Machine. *arXiv* **2018**, arXiv:1806.02502.

63. Stijven, S.; Vladislavleva, E.; Kordon, A.; Willem, L.; Kotanchek, M.E. Prime-Time: Symbolic Regression Takes Its Place in the Real World. In *Genetic Programming Theory and Practice XIII*; Springer: Berlin/Heidelberg, Germany, 2016; pp. 241–260.

64. Reddy, C.S.; Pattanaik, C.; Dhal, N.K.; Biswal, A.K. Vegetation and Floristic Diversity of Bhitarkanika National Park, Orissa, India. *Indian For.* **2006**, *132*, 664–680.

65. Forest Survey of India (FSI). *State of Forest Report*; Forest Survey of India (FSI): Dehradun, India, 2017.

66. Reddy, C.S. *Field Identification Guide for Indian Mangroves*; Bishen Singh Mahendra Pal Singh: Dehradun, India, 2008; Volume 001.

67. Pattanaik, C.; Reddy, C.S.; Dhal, N.K.; Das, R. Utilisation of mangrove forests in Bhitarkanika wildlife sanctuary, Orissa. *Indian J. Tradit. Knowl.* **2008**, *7*, 598–603.

68. Sullivan, M.J.P.; Lewis, S.L.; Hubau, W.; Qie, L.; Baker, T.R.; Banin, L.F.; Chave, J.; Cuni-Sanchez, A.; Feldpausch, T.R.; Lopez-Gonzalez, G.; et al. Field methods for sampling tree height for tropical forest biomass estimation. *Methods Ecol. Evol.* **2018**, *9*, 1179–1189. [CrossRef] [PubMed]

69. Yague-Martinez, N.; Prats-Iraola, P.; Gonzalez, F.R.; Brcic, R.; Shau, R.; Geudtner, D.; Eineder, M.; Bamler, R. Interferometric Processing of Sentinel-1 TOPS Data. *IEEE Trans. Geosci. Remote Sens.* **2016**, *54*, 2220–2234. [CrossRef]

70. Bickel, D.L. *SAR Image Effects on Coherence and Coherence Estimation*; Sandia National Laboratories: Albuquerque, NM, USA, 2014.

71. Louis, J.; Debaecker, V.; Pflug, B.; Main-Knorn, M.; Bieniarz, J.; Mueller-Wilm, U.; Cadau, E.; Gascon, F. Sentinel-2 SEN2COR: L2A processor for users. In Proceedings of the Living Planet Symposium 2016, Prague, Czech Republic, 13 May 2016; Volume ESA SP-740.

72. Markus, T.; Neumann, T.; Martino, A.; Abdalati, W.; Brunt, K.; Csatho, B.; Farrell, S.; Fricker, H.; Gardner, A.; Harding, D.; et al. The Ice, Cloud, and land Elevation Satellite-2 (ICESat-2): Science requirements, concept, and implementation. *Remote Sens. Environ.* **2017**, *190*, 260–273. [CrossRef]

73. Neuenschwander, A.L.; Popescu, S.C.; Nelson, R.F.; Harding, D.; Pitts, K.L.; Robbins, J. ATLAS/ICESat-2 L3A Land and Vegetation Height, Version 1. Available online: https://doi.org/10.5067/ATLAS/ATL08.001 (accessed on 10 August 2019).

74. Ferretti, A.; Monti-guarnieri, A.; Prati, C.; Rocca, F. *InSAR Principles: Guidelines for SAR Interferometry Processing and Interpretation, Part C. InSAR Processing: A Mathematical Approach*; ESA Publications: Auckland, New Zealand, 2007.

75. Geudtner, D.; Torres, R.; Snoeij, P.; Ostergaard, A.; Navas-Traver, I.; Rommen, B.; Brown, M. Sentinel-1 system overview and performance. In Proceedings of the ESA Living Planet Symposium, Edinburgh, UK, 13 September 2013; pp. 1719–1721.

76. Weiss, M.; Baret, F. *S2ToolBox Level 2 Products: LAI, FAPAR, FCOVER*; Institut National de la Recherche Agronomique (INRA): Avignon, France, 2016.

77. Max, K.; Kuhn, M. Building Predictive Models in R Using the caret Package. *J. Stat. Softw.* **2008**, *28*, 1–26.

78. Liaw, A.; Wiener, M. Classification and Regression by randomForest. *R News* **2002**, *2*, 18–22.

79. Aryadoust, V. Application of Evolutionary Algorithm-Based Symbolic Regression to Language Assessment: Toward Nonlinear Modeling. *Psychol. Test Assess. Model.* **2015**, *57*, 301.

80. Schmidt, M.; Lipson, H. Distilling Free-Form Natural Laws from Experimental Data. *Science* **2009**, *324*, 81–85. [CrossRef]

81. Dyk, M. Van Identifying Patterns in Course-Leaving That Predict Student Leaving—A Comparison of Different Predictive Algorithms. Master's Thesis, University of Oklahoma, Norman, OK, USA, 2018.

82. Strobl, C.; Boulesteix, A.-L.; Kneib, T.; Augustin, T.; Zeileis, A. Conditional variable importance for random forests. *BMC Bioinf.* **2008**, *9*, 307. [CrossRef]

83. Dube, T.; Mutanga, O.; Elhadi, A.; Ismail, R. Intra-and-inter species biomass prediction in a plantation forest: Testing the utility of high spatial resolution spaceborne multispectral RapidEye sensor and advanced machine learning algorithms. *Sensors* **2014**, *14*, 15348–15370. [CrossRef]

84. Stijven, S.; Minnebo, W.; Vladislavleva, K. Separating the Wheat from the Chaff: On Feature Selection and Feature Importance in Regression Random Forests and Symbolic Regression. In Proceedings of the 13th annual conference companion on genetic and evolutionary computation, Dublin, Ireland, 12 July 2011; pp. 623–630.

85. Chen, Q.; Zhang, M.; Xue, B. Feature selection to improve generalization of genetic programming for high-dimensional symbolic regression. *IEEE Trans. Evol. Comput.* **2017**, *21*, 792–806. [CrossRef]

86. Simard, M.; Pinto, N.; Fisher, J.B.; Baccini, A. Mapping forest canopy height globally with spaceborne lidar. *J. Geophys. Res. Biogeosci.* **2011**, *116*, 1–12. [CrossRef]

87. Korhonen, L.; Korhonen, K.T.; Stenberg, P.; Maltamo, M.; Rautiainen, M. Local models for forest canopy cover with beta regression. *Silva Fenn.* **2007**, *41*, 671–685. [CrossRef]

88. Schlund, M.; von Poncet, F.; Hoekman, D.H.; Kuntz, S.; Schmullius, C. Importance of bistatic SAR features from TanDEM-X for forest mapping and monitoring. *Remote Sens. Environ.* **2014**, *151*, 16–26. [CrossRef]

89. Jaimez, R.E.; Araque, O.; Guzman, D.; Mora, A.; Espinoza, W.; Tezara, W. Agroforestry systems of timber species and cacao: Survival and growth during the early stages. *J. Agric. Rural Dev. Trop. Subtrop.* **2013**, *114*, 1–11.

90. Ali, I.; Greifeneder, F.; Stamenkovic, J.; Neumann, M.; Notarnicola, C. Review of machine learning approaches for biomass and soil moisture retrievals from remote sensing data. *Remote Sens.* **2015**, *7*, 16398–16421. [CrossRef]

91. Hengl, T.; Nussbaum, M.; Wright, M.N.; Heuvelink, G.B.M.; Gräler, B. Random forest as a generic framework for predictive modeling of spatial and spatio-temporal variables. *PeerJ* **2018**, *2018*, e5518. [CrossRef]

92. Castillo, F.; Marshall, K.; Green, J.; Kordon, A. A methodology for combining symbolic regression and design of experiments to improve empirical model building. *Lect. Notes Comput. Sci.* **2003**, *2724*, 1975–1985.

93. Neuenschwander, A.; Klotz, B.; Jelley, B. ATL08 Known Issues—Release 001. 2019. Available online: https://nsidc.org/sites/nsidc.org/files/technical-references/ATL08_Release001_Known%20Issues.2.pdf (accessed on 26 February 2020).

94. Neuenschwander, A.L.; Magruder, L.A. Canopy and Terrain Height Retrievals with ICESat-2: A First Look. *Remote Sens.* **2019**, *11*, 1721. [CrossRef]

95. Wannasiri, W.; Nagai, M.; Honda, K.; Santitamnont, P.; Miphokasap, P. Extraction of Mangrove Biophysical Parameters Using Airborne LiDAR. *Remote Sens.* **2013**, *5*, 1787–1808. [CrossRef]

Article

Forest Height Estimation Based on P-Band Pol-InSAR Modeling and Multi-Baseline Inversion

Xiaofan Sun [1,2], **Bingnan Wang** [1,*], **Maosheng Xiang** [1,2], **Liangjiang Zhou** [1] **and Shuai Jiang** [3]

[1] National Key Lab of Microwave Imaging Technology, Aerospace Information Research Institute, Chinese Academy of Sciences, Beijing 100190, China; sunxiaofan15@mails.ucas.ac.cn (X.S.); xms@mail.ie.ac.cn (M.X.); ljzhou@mail.ie.ac.cn (L.Z.)

[2] School of Electronic, Electrical and Communication Engineering, University of Chinese Academy of Sciences, Beijing 100049, China

[3] Beijing Institute of Spacecraft System Engineering, China Academy of Space Technology, Beijing 100094, China; jiangshuai13@mails.ucas.ac.cn

*   Correspondence: wbn@mail.ie.ac.cn

Received: 11 March 2020; Accepted: 13 April 2020; Published: 22 April 2020

Abstract: The Gaussian vertical backscatter (GVB) model has a pivotal role in describing the forest vertical structure more accurately, which is reflected by P-band polarimetric interferometric synthetic aperture radar (Pol-InSAR) with strong penetrability. The model uses a three-dimensional parameter space (forest height, Gaussian mean representing the strongest backscattered power elevation, and the corresponding standard deviation) to interpret the forest vertical structure. This paper establishes a two-dimensional GVB model by simplifying the three-dimensional one. Specifically, the two-dimensional GVB model includes the following three cases: the Gaussian mean is located at the bottom of the canopy, the Gaussian mean is located at the top of the canopy, as well as a constant volume profile. In the first two cases, only the forest height and the Gaussian standard deviation are variable. The above approximation operation generates a two-dimensional volume only coherence solution space on the complex plane. Based on the established two-dimensional GVB model, the three-baseline inversion is achieved without the null ground-to-volume ratio assumption. The proposed method improves the performance by 18.62% compared to the three-baseline Random Volume over Ground (RVoG) model inversion. In particular, in the area where the radar incidence angle is less than 0.6 rad, the proposed method improves the inversion accuracy by 34.71%. It suggests that the two-dimensional GVB model reduces the GVB model complexity while maintaining a strong description ability.

Keywords: P-band polarimetric interferometric synthetic aperture radar (Pol-InSAR); forest vertical structure; Gaussian vertical backscatter (GVB) volume; multi-baseline optimization

---

1. Introduction

The essence of the model-based polarimetric interferometric synthetic aperture radar (Pol-InSAR) parameter inversion is to solve the model equations established between the synthetic aperture radar (SAR) observations and the theoretical forest vertical structure [1–7]. The Random Volume over Ground (RVoG) model based on the assumption of vertical homogeneous random volume was widely verified at relatively high frequencies (e.g., X-, C- and L-band) [5–14].

In the process of the Pol-InSAR model inversion, a non-negligible issue is whether the equations has a unique solution. An effective precondition for single-baseline Quad-pol configuration to make the RVoG model solution unique is the null ground-to-volume ratio assumption (commonly less than −10 dB) [6]. Nevertheless, this assumption is no longer valid for P-band observations in which all polarization channels generally contain significant ground responses [15–19].

A common solution to this case is to fix the extinction coefficient in the RVoG model [15,16,20–22]. However, as the inversion performance depends heavily on the selection of extinction coefficient, this technique is restricted in complex actual scene, in which it is unreasonable to apply only a single extinction coefficient to the parameter estimation of whole scene. An alternative is to increase the observation space through the multi-baseline configuration [8,17,23–26], and yet in this case, the inversion based on RVoG model is still obviously underestimated, indicating that multi-baseline is not enough to make up for the lack of the RVoG model [17,23]. Actually, when P-band SAR systems are used to observe the low-density forest, the volume scattering layer exhibits a vertical heterogeneous distribution, leading to the ineffectiveness of RVoG model [17].

To cope with this issue, the Gaussian vertical backscatter (GVB) model with a stronger expression ability was proposed [27,28]. The GVB model is mainly used to describe the vertical heterogeneity of volume scattering, which is often effective for the Pol-InSAR inversion where most backscatters are concentrated in the middle or lower part of the canopy. Compared with the RVoG model which has only two parameters (forest height and extinction coefficient), the GVB model describes the forest vertical structure with three parameters (forest height, Gaussian mean and standard deviation), which effectively improves the description ability of the model, yet brings serious inversion complexity simultaneously [24,25].

In view of the characteristics of P-band Pol-InSAR observations and the limitations of the existing methods, this paper proposes the two-dimensional GVB model and the corresponding multi-baseline inversion. Concretely, in order to apply to the heterogeneity of the forest vertical volume under P-band observations, the GVB model framework is used to interpret the forest vertical structure. Concurrently, to alleviate the high complexity of the three-dimensional GVB model, the two-dimensional GVB model is developed. As for the the model inversion, considering the assumption of null ground-to-volume ratio is no longer tenable, the multi-baseline configuration is adoped to optimize the model equations.

The remaining part of the paper proceeds as follows. Section 2 retrospects the model-based Pol-InSAR inversion theory. Section 3 presents the inversion based on the two-dimensional GVB model, while Section 4 validates the proposed approach with the actual E-SAR data. Finally, the discussions and conclusions are drawn in Section 5.

2. Model-Based Pol-InSAR Inversion Theory

Once temporal decorrelation and noise decorrelation are ignored, the interferometric coherence in Pol-InSAR observation mainly includes the volume decorrelation [5–7,29,30]

$$\gamma = \frac{\int_{z_0}^{z_0+h_v} F(z) e^{jk_z z} dz}{\int_{z_0}^{z_0+h_v} F(z) dz} \tag{1}$$

where $F(z)$ is the structure function related to the vertical coordinate z, h_v and z_0 refer to the forest height and the ground starting coordinate respectively. The interferometric vertical wavenumber k_z is related to the radar wavelength λ, incidence angle θ and the incidence angle difference between two interferometric antennas $\delta\theta$ [31]

$$k_z = \frac{4\pi}{\lambda \sin\theta} \delta\theta \tag{2}$$

2.1. RVoG Model and Inversion

As shown in Figure 1, the vertical structure function based on the RVoG model is given [6,7,29,30,32]

$$F\left(z\right) = \rho_v e^{\left(2\sigma\left(z-z_0-h_v\right)/\cos\theta\right)} + \rho_g e^{\left(-2\sigma h_v/\cos\theta\right)} \delta\left(z-z_0\right) \tag{3}$$

where σ indicates the extinction coefficient independent of polarization, $\delta\left(\bullet\right)$ refers to the impulse function, ρ_v and ρ_g related to the specific polarization state represent the backscatter intensities per unit length corresponding to the volume and ground, respectively.

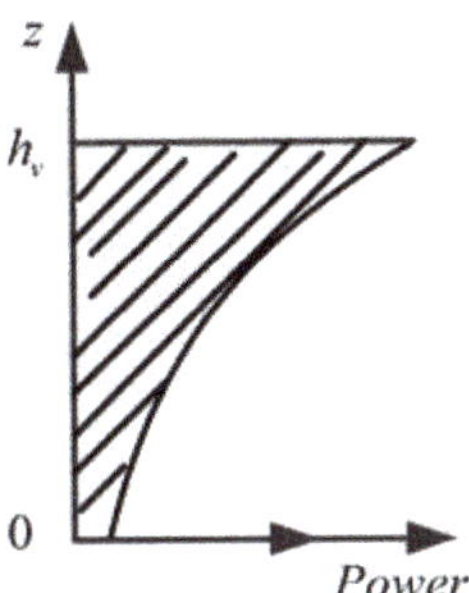

Figure 1. The profile of Random Volume over Ground model.

Substituting Equation (3) into Equation (1) gives the volume decorrelation under the assumption of the RVoG model

$$\gamma\left(\omega\right) = e^{j\phi_0} \frac{\gamma_v + \mu\left(\omega\right)}{1 + \mu\left(\omega\right)} \tag{4}$$

where $\phi_0 = k_z z_0$ is the ground interferometric phase, ω represents the unit vector characterizing polarization, γ_v and $\mu\left(\omega\right)$ indicate the volume only coherence and the ground-to-volume power ratio [6,7,29,30,32]

$$\gamma_v\left(h_v,\sigma\right) = \frac{\int_0^{h_v} e^{\left(2\sigma z/\cos\theta\right)} e^{jk_z z}dz}{\int_0^{h_v} e^{\left(2\sigma z/\cos\theta\right)}dz} = \frac{2\sigma}{\cos\theta\left(e^{\left(2\sigma h_v/\cos\theta\right)} - 1\right)} \int_0^{h_v} e^{\left(2\sigma z/\cos\theta\right)} e^{jk_z z}dz \tag{5}$$

$$\mu\left(\omega\right) = \frac{2\sigma}{\cos\theta\left(e^{\left(2\sigma h_v/\cos\theta\right)} - 1\right)} \frac{\rho_g}{\rho_v} \tag{6}$$

According to the random volume hypothesis, the extinction coefficient σ is independent of polarization, while ϕ_0 depends on the baseline. Therefore, once the baseline is determined, h_v, σ and ϕ_0 in Equations (4) and (5) remain unchanged, while only $\mu\left(\omega\right)$ varies with polarization state. Since the single-baseline Quad-pol provides three measured complex coherences corresponding to independent polarization channels, the parameter inversion can be achieved by a six-dimensional optimization process [5,6,15,17,19]

$$\begin{bmatrix} \hat{h}_v, & \hat{\sigma}, & \hat{\phi}_0, & \hat{\mu}_i \end{bmatrix}^T = \underset{[h_v,\sigma,\phi_0,\mu_i]}{\arg\min} \left\{ \left\| \begin{bmatrix} \hat{\gamma}\left(\omega_1\right) \\ \hat{\gamma}\left(\omega_2\right) \\ \hat{\gamma}\left(\omega_3\right) \end{bmatrix} - \begin{bmatrix} \gamma\left(h_v,\sigma,\phi_0,\mu_1\right) \\ \gamma\left(h_v,\sigma,\phi_0,\mu_2\right) \\ \gamma\left(h_v,\sigma,\phi_0,\mu_3\right) \end{bmatrix} \right\| \right\} \tag{7}$$

where $\left[\hat{h}_v, \ \hat{\sigma}, \ \hat{\phi}_0, \ \hat{\mu}_i \right]^T$ is the parameter estimation and subscript $i = 1, 2, 3$ represent any three independent polarization channels. $\|\bullet\|$ denotes the matrix 2-norm operation, $\hat{\gamma}$ and γ represent the radar observation and the theoretical prediction respectively. Although the optimization process consists of six unknowns and six independent equations, the inversion cannot get a unique solution due to the nonlinear form of the equation [17].

One way to guarantee a unique solution to the single-baseline configuration is to assume that there is at least one polarization channel with little ground response [6,7,17]. Under this assumption, $\mu_3 = 0$ can be set directly, and the optimization expression at this time is as follows

$$\left[\hat{h}_v, \ \hat{\sigma}, \ \hat{\phi}_0, \ \hat{\mu}_i \right]^T = \underset{[h_v,\sigma,\phi_0,\mu_i]}{\arg\min} \left\{ \left\| \begin{bmatrix} \hat{\gamma}(\omega_1) \\ \hat{\gamma}(\omega_2) \\ \hat{\gamma}(\omega_3) \end{bmatrix} - \begin{bmatrix} \gamma(h_v,\sigma,\phi_0,\mu_1) \\ \gamma(h_v,\sigma,\phi_0,\mu_2) \\ \gamma(h_v,\sigma,\phi_0,\mu_3 = 0) \end{bmatrix} \right\| \right\} \tag{8}$$

As mentioned before, the null ground-to-volume ratio assumption is no longer effective for P-band SAR observations. In this case, a common single-baseline inversion method for P-band observations is to fix the extinction coefficient [15,16,20–22]

$$\left[\hat{h}_v, \ \hat{\phi}_0, \ \hat{\mu}_i \right]^T = \underset{[h_v,\phi_0,\mu_i]}{\arg\min} \left\{ \left\| \begin{bmatrix} \hat{\gamma}(\omega_1) \\ \hat{\gamma}(\omega_2) \\ \hat{\gamma}(\omega_3) \end{bmatrix} - \begin{bmatrix} \gamma(h_v,\sigma = \sigma_{pre},\phi_0,\mu_1) \\ \gamma(h_v,\sigma = \sigma_{pre},\phi_0,\mu_2) \\ \gamma(h_v,\sigma = \sigma_{pre},\phi_0,\mu_3) \end{bmatrix} \right\| \right\} \tag{9}$$

where σ_{pre} represents the extinction coefficient that is determined by prior or empirical information.

Another method to ensure that the inversion has a unique solution while not requiring the null ground-to-volume ratio assumption is the multi-baseline optimization [17,23]

$$\left[\hat{h}_v, \ \hat{\sigma}, \ \hat{\phi}_0^j, \ \hat{\mu}_i \right]^T = \underset{[h_v,\sigma,\phi_0^j,\mu_i]}{\arg\min} \left\{ \left\| \begin{bmatrix} \hat{\gamma}^1(\omega_1) \\ \hat{\gamma}^1(\omega_2) \\ \hat{\gamma}^1(\omega_3) \\ \dots \\ \hat{\gamma}^n(\omega_1) \\ \hat{\gamma}^n(\omega_2) \\ \hat{\gamma}^n(\omega_3) \end{bmatrix} - \begin{bmatrix} \gamma^1(h_v,\sigma,\phi_0^1,\mu_1) \\ \gamma^1(h_v,\sigma,\phi_0^1,\mu_2) \\ \gamma^1(h_v,\sigma,\phi_0^1,\mu_3) \\ \dots \\ \gamma^n(h_v,\sigma,\phi_0^n,\mu_1) \\ \gamma^n(h_v,\sigma,\phi_0^n,\mu_2) \\ \gamma^n(h_v,\sigma,\phi_0^n,\mu_3) \end{bmatrix} \right\| \right\} \tag{10}$$

where the superscript $j = 1, 2, ..., n$ represent different baselines, among which n is the total number of baselines involved in the optimization. Under similar experimental environments (radar system parameters, weather conditions, etc.), μ_i of different interferometric pairs are assumed to be constant in the same polarization [24,25,33]. Consequently, for each additional interferometric pair, the corresponding optimization increases three observational complex coherences and produces an unknown variable ϕ_0 simultaneously. In other words, when n interferometric pairs are involved in inversion, the optimization process includes $6n$ independent equations and $n + 5$ unknowns concurrently.

2.2. GVB Model and Inversion

As presented in Figure 2, the GVB model can be divided into three situations in line with the position of δ. The first case in Figure 2a is similar to the RVoG model, which assumes that the strongest backscatter power is located at the top of the canopy, while Figure 2b,c are supplements of GVB model for the heterogeneity of the canopy under P-band observations [25]. The volume only coherence based on the GVB model is [27,28]

$$\gamma_v\left(h_v,\delta,\chi\right)=\frac{\int_0^{h_v}e^{-\frac{(z-\delta)^2}{2\chi^2}+jk_zz}dz}{\int_0^{h_v}e^{-\frac{(z-\delta)^2}{2\chi^2}}dz}=e^{-\frac{\chi^2k_z^2}{2}+j\delta k_z}\frac{erf\left(\frac{1}{\sqrt{2}}\left(j\chi k_z+\frac{\delta}{\chi}\right)\right)-erf\left(\frac{1}{\sqrt{2}}\left(j\chi k_z+\frac{\delta-h_v}{\chi}\right)\right)}{erf\left(\frac{h_v-\delta}{\sqrt{2}\chi}\right)+erf\left(\frac{\delta}{\sqrt{2}\chi}\right)} \tag{11}$$

where δ and χ represent the backscatter mean elevation and the corresponding standard deviation respectively, and $erf\left(\bullet\right)$ indicates the error function.

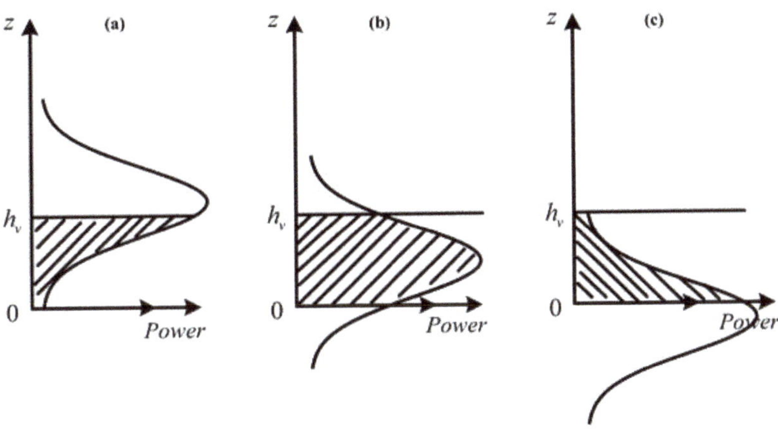

Figure 2. The profile of Gaussian vertical backscatter volume over ground. (**a**): $\delta\geq h_v$; (**b**): $\delta\in(0,\ h_v)$; (**c**): $\delta\leq 0$.

When δ or χ is fixed, the corresponding solution spaces of GVB model are shown in Figure 3, from which it can be found that there is an overlapping area between the two solution spaces. It illustrates that once δ and χ both can take arbitrary values, there are some identical elements in the corresponding three-dimensional volume only coherence solution space. This will lead to multiple solutions [24,25].

Under the assumption of the two-layer volume over ground model, the interferometric coherence based on the GVB model also conforms to Equation (4). As more model parameters are included as well as the Gaussian form of vertical structure function (resulting in multiple solutions as presented in Figure 3), the GVB model inversion theoretically requires multi-baseline observations to ensure that the forest vertical structure with reasonable physical significance can be obtained [24,25]

$$\left[\hat{h}_v,\ \hat{\delta},\ \hat{\chi},\ \hat{\phi}_0^j,\ \hat{\mu}_i\right]^T=\underset{\left[h_v,\delta,\chi,\phi_0^j,\mu_i\right]}{arg\ min}\left\{\left\|\begin{bmatrix}\hat{\gamma}^1\left(\omega_1\right)\\\hat{\gamma}^1\left(\omega_2\right)\\\hat{\gamma}^1\left(\omega_3\right)\\\cdots\\\hat{\gamma}^n\left(\omega_1\right)\\\hat{\gamma}^n\left(\omega_2\right)\\\hat{\gamma}^n\left(\omega_3\right)\end{bmatrix}-\begin{bmatrix}\gamma^1\left(h_v,\delta,\chi,\phi_0^1,\mu_1\right)\\\gamma^1\left(h_v,\delta,\chi,\phi_0^1,\mu_2\right)\\\gamma^1\left(h_v,\delta,\chi,\phi_0^1,\mu_3\right)\\\cdots\\\gamma^n\left(h_v,\delta,\chi,\phi_0^n,\mu_1\right)\\\gamma^n\left(h_v,\delta,\chi,\phi_0^n,\mu_2\right)\\\gamma^n\left(h_v,\delta,\chi,\phi_0^n,\mu_3\right)\end{bmatrix}\right\|\right\} \tag{12}$$

where $\left[\hat{h}_v,\ \hat{\delta},\ \hat{\chi},\ \hat{\phi}_0^j,\ \hat{\mu}_i\right]^T$ is the the parameter estimation. Similarly, it is assumed that δ and χ are independent of baseline and polarization. When n interferometric pairs are involved in inversion, the optimization process will include $6n$ independent equations and $n+6$ unknowns concurrently.

In addition, in order to ensure the physical significance of parameter estimation, some prior or empirical knowledge can be used to calibrate the initial values and constraints in the optimization Equation (12) [24,25].

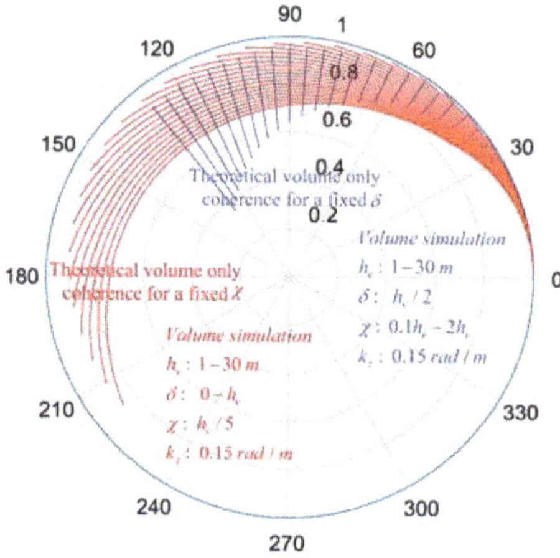

Figure 3. The solution space of Gaussian vertical backscatter model when δ or χ is fixed. Each blue or red curve represents the volume only coherence for a given forest height at an interval of 1 m. The terrain slope is set to 0.

3. Two-Dimensional GVB Model and Inversion

3.1. Two-Dimensional GVB Model

Corresponding to the three cases in Figure 2, the two-dimensional GVB model is constructed by simplifying the original three-dimensional GVB model, as illustrated in Figure 4. Cases $\delta \geq h_v$ and $\delta \leq 0$ are approximated to $\delta = h_v$ and $\delta = 0$ respectively, while case $\delta \in (0, h_v)$ is approximated to a constant volume profile. In these three cases, h_v and χ are varied, while δ remains stationary. Since the constant profile is equivalent to the case that χ is infinite when $\delta = h_v$ or $\delta = 0$, only the following two cases are considered in the modeling of volume only coherence

$$\gamma_v\left(h_v, \delta, \chi\right)\big|_{\delta=0} = \frac{\int_0^{h_v} e^{-\frac{z^2}{2\chi^2}+jk_z z}\,dz}{\int_0^{h_v} e^{-\frac{z^2}{2\chi^2}}\,dz} = e^{-\frac{\chi^2 k_z^2}{2}}\frac{erf\left(\frac{1}{\sqrt{2}}j\chi k_z\right) - erf\left(\frac{1}{\sqrt{2}}\left(j\chi k_z - \frac{h_v}{\chi}\right)\right)}{erf\left(\frac{h_v}{\sqrt{2}\chi}\right)} \tag{13}$$

$$\gamma_v\left(h_v, \delta, \chi\right)\big|_{\delta=h_v} = \frac{\int_0^{h_v} e^{-\frac{(z-h_v)^2}{2\chi^2}+jk_z z}\,dz}{\int_0^{h_v} e^{-\frac{(z-h_v)^2}{2\chi^2}}\,dz} = e^{-\frac{\chi^2 k_z^2}{2}+jh_v k_z}\frac{erf\left(\frac{1}{\sqrt{2}}\left(j\chi k_z + \frac{h_v}{\chi}\right)\right) - erf\left(\frac{1}{\sqrt{2}}j\chi k_z\right)}{erf\left(\frac{h_v}{\sqrt{2}\chi}\right)} \tag{14}$$

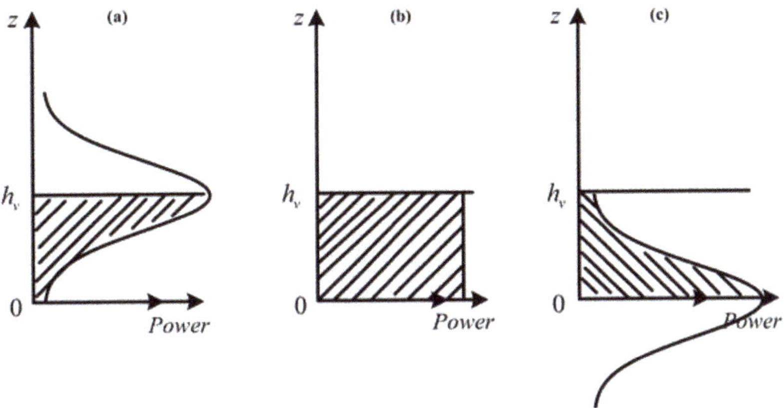

Figure 4. The profile of two-dimensional Gaussian vertical backscatter volume over ground. (**a**): $\delta = h_v$; (**b**): constant volume profile; (**c**): $\delta = 0$.

In line with Equations (13) and (14), the solution space of the two-dimensional GVB model $\gamma_v\left(h_v, \delta = 0/h_v, \chi\right)$ on the complex plane can be drawn, as shown in Figure 5. Although δ may be 0 or h_v, the two cases correspond to two completely non-overlapping regions (blue and green solution spaces in Figure 5, respectively) on the complex plane. According to the setting of volume simulation parameters, with the same constant profile, the forest height corresponding to P_1 is smaller than that corresponding to P_2. Simultaneously, it can be seen that the solution volume only coherence and P_2 are on the same forest height curve. Hence, when the solution volume only coherence lies in the blue solution space, the solution space containing only $\delta = h_v$ will cause underestimation, which can be ameliorated by $\delta = 0$ solution space. Since the null extinction (the extinction coefficient is zero) in RVoG model has the same solution space curve as the constant profile of the two-dimensional GVB model, and considering that the vertical structure described in Figures 1 and 4a are similar, the solution space of RVoG model can be approximately equivalent to the case of $\delta = h_v$ in two-dimensional GVB model. In a word, the added $\delta = 0$ solution space is the reason the two-dimensional GVB model is superior to the RVoG model [25].

In accordance with Figure 5, the solution space relying on the two-dimensional GVB model is a two-dimensional complex space. Therefore, each element in the solution space can be uniquely identified

$$\gamma_v\left(h_v, \delta = 0/h_v, \chi\right) = \gamma_v{}^{re} + j\gamma_v{}^{im} \tag{15}$$

where $\gamma_v{}^{re}$ and $\gamma_v{}^{im}$ are the real part and the imaginary part of the homologous volume only coherence, respectively.

In line with Figures 4 and 5 and Equations (13) and (14), the solution spaces of $\delta = h_v$ and $\delta = 0$ are not overlapped, so $(\delta = 0/h_v, \chi)$ can be marked with a factor $\eta\ (\delta = 0/h_v, \chi)$. The physical meaning of (h_v, η) represents the vertical structure profile with a unique volume only coherence. In this case, Equation (15) can be rewritten as

$$\gamma_v\left(h_v, \eta\right) = \gamma_v{}^{re} + j\gamma_v{}^{im} \tag{16}$$

where η can be any marking mode in engineering implementation.

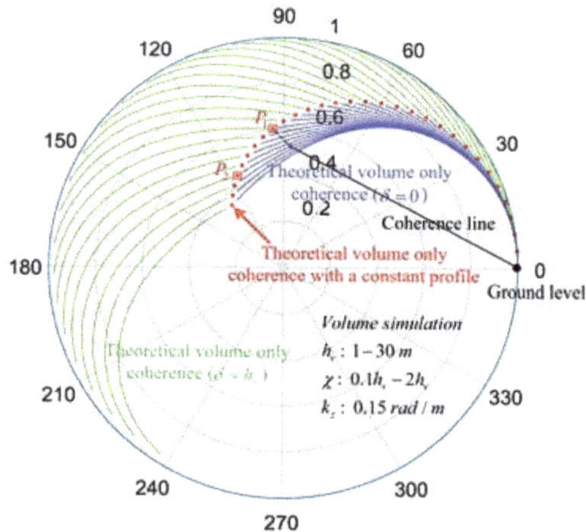

Figure 5. The solution space of two-dimensional Gaussian vertical backscatter model. Each blue or green curve, or red dot represents the volume only coherence for a given forest height at an interval of 1 m. The terrain slope is set to 0. The blue dot refers to the solution volume only coherence, and the minimum distance between it and the $\delta = h_v$ solution space is indicated by the black dotted line. P_1 is the final solution in $\delta = h_v$ solution space, while P_2 corresponds to the solution forest height in $\delta = 0$ solution space. Both P_1 and P_1 lie on the solution space curve of the constant profile.

3.2. Pol-InSAR Inversion Based on Two-Dimensional GVB Model

The inversion conditions of the two-dimensional GVB volume over ground model and the two-parameter (h_v and σ) RVoG model are the same, i.e., single-baseline Quad-pol configuration constructs six independent equations and six unknowns simultaneously. Therefore, in order to ensure a unique solution and avoid the null ground-to-volume ratio assumption, the multi-baseline configuration is used to carry out the two-dimensional GVB model inversion

$$\begin{bmatrix} \hat{h}_v, & \hat{\eta}, & \hat{\phi}_0^j, & \hat{\mu}_i \end{bmatrix}^T = \underset{\left[h_v, \eta, \phi_0^j, \mu_i\right]}{arg\,min} \left\{ \left\| \begin{bmatrix} \hat{\gamma}^1(\omega_1) \\ \hat{\gamma}^1(\omega_2) \\ \hat{\gamma}^1(\omega_3) \\ \dots \\ \hat{\gamma}^n(\omega_1) \\ \hat{\gamma}^n(\omega_2) \\ \hat{\gamma}^n(\omega_3) \end{bmatrix} - \begin{bmatrix} \gamma^1\left(h_v, \eta, \phi_0^1, \mu_1\right) \\ \gamma^1\left(h_v, \eta, \phi_0^1, \mu_2\right) \\ \gamma^1\left(h_v, \eta, \phi_0^1, \mu_3\right) \\ \dots \\ \gamma^n\left(h_v, \eta, \phi_0^n, \mu_1\right) \\ \gamma^n\left(h_v, \eta, \phi_0^n, \mu_2\right) \\ \gamma^n\left(h_v, \eta, \phi_0^n, \mu_3\right) \end{bmatrix} \right\| \right\} \tag{17}$$

where $\begin{bmatrix} \hat{h}_v, & \hat{\eta}, & \hat{\phi}_0^j, & \hat{\mu}_i \end{bmatrix}^T$ is the parameter estimation. Please note that since δ and χ are assumed to be independent of baseline and polarization, the factor η is also independent of these two system observation variables. In Equation (17), n baselines can provide $6n$ independent equations, while there are $n + 5$ unknowns to be solved.

Figure 6 presents the flowchart of the proposed two-dimensional GVB modeling and multi-baseline inversion. The flowchart begins by the conventional Pol-InSAR processing (including imaging, coregistration, interferometry, flat earth removal, multilooking, filtering and coherence estimation). It then goes on to the modeling of the two-dimensional GVB model and the corresponding multi-baseline inversion, which is also the primary contribution made by this paper. First of all, the original three-dimensional GVB model is approximate to the two-dimensional GVB model. Although δ in two-dimensional GVB model still exists two possible values (0 or h_v), these two situations constitute a two-dimensional solution space of the new model without overlapping. Therefore, two parameters (h_v and η in Equation (16)) can be used to mark every possible volume only coherence in the solution space. In model inversion stage, the solution space corresponding to each baseline is different, which is related to the interferometric vertical wavenumber k_z, while (h_v, $\delta = 0/h_v$, χ) is independent of the baseline. Therefore, in line with the one-to-one mapping relationship between (h_v, η) and (h_v, $\delta = 0/h_v$, χ), forest vertical structure estimates are ultimately available by using the genetic algorithm to optimize the inversion consisting of no less than two baselines [17].

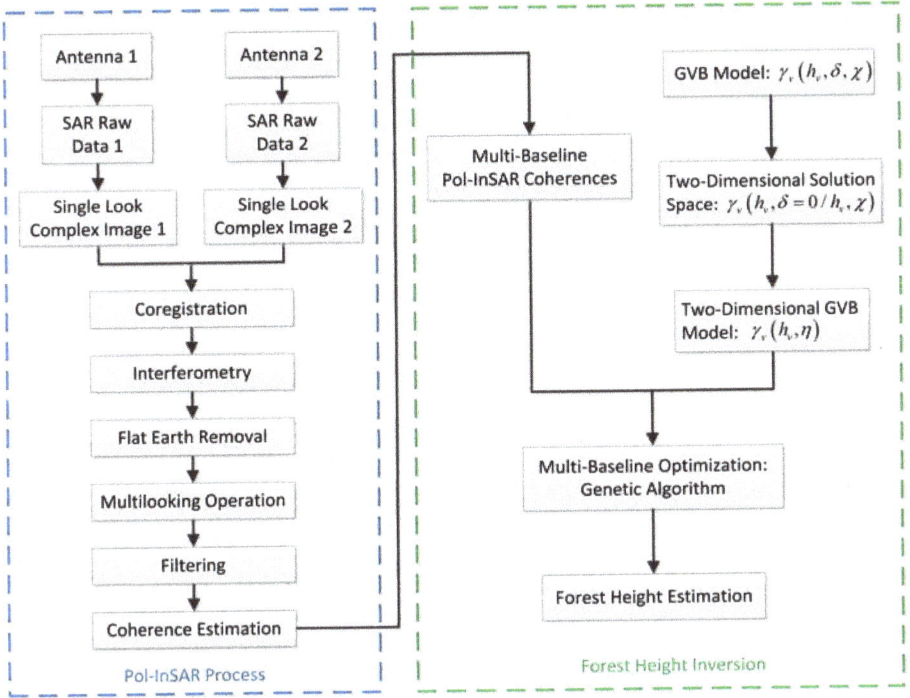

Figure 6. Flowchart of the proposed two-dimensional GVB modeling and multi-baseline inversion.

4. Validation with P-Band Pol-InSAR Data

4.1. BIOSAR 2008 Data Introduction

The data set applied in this paper is from European Space Agency (ESA) BIOSAR 2008 campaign, under which the P-band Pol-InSAR experiment was conducted by German Aerospace Center (DLR) E-SAR system. The experimental site is located in Krycklan catchment, Northern Sweden, mainly covered by boreal forest (spruce, pine, and birch), and the Pauli-basis polarization composite map is presented in Figure 7 [23]. On account of the significant topographic fluctuations in the test area, the external Digital Elevation Model (DEM) data is used to generate terrain slope to compensate the inversion model, without which there would be a certain degree of inversion error [23,26,29,30,34–37]. In addition, the LiDAR-measured forest height provided in the data set is used as the true value to evaluate the inversion performance. To carry out the multi-baseline optimization, three interferometric pairs with baseline lengths of 16 m, 24 m and 32 m are selected in this paper, and the corresponding parameter information is illustrated in Table 1. Please note that the incidence angle range in Table 1 is not corrected for terrain slope.

Figure 7. Pauli-basis polarization composite map.

Table 1. Pol-InSAR data parameters.

Track	Baseline (m)	k_z Range (rad/m)	θ Range (rad)	Band	Polarization
Master	Master	Master	0.44–0.96	P	Quad
Slave	16	0.014–0.135	Slave	P	Quad
Slave	24	0.021–0.181	Slave	P	Quad
Slave	32	0.026–0.245	Slave	P	Quad

4.2. Experimental Results and Analysis

In line with the flowchart in Figure 6, the Pol-InSAR process is first implemented to acquire the multi-polarization coherences. In the process of coherence estimation, the coherence optimization is used to suppress the influence of noise and other interferences [38]. The forest height inversion is completed by inputting the estimated Pol-InSAR coherences into Equation (17), which is an optimization process weighted by multi-baseline observations. In this paper, three P-band Quad-pol interferometric pairs with uniform spatial baseline sampling are applied, as shown in Table 1. 200 verification stands are randomly selected to assess the inversion performance. The pixels in the same stand have similar LiDAR forest heights.

Besides the two-dimensional GVB inversion (consisting of $\delta = h_v$ and $\delta = 0$), the two-dimensional GVB with only $\delta = h_v$ inversion and the RVoG inversion are carried out as well. Based on Equations (10) and (17), all these three inversions use a three-baseline configuration to increase observation spaces, thereby avoiding the assumption of null ground-to-volume ratio. In line with the previous analysis, the advantage of two-dimensional GVB model over RVoG model is that the former not only has the solution space corresponding to $\delta = h_v$ (the green curves in Figure 5), but also has the solution space corresponding to $\delta = 0$ (the blue curves in Figure 5) which is actually a complement to the RVoG model. Therefore, theoretically, the inversion performance of the two-dimensional GVB model with only $\delta = h_v$ should be very similar to that of the RVoG model, and the performance of these two inversions should be inferior to that of the two-dimensional GVB model containing both $\delta = h_v$ and $\delta = 0$.

The inversions corresponding to the two-dimensional GVB model (Figure 8a), the two-dimensional GVB model with only $\delta = h_v$ (Figure 8b) and the RVoG model (Figure 8c) are presented in Figure 8. The root mean square error (RMSE) of the three inversions is 3.19 m, 3.90 m, and 3.92 m in turn. The homologous correlation coefficients R^2 are 0.61, 0.62 and 0.62. Consistent with the theoretical analysis, the performance of the two-dimensional GVB with only $\delta = h_v$ inversion and the RVoG inversion are very close. Simultaneously, it can be seen that these two inversions still present a certain degree of underestimation, which cannot be compensated by a multi-baseline configuration. After adding the case $\delta = 0$ to the two-dimensional GVB solution space, the underestimation can be alleviated. Compared with the two-dimensional GVB with only $\delta = h_v$ inversion and the RVoG inversion, the two-dimensional GVB inversion improves the performance by 18.21% and 18.62%

respectively, which indicates the superiority of the proposed two-dimensional GVB model containing both $\delta = h_v$ and $\delta = 0$. Since the three correlation coefficients R^2 are very close, this indicator cannot reflect the performance difference of different inversions. In addition, the statistical histogram of the differences between the above three inversions and the LiDAR measurement is shown in Figure 8d, which clearly reflects the underestimation of the two-dimensional GVB with only $\delta = h_v$ inversion and the RVoG inversion, and the improvement of the two-dimensional GVB inversion for this underestimation.

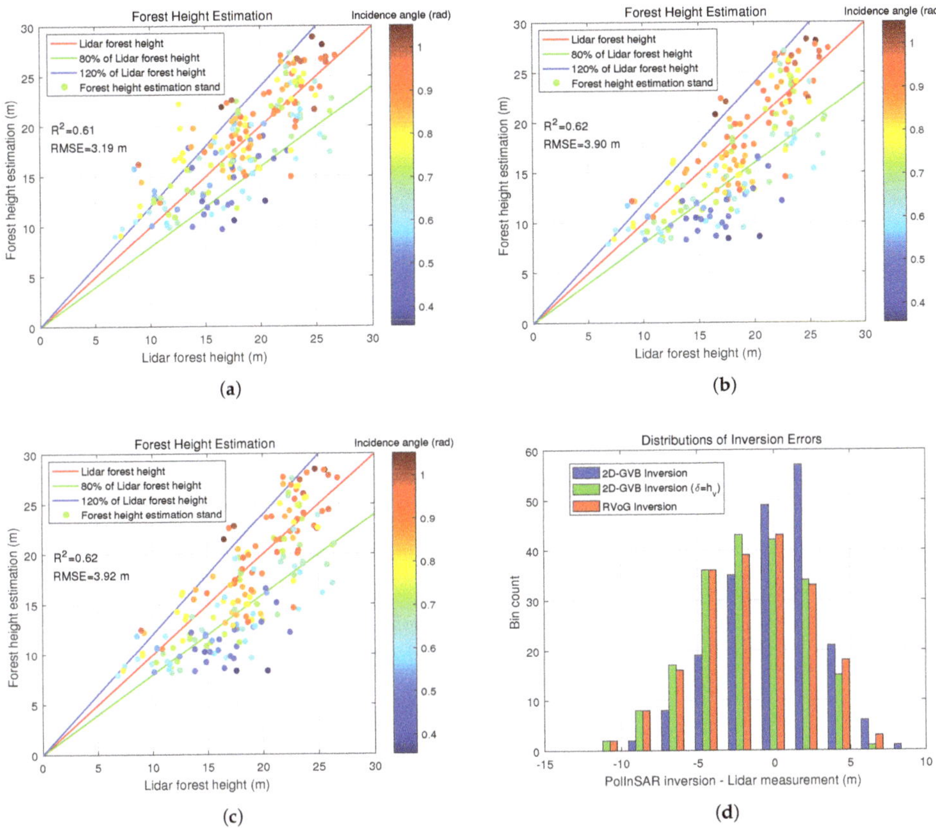

Figure 8. The forest height estimation of validation stands with (**a**) two-dimensional GVB model, (**b**) two-dimensional GVB model with only $\delta = h_v$, and (**c**) RVoG model, compared with the LiDAR measurement. The histogram (**d**) shows the differences between the Pol-InSAR inversion and the LiDAR measurement with the two-dimensional GVB model, two-dimensional GVB model with only $\delta = h_v$, and RVoG model, respectively. The stands that are counted in (**d**) correspond to the stands in (**a**–**c**).

In line with the previous work [17,23,25], the multi-baseline RVoG inversion has a relatively satisfactory performance in the region with a large incidence angle, while the underestimation mainly exists in the region with a small incidence angle. As a result, the following analysis will focus on the inversion performance in the small incidence angle region. The stands with an incidence angle less than 0.6 rad (the lower the threshold is set, the more significant the performance difference) in Figure 8 are selected for quantitative evaluation, as shown in Figure 9. The RMSEs of the three inversions are 3.48 m, 5.31 m and 5.33 m, and the homologous correlation coefficients R^2 are 0.41, 0.41 and 0.41. Compared with Figure 8a, the two-dimensional GVB inversion in Figure 9a maintains almost the same RMSE, which indicates that the inversion has a stable performance over the entire incidence angle range. Nevertheless, compared with Figure 8b,c, the two-dimensional GVB with only $\delta = h_v$ inversion (Figure 9b) and the RVoG inversion (Figure 9c) deteriorated in the region with a small incidence angle. Specifically, the performance of these two inversions is reduced by 36.15% and 35.97%, respectively. This is because the two-dimensional GVB model with only $\delta = h_v$ and the RVoG model are more effective in the region with a large incidence angle , while the two-dimensional GVB model with only $\delta = 0$ mainly works in the region with a small incidence angle. Therefore, the performance improvement of the proposed two-dimensional GVB inversion is more obvious in the area of small incidence angle, whose accuracy is improved by 34.46% and 34.71% compared to Figure 9b,c, respectively. Similarly, from the histogram in Figure 9d, it is apparent that the two-dimensional GVB inversion has a significant performance improvement over the two-dimensional GVB with only $\delta = h_v$ inversion and the RVoG inversion.

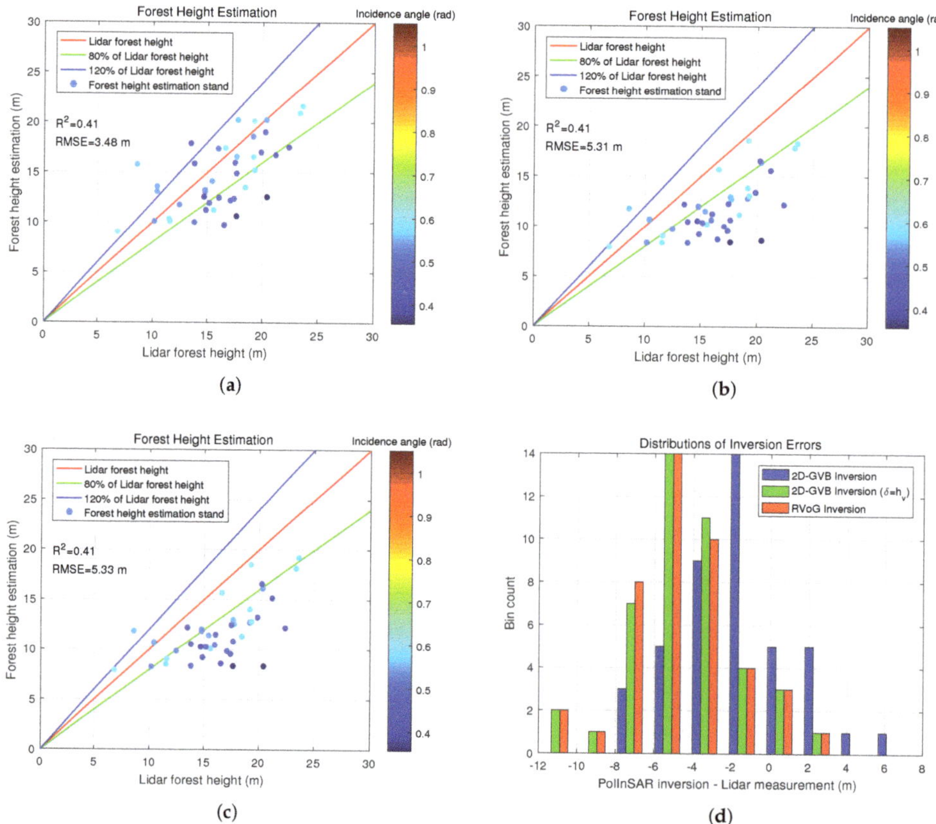

Figure 9. The forest height estimation of validation stands with (**a**) two-dimensional GVB model, (**b**) two-dimensional GVB model with only $\delta = h_v$, and (**c**) RVoG model, compared with the LiDAR measurement. These validation stands correspond to those in Figure 8 with an incidence angle less than 0.6 rad. The histogram (**d**) shows the differences between the Pol-InSAR inversion and the LiDAR measurement with the two-dimensional GVB model, two-dimensional GVB model with only $\delta = h_v$, and RVoG model, respectively. The stands that are counted in (**d**) correspond to the stands in (**a**–**c**).

5. Discussions and Conclusions

Since the research in this paper is based on P-band Pol-InSAR observations, the characteristics of P-band SAR systems were consistently the focus of the entire research. This paper analyzes the relationships between model parameters and observation spaces (observation baselines). To ensure that the RVoG optimization has a unique solution, the single-baseline inversion must be combined with certain rules (such as the null ground-to-volume ratio assumption or guessing a model parameter based on prior knowledge in advance) [17]. The multi-baseline inversion becomes an effective choice when the null ground-to-volume ratio assumption is no longer valid (such as the sparse boreal forest under P-band observations) and a parameter in the model cannot be accurately predetermined. Furthermore, under P-band observations, the forest canopy exhibits vertical structure heterogeneity, which makes it necessary to adjust the traditional RVoG model that uses the exponential function to describe the forest vertical structure. This paper approximates $\delta \geq h_v$, $\delta \leq 0$ and $\delta \in (0, h_v)$ in GVB model as $\delta = h_v$, $\delta = 0$ and a constant volume profile, thereby simplifying the three-dimensional GVB model into a two-dimensional GVB model. The two-dimensional GVB model achieves dimensionality reduction while retaining the advantage of the GVB model over the RVoG model (that is, increasing the solution space corresponding to $\delta = 0$ in the GVB model). Due to the same mathematical conditions as the RVoG model, the two-dimensional GVB model optimization also requires a multi-baseline configuration.

The experimental results illustrate that there is clear dependence of the RVoG inversion accuracy from the incidence angle. A possible reasonable explanation is that the penetration depth of the electromagnetic wave into the canopy is negatively related to the incidence angle. Specifically, the RVoG model presents an underestimation in the region with a small incidence angle, while this underestimation can be improved by the proposed two-dimensional GVB model. The reason for the underestimation improvement is that the vertical backscatter profile is closer to Figure 4c, which cannot be expressed by the RVoG model. The corresponding solution space is represented by the blue curves in Figure 5. When the solution volume only coherence is located in this supplementary solution space (e.g., the blue dot in Figure 5), the inversion performance based on the two-dimensional GVB model will be better than that based on the RVoG model.

Compared with the previous work [25], the proposed method achieves similar accuracy while simplifies the inversion process. Furthermore, the regularity among the parameters used in reference [25] depends on the specific experimental data, which may not be universal. In contrast, the method proposed in this paper has more advantages in general adaptability.

Some limitations are also noteworthy. Compared to the single-baseline, the multi-baseline configuration increases the complexity of the optimization process. In practice, since it may fall into an optimization without physical significance, the multi-baseline inversion depends to some extent on optimization initial values and constraints [24,25]. Therefore, a possible solution is to substitute with the single-baseline configuration to optimize the two-dimensional GVB model to avoid multi-baseline instability. According to the relationships between independent observation equations and unknown parameters based on the single-baseline configuration, the realization of the single-baseline inversion depends on certain prior knowledge, which may be feasible due to a large number of data sources obtained by multi-sensor at present. Furthermore, the reason for simplifying the three-dimensional GVB model is that it is too complicated and causes great difficulties in the inversion process, and yet this simplification may introduce some errors. Although the appearance of the errors may rely on specific experimental data, it is meaningful to carry out an analysis on the errors introduced by the simplification operation, which may become a key aspect of future research.

Author Contributions: Conceptualization, X.S.; formal analysis, X.S. and S.J.; funding acquisition, B.W., M.X. and L.Z.; investigation, X.S.; methodology, X.S.; project administration, B.W., M.X. and L.Z.; resources, B.W., M.X. and L.Z.; software, X.S.; supervision, B.W. and M.X.; writing—original draft, X.S.; writing—review and editing, X.S., B.W., M.X. and S.J. All authors have read and agreed to the published version of the manuscript.

Funding: This work was supported in part by the Equipment Development Department Pre-Research Fund under Grant 61404130308.

Acknowledgments: The E-SAR data (© ESA 2017) was provided by the European Space Agency (ESA) under the BIOSAR 2008 campaign (ESA EO Project Campaign ID 37708). The ancillary Digital Elevation Model (DEM) data (© Lantmäteriet) was provided by the Lantmäteriet.

Conflicts of Interest: The authors declare no conflict of interest.

References

1. Treuhaft, R.N.; Madsen, S.N.; Moghaddam, M.; Van Zyl, J.J. Vegetation characteristics and underlying topography from interferometric radar. *Radio Sci.* **1996**, *31*, 1449–1485. [CrossRef]
2. Cloude, S.R.; Papathanassiou, K.P. Polarimetric SAR interferometry. *IEEE Trans. Geosci. Remote Sens.* **1998**, *36*, 1551–1565. [CrossRef]
3. Treuhaft, R.N.; Cloude, S.R. The structure of oriented vegetation from polarimetric interferometry. *IEEE Trans. Geosci. Remote Sens.* **1999**, *37*, 2620–2624. [CrossRef]
4. Treuhaft, R.N.; Siqueira, P.R. Vertical structure of vegetated land surfaces from interferometric and polarimetric radar. *Radio Sci.* **2000**, *35*, 141–177. [CrossRef]
5. Papathanassiou, K.P.; Cloude, S.R. Single-baseline polarimetric SAR interferometry. *IEEE Trans. Geosci. Remote Sens.* **2001**, *39*, 2352–2363. [CrossRef]
6. Cloude, S.R.; Papathanassiou, K.P. Three-stage inversion process for polarimetric SAR interferometry. *Proc. Inst. Elect. Eng. Radar Sonar Navig.* **2003**, *150*, 125–134. [CrossRef]
7. Cloude, S.R. *Polarisation: Applications in Remote Sensing*; Oxford University Press: Oxford, UK, 2009.
8. Cloude, S.R. Robust parameter estimation using dual baseline polarimetric SAR interferometry. In Proceedings of the IEEE International Geoscience and Remote Sensing Symposium, Toronto, ON, Canada, 24–28 June 2002; pp. 838–840.
9. Mette, T.; Kugler, F.; Papathanassiou, K.P.; Hajnsek, I. Forest and the random volume over ground—Nature and effect of 3 possible error types. In Proceedings of the European Conference on Synthetic Aperture Radar, Dresden, Germany, 16–18 May 2006.
10. Garestier, F.; Dubois-Fernandez, P.C.; Papathanassiou, K.P. Pine forest height inversion using single-pass X-band PolInSAR data. *IEEE Trans. Geosci. Remote Sens.* **2008**, *46*, 59–68. [CrossRef]
11. Neumann, M.; Saatchi, S.S.; Ulander, L.M.H.; Fransson, J.E.S. Assessing performance of L-and P-band polarimetric interferometric SAR data in estimating boreal forest above-ground biomass. *IEEE Trans. Geosci. Remote Sens.* **2012**, *50*, 714–726. [CrossRef]
12. López-Martínez, C.; Alonso-González, A. Assessment and estimation of the RVoG model in polarimetric SAR interferometry. *IEEE Trans. Geosci. Remote Sens.* **2014**, *52*, 3091–3106. [CrossRef]
13. Lei, Y.; Siqueira, P. Estimation of forest height using spaceborne repeat-pass L-band InSAR correlation magnitude over the US state of Maine. *Remote Sens.* **2014**, *6*, 10252-10285. [CrossRef]
14. Ballester-Berman, J.D.; Vicente-Guijalba, F.; Lopez-Sanchez, J.M. A simple RVoG test for PolInSAR data. *IEEE J. Sel. Top. Appl. Earth Obs. Remote Sens.* **2015**, *8*, 1028–1040. [CrossRef]
15. Kugler, F.; Koudogbo, F.N.; Gutjahr, K.; Papathanassiou, K.P. Frequency effects in Pol-InSAR forest height estimation. In Proceedings of the European Conference on Synthetic Aperture Radar, Dresden, Germany, 16–18 May 2006.
16. Garestier, F.; Dubois-Fernandez, P.C.; Champion, I. Forest height inversion using high-resolution P-band Pol-InSAR data. *IEEE Trans. Geosci. Remote Sens.* **2008**, *46*, 3544–3559. [CrossRef]
17. Kugler, F.; Lee, S.-K.; Papathanassiou, K.P. Estimation of forest vertical structure parameter by means of multi-baseline Pol-InSAR. In Proceedings of the IEEE International Geoscience and Remote Sensing Symposium, Cape Town, South Africa, 12–17 July 2009; pp. IV-721–IV-724.
18. Garestier, F.; Dubois-Fernandez, P.C.; Guyon, D.; Le Toan, T. Forest biophysical parameter estimation using L- and P-band polarimetric SAR data. *IEEE Trans. Geosci. Remote Sens.* **2009**, *47*, 3379–3388. [CrossRef]
19. Lee, S.-K.; Kugler, F.; Hajnsek, I.; Papathanassiou, K.P. The potential and challenges of polarimetric SAR interferometry techniques for forest parameter estimation at P-band. In Proceedings of the European Conference on Synthetic Aperture Radar, Aachen, Germany, 7–10 June 2010; pp. 503–505.
20. Dubois-Fernandez, P.C.; Souyris, J.-C.; Angelliaume, S.; Garestier, F. The compact polarimetry alternative for spaceborne SAR at low frequency. *IEEE Trans. Geosci. Remote Sens.* **2008**, *46*, 3208–3222. [CrossRef]

21. Roueff, A.; Arnaubec, A.; Dubois-Fernandez, P.C.; Refregier, P. Cramer-rao lower bound analysis of vegetation height estimation with random volume over ground model and polarimetric SAR interferometry. *IEEE Geosci. Remote Sens. Lett.* **2011**, *8*, 1115–1119. [CrossRef]

22. Arnaubec, A.; Roueff, A.; Dubois-Fernandez, P.C.; Réfrégier, P. Vegetation height estimation precision with compact PolInSAR and homogeneous random volume over ground model. *IEEE Trans. Geosci. Remote Sens.* **2014**, *52*, 1879–1891. [CrossRef]

23. Hajnsek, I.; Scheiber, R.; Keller, M.; Horn, R.; Lee, S.-K.; Ulander, L.; Gustavsson, A.; Sandberg, G.; Le Toan, T.; Tebaldini, S.; et al. BIOSAR 2008: Final Report. 2009. Available online: https://earth.esa.int/c/document_library/get_file?folderId=21020&name=DLFE-903.pdf (accessed on 21 November 2018).

24. Fu, H.; Wang, C.; Zhu, J.; Xie, Q.; Zhang, B. Estimation of pine forest height and underlying DEM using multi-baseline P-band PolInSAR data. *Remote Sens.* **2016**, *8*, 820. [CrossRef]

25. Sun, X.; Wang, B.; Xiang, M.; Jiang, S.; Fu, X. Forest height estimation based on constrained Gaussian vertical backscatter model using multi-baseline P-band Pol-InSAR data. *Remote Sens.* **2019**, *11*, 42. [CrossRef]

26. Xie, Q.; Zhu, J.; Wang, C.; Fu, H.; Lopez-Sanchez, J.M.; Ballester-Berman, J.D. A modified dual-baseline PolInSAR method for forest height estimation. *Remote Sens.* **2017**, *9*, 819. [CrossRef]

27. Garestier, F.; Le Toan, T. Forest modeling for height inversion using single-baseline InSAR/Pol-InSAR data. *IEEE Trans. Geosci. Remote Sens.* **2010**, *48*, 1528–1539. [CrossRef]

28. Garestier, F.; Le Toan, T. Estimation of the backscatter vertical profile of a pine forest using single baseline P-band (Pol-)InSAR data. *IEEE Trans. Geosci. Remote Sens.* **2010**, *48*, 3340–3348. [CrossRef]

29. Hajnsek, I.; Kugler, F.; Lee, S.-K.; Papathanassiou, K.P. Tropical-forest-parameter estimation by means of Pol-InSAR: The INDREX-II campaign. *IEEE Trans. Geosci. Remote Sens.* **2009**, *47*, 481–493. [CrossRef]

30. Kugler, F.; Lee, S.-K.; Hajnsek, I.; Papathanassiou, K.P. Forest height estimation by means of Pol-InSAR data inversion: The role of the vertical wavenumber. *IEEE Trans. Geosci. Remote Sens.* **2015**, *53*, 5294–5311. [CrossRef]

31. Bamler, R.; Hartl, P. Synthetic aperture radar interferometry. *Inverse Probl.* **1998**, *14*, R1–R54. [CrossRef]

32. Lavalle, M.; Simard, M.; Hensley, S. A temporal decorrelation model for polarimetric radar interferometers. *IEEE Trans. Geosci. Remote Sens.* **2012**, *50*, 2880–2888. [CrossRef]

33. Lavalle, M.; Khun, K. Three-baseline InSAR estimation of forest height. *IEEE Geosci. Remote Sens. Lett.* **2014**, *11*, 1737–1741. [CrossRef]

34. Park, S.-E.; Moon, W.M.; Pottier, E. Assessment of scattering mechanism of polarimetric SAR signal from mountainous forest areas. *IEEE Trans. Geosci. Remote Sens.* **2012**, *50*, 4711–4719. [CrossRef]

35. Lu, H.; Suo, Z.; Guo, R.; Bao, Z. S-RVoG model for forest parameters inversion over underlying topography. *Electron. Lett.* **2013**, *49*, 618–620. [CrossRef]

36. Zhang, Q.; Liu, T.; Ding, Z.; Zeng, T.; Long, T. A modified three-stage inversion algorithm based on R-RVoG model for Pol-InSAR data. *Remote Sens.* **2016**, *8*, 861. [CrossRef]

37. Sun, X.; Wang, B.; Xiang, M.; Fu, X.; Zhou, L.; Li, Y. S-RVoG model inversion based on time-frequency optimization for P-band polarimetric SAR interferometry. *Remote Sens.* **2019**, *11*, 1033. [CrossRef]

38. Tabb, M.; Orrey, J.; Flymn, T.; Carande, R. Phase diversity: A decomposition for vegetation parameter estimation using polarimetric SAR interferometry. In Proceedings of the European Conference on Synthetic Aperture Radar, Cologne, Germany, 4–6 June 2002; pp.721–724.

Article

Using Field Spectroradiometer to Estimate the Leaf N/P Ratio of Mixed Forest in a Karst Area of Southern China: A Combined Model to Overcome Overfitting

Wen He [1,2], Yanqiong Li [3], Jinye Wang [1,*], Yuefeng Yao [2], Ling Yu [4], Daxing Gu [2] and Longkang Ni [2]

1   College of Environmental Science and Engineering, Guilin University of Technology, Guilin 541006, China; hw@gxib.cn
2   Guangxi Key Laboratory of Plant Conservation and Restoration Ecology in Karst Terrain, Guangxi Institute of Botany, Guangxi Zhuang Autonomous Region and Chinese Academy of Sciences, Guilin 541006, China; yf.yao@gxib.cn (Y.Y.); GIB@gxib.cn (D.G.); nlk@gxib.cn (L.N.)
3   Key Laboratory of Vegetation Restoration and Management of Degraded Ecosystems, South China Botanical Garden, Chinese Academy of Sciences, Guangzhou 510650, China; liyq2018@scbg.ac.cn
4   School of Geographical Sciences, Southwest University, Chongqing 400715, China; yl0435@email.swu.edu.cn
*   Correspondence: 2005010@glut.edu.cn

Abstract: The ratio between nitrogen and phosphorus (N/P) in plant leaves has been widely used to assess the availability of nutrients. However, it is challenging to rapidly and accurately estimate the leaf N/P ratio, especially for mixed forest. In this study, we collected 301 samples from nine typical karst areas in Guangxi Province during the growing season of 2018 to 2020. We then utilized five models (partial least squares regression (PLSR), backpropagation neural network (BPNN), general regression neural network (GRNN), PLSR+BPNN, and PLSR+GRNN) to estimate the leaf N/P ratio of plants based on these samples. We also applied the fractional differentiation to extract additional information from the original spectra of each sample. The results showed that the average leaf N/P ratio of plants was 17.97. Plant growth was primarily limited by phosphorus in these karst areas. The sensitive spectra to estimate leaf N/P ratio had wavelengths ranging from 400–730 nm. The prediction capabilities of these five models can be ranked in descending order as PLSR+GRNN, PLSR+BPNN, PLSR, GRNN, and BPNN when considering both accuracy and robustness. The PLSR+GRNN model yielded high R^2 and performance to deviation (RPD), and low root mean squared error (RMSE) with values of 0.91, 3.15, and 1.98, respectively, for the training test and 0.81, 2.25, and 2.46, respectively, for validation test. Compared with the PLSR model, both PLSR+BPNN and PLSR+GRNN models had higher accuracy and were more stable. Moreover, both PLSR+BPNN and PLSR+GRNN models overcame the issue of overfitting, which occurs when a single model is used to predict leaf N/P ratio. Therefore, both PLSR+BPNN and PLSR+GRNN models can be used to predict the leaf N/P ratio of plants in karst areas. Fractional differentiation is a promising spectral preprocessing technique that can improve the accuracy of models. We conclude that the leaf N/P ratio of mixed forest can be effectively estimated using combined models based on field spectroradiometer data in karst areas.

Keywords: mixed forest in karst areas; leaf N/P ratio; fractional differentiation; combined model; overcome overfitting; field spectroradiometer

Citation: He, W.; Li, Y.; Wang, J.; Yao, Y.; Yu, L.; Gu, D.; Ni, L. Using Field Spectroradiometer to Estimate the Leaf N/P Ratio of Mixed Forest in a Karst Area of Southern China: A Combined Model to Overcome Overfitting. *Remote Sens.* **2021**, *13*, 3368. https://doi.org/10.3390/rs13173368

Academic Editor: James Cleverly

Received: 5 July 2021
Accepted: 22 August 2021
Published: 25 August 2021

Publisher's Note: MDPI stays neutral with regard to jurisdictional claims in published maps and institutional affiliations.

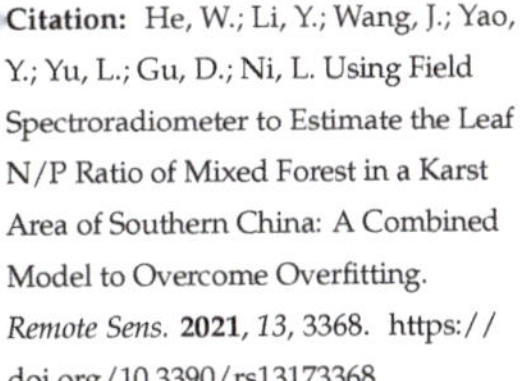

1. Introduction

Nitrogen (N) and phosphorus (P) are crucial functional elements for organisms [1] and play a vital role in plant physiological processes [2–4]. Changes of the N and P concentration are essential for the growth of plants, as they are closely related to photosynthesis, respiration, N_2 fixation, and organic matter mineralization [5]. N and P are the primary limiting nutrients for plant growth in most natural systems. An N/P ratio greater than 16 indicates that plant growth is limited by P, while an N/P ratio of less than 14 is limited by N. Values between 14 and 16 suggest either N or P can be limited, or plant growth is co-limited by both N and P [6,7], although there is no general consensus about an N/P ratio which is not limiting in either or both nutrients. Therefore, exploring the N and P stoichiometry of leaves plays a critical role in better understanding the survival and adaptation strategies of plants. However, current methods for monitoring N and P content mainly rely on laboratory analysis. The precision of these traditional analytical methods is high, but they are time-consuming, complex, and require difficult storage and transportation logistics of field samples, which limits rapid on-site and non-destructive detection [8]. As a real-time, fast, and non-destructive technology, spectral technology plays an increasingly important role in detecting nutrient acquisition and growth of plants.

The mechanism and methods of estimating leaf biochemical parameters with remote sensing techniques have been advanced in recent years, especially ground-based hyperspectral remote sensing. For example, Sonobe and Wang [9] used hyperspectral techniques to estimate the chlorophyll content in leaves of a deciduous forest, and found that the normalized difference spectral index using the first-order differentiation of reflectance at 522 and 728 nm was the best index. Zhao et al. [10] used a hyperspectral stepwise regression analysis to estimate the water content of apple tree canopies and found that an equidistant sampling method improved the prediction accuracy. Yamashita et al. found that machine learning methods can predict chlorophyll and N content based on field spectroradiometer [11]. However, most previous research has focused on estimating chlorophyll, water, organic carbon (C), and N. Few studies have investigated N/P and other stoichiometric ratios that are significant indicators of plant growth. The patterns found for N and P cannot be directly applied to N/P. For example, Cui et al. [8] proposed that N and P display a strong positive correlation with the spectral reflectance at 650 nm, while the N/P ratio displays a strong negative correlation with the spectral reflectance at 650 nm. Therefore, further studies of stoichiometric ratio estimation using field spectroradiometer are needed.

Previous studies on the spectral inversion of biochemical parameters focused on single species, such as rice [12], wheat [13], tobacco [14], apple [10], and tea [15]. As many species co-exist within natural habitats, results derived for single species or cropland is not applicable to investigate patterns in complex mixed forests. Therefore, establishing a database with multiple plants is critical to estimate plant biochemical parameters through remote sensing methods. However, mixed species samples are usually more difficult to collect, and it is more complex to predict biochemical parameters. In contrast, single plant species are easier to predict with high accuracy. For example, the best coefficient of determination (R^2) for the estimation of leaf mass per area (LMA) in mixed-species by Cheng et al. [16] was 0.74, while Inoue et al. [17] produced an R^2 above 0.9 in rice. Further research is required to improve model accuracy to estimate biochemical parameters in mixed-species ecosystems.

In terms of the algorithm employed in previous studies, mainstream empirical methods can be summarized into three categories; spectral index methods [9], linear regression methods [13], and machine learning methods [11]. Both the spectral index and the linear regression methods focus on finding the linear relationship between spectral reflectance and plant biochemical parameters. By contrast, machine learning methods can calculate nonlinear relationships between spectral reflectance and plant biochemical parameters. Each method has limitations and advantages. Linear models use simple principles, are computationally efficient, and more robust in prediction accuracy, but are often inferior to machine learning methods in terms of prediction capability. Theoretically, there will be both linear and nonlinear relationships between spectral reflectance and biochemical parameters of plant leaves. Therefore, the use of a combined model including both linear and nonlinear methods may improve prediction capability.

Overfitting is a common problem in many empirical models, especially for machine learning methods [18], and can cause deceptive diagnostic results and reduce the transferability of the model. A small number of samples, a large number of variables, and high-dimensionality may lead to overfitting of the model [19]. Field spectroradiometer data are characterized by high-dimensionality and a large number of bands (variables), meaning that overfitting is a critical issue that has to be considered in the estimation of biochemical parameters. It has been proven that increasing data samples, cross-validation [20], regularization [19], noise removal [21], and integration of multiple models [22] are effective ways to overcome overfitting [23,24]. Therefore, combining linear and nonlinear models may be a way to reduce overfitting.

Spectral differentiation transform methods play an essential role in estimating plant parameters, and the most commonly used are spectral first-order and second-order differentiation [25]. It has been demonstrated that differential transformation of spectra can improve model performance for estimating plant water [10], chlorophyll [9], and N content [17,26]. However, the application of integer-order differentiation is not always sufficient, as the spectral curve shifts in shape from n-order to n + 1-order in a sharply fluctuating way, and there is no smooth transition between the intermediate stages. Fractional order transform methods allow differentiation from zero to arbitrary real numbers [27]. Using fractional order differentiation transforms spectra more continuously and produces more detailed information about the spectra.

The prediction of leaf N/P ratio of a particular species by field spectroradiometer has been reported [8,28], but directly estimating the leaf N/P ratio of all plants worldwide is difficult to achieve with current methods. Therefore, improving model performance and applying these methods to estimate leaf N/P ratio of regional mixed-species ecosystems is a pressing issue. Karst landscapes are one of the most crucial landform types globally, accounting for roughly 15% of the Earth's total land area and inhabited by about 1 billion people [29]. Southwestern China has the largest karst area on the planet [30] and is one of 25 global biodiversity hotspots that contain many endemic and threatened species [31]. The karst areas of Guangxi Province are an important part of the southwest karst region, containing diverse landscape types, a wide variety of plants, and representing a key area for biodiversity conservation. However, anthropogenic disturbances have led to species loss in karst areas [32]. Non-destructive and rapid estimation of leaf N/P ratio of plants is required for ecological restoration and conservation in karst areas.

In this paper, we simulated the leaf N/P ratio of mixed forest in karst areas of Guangxi Province using fractional spectral differentiation and multiple models, combined with field spectroradiometer data. The primary objectives of this study are to (1) explore the predictive capabilities of linear regression models (partial least squares regression, PLSR) and nonlinear regression models (backpropagation neural network (BPNN), generalized neural network (GRNN)) to estimate leaf N/P ratio of mixed species based on field spectroradiometer data, (2) evaluate the contribution of fractional differentiation in improving these linear and nonlinear regression models' performance, and (3) propose the best models that can overcome the insufficient accuracy of the PLSR model and overfitting of the BPNN and GRNN models to estimate leaf N/P ratio of plants in karst areas.

2. Materials and Methods

2.1. Study Area

The study area is located in Guangxi Province (20°54′–26°24′N, 104°28′–112°04′E) in south China (Figure 1). The elevation of the study area ranges from 0 to 2141.50 m. This area borders the South China Sea and has a tropical and subtropical climate. The average annual temperature ranges from 17.30 to 23.80 °C, and annual precipitation ranges from 1024.60 to 2358.60 mm. Karst landforms are distributed widely throughout Guangxi and cover about 97,000 km^2, accounting for 41% of the total area of the province. There are more than 4000 species of vascular plants, including more than 2000 species of medicinal plants, in the karst areas of Guangxi Province [33]. Nine typical karst experimental plots were selected, containing primary forests, secondary forests, and shrubs that represent the vegetation succession in karst areas. The area of each plot was about 200 m^2. Detailed information on each plot is described in Table 1.

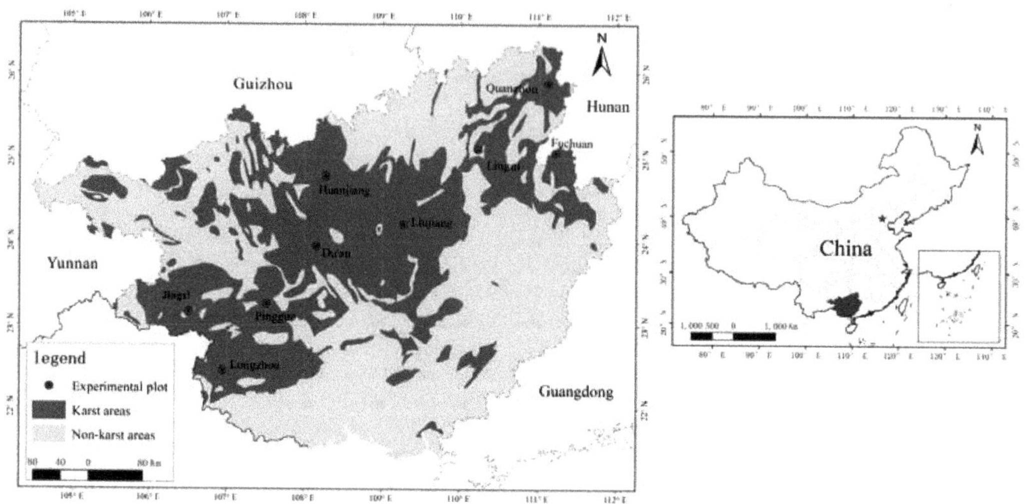

Figure 1. Location of the nine study plots.

Table 1. Brief description of the experimental plots.

No.	Name of the Experimental Plot	Successional Stages of the Plant Community	Name of Dominant Species	Average Annual Temperature	Average Annual Rainfall
1	Jingxi	Secondary forest	*Cladrastis platycarpa (Maxim.) Makino, Bruguiera gymnorhiza (L.) Lam., Buddleja officinalis, Abelia biflora Turcz.*	21.68	1621.92
2	Longzhou	Primary forest	*Canthium dicoccum, Memecylon scutellatum, Pistacia weinmanniifolia, Boniodendron minus, Excentrodendron hsienmu*	23.28	1272.72
3	Pingguo	Shrubs	*Rhus chinensis Mill., Cipadessa baccifera (Roth.) Miq., Vitex negundo L., Alchornea trewioides*	22.03	1328.63
4	Du'an	Shrubs	*Psidium guajava, Vitis heyneana, Buddleja officinalis, Serissa japonica*	22.03	1733.37
5	Huanjiang	Secondary forest	*Solanum indicum L., Ficus tinctoria Forst. F. subsp. gibbosa (Bl.) Corner, Albizia lebbeck (Linn.) Benth., Vitex negundo L.*	22.50	1392.50
6	Liujiang	Secondary forest	*Alchornea trewioides, Litsea glutinosa, Maclura cochinchinensis, Vitex negundo L.*	21.57	1433.62
7	Lingui	Scrubs	*Bauhinia championii, Zanthoxylum bungeanum, Sageretia thea, Rosa cymosa*	21.56	1891.94
8	Quanzhou	Scrubs	*Paliurus ramosissimus, Ilex corallina var. loeseneri, Bauhinia championii, Sageretia thea*	21.65	1529.96
9	Fuchuan	Secondary forest	*Albizia kalkora, Pistacia chinensis Bunge, Sapium sebiferum (L.) Roxb., Vitex negundo L.*	19.47	1685.97

2.2. Data Collection

The leaves were sampled from July 2018 to September 2020. In each experimental plot, leaf samples from 8–15 plants of locally dominant species were collected. In total, the database includes 301 samples covering 37 families, 59 genera, and 70 species. As plants are susceptible to light conditions [34], we collected leaves from three directions for each plant (0–120°, 120–240°, and 240–360°, with 0° due north) to reduce random errors in samples.

Leaf spectral reflectance was measured in attached leaves using a spectroradiometer (Fieldspec 4, Analytical Spectral Devices, ASD, Boulder, CO, USA), with a spectral resolution of 3 nm in the visible and near-infrared (NIR) (350–1000 nm) and 8 nm in shortwave-infrared (SWIR, 1000–2500 nm) [14]. Reference plate (white reference) calibration was performed every 10 min during the measurement. Three branches of each tree were selected for measurement. As the instrument battery only has a continuous operating time of about 4 h in the field, only two mature and healthy leaves per branch were taken for spectral scanning due to time limitations. Finally, the scanned spectral reflectance of all leaves on each branch were arithmetically averaged, and the average value was taken as the spectral sample of each tree.

After the spectral measurements, the healthy and mature leaves on the branches were collected. It was then kept intact in a self-sealing bag and immediately placed in an incubator (ICERSICE940). Leaf samples were transported back to the laboratory within 24 h and dried at 75 °C. The dry samples were entirely sieved through a 100 mesh sieve for physicochemical analysis. Finally, the total nitrogen (TN) content of plant leaves was measured using the Kjeldahl method [35], and the total phosphorus (TP) content of plant leaves was measured by the phosphomolybdate blue spectrophotometry method [36]. The ratio of TN to TP was determined as the leaf N/P ratio.

2.3. Methodology

2.3.1. Fractional Differentiation (FD)

Fractional differentiation is an extension of integer differentiation to arbitrary differentiation [37] and is widely applied in electromagnetic field theory, control systems, nonlinear dynamics, biomedicine, and digital signal processing [38]. The most common method of fractional differentiation is mainly in the form of Riemann–Liouville, Grünwald-Letnikov, and Caputo [25,35]. The Grünwald–Letnikov is a finite-difference expression:

$$d^v f(x) = \lim_{h \to \infty} \frac{1}{h^v} \sum_{m=0}^{\frac{t-a}{h}} (-1)^m \frac{\Gamma(v+1)}{m! \Gamma(v-m+1)} f(x - mh) \tag{1}$$

where v is the order of differentiation, h is the step size, and t and a are the upper and lower limits of differentiation, respectively. $\Gamma(\cdot)$ is the Gamma function:

$$\Gamma(\beta) = \int_0^\infty e^{-t} t^{\beta-1} dt = (\beta - 1)! \tag{2}$$

where β is an arbitrary variable (we defined it as the order of differentiation in this study). In this paper, the plant leaf spectra were differentiated in the range between 0 to 3 orders (at an interval of 0.1 order). The integer order refers to zero, first, second, and third orders, while the other values are fractional orders.

2.3.2. Partial Least Squares Regression (PLSR)

The Partial Least Squares Regression (PLSR) model (Höskuldsson 1988) combines the merits of principal components, typical correlation, and multiple linear regression analysis. This method is essentially based on the assumption that the sample size is n and the data sets for the independent and dependent variables are $Z = [z_1, z_2, \cdots z_k]_{n \times k}$, $Q = [q]_{n \times 1}$, respectively. The first component f_1 is extracted from Z. f_1 is a linear combination with $z_1, z_2, \cdots z_k$ that carries the maximum variance information in Z and reaches the maximum correlation with q. If the accuracy of the model is satisfied, the component extraction is stopped. Otherwise, the next principal component is extracted until the requirement is satisfied.

$$q = f_1 a_1 + f_2 a_2 + \cdots + f_k a_k \tag{3}$$

$$f_m = w_{m1}z_1 + w_{m2}z_2 + \cdots + w_{mk}z_k \tag{4}$$

where m is the number of principal components, k is the number of independent variables a is the regression coefficients of y on f, and w is the linear coefficient of f on z. In this study, the fractional differentiation spectral reflectance of each order that had a significant correlation ($p < 0.05$) on the leaf N/P ratio of karst plants was used as the independent variable Z. This method is the same as used for the BPNN and GRNN models described later.

2.3.3. Back Propagation Neural Network (BPNN)

Back Propagation Neural Network (BPNN) is a multilayer feed-forward neural network [39]. The key traits of this method are the forward transmission of signals and the backpropagation of errors [39]. During forward transmission, the input signal is transferred from the input layer through the hidden layer and is then output. The neuron state of each layer only affects the neuron state of the next layer. If the output layer does not return the expected results, it transfers to backpropagation. The weights and thresholds of the network are adjusted according to the prediction error, resulting in the predicted output of the BPNN continuously approximating the expected results.

In this study, a one-hidden layer with the tansig function and an output layer with the purelin function neural network was built. The number of nodes in the hidden layer significantly impacts the output result [40], so 5-fold cross-validation was used to select the optimal number of hidden layer nodes (from 4–20). We used the arithmetic mean value of 10 consecutive operations of the BPNN model as the final results to eliminate fluctuations of the neural network operation. This threshold was also applied to the GRNN, PLSR+BPNN, and PLSR+GRNN models.

2.3.4. Generalized Regression Neural Network (GRNN)

Generalized regression neural network (GRNN) is a radial basis function neural network model proposed by Specht (1991) [41]. GRNN essentially derives the maximum probability estimate from the training data, which can be considered as an arbitrary function between input and output vectors. Unlike BPNN, GRNN does not require an iterative training procedure, making it significantly faster than BPNN in terms of computational efficiency [42]. This method displays greater prediction capability for nonlinear estimation. The prediction function can be expressed as:

$$\hat{Y}(X) = \frac{\sum\limits_{i=1}^{j} Y_i \exp\left[-\frac{(X-X_i)^T(X-X_i)}{2\delta^2}\right]}{\sum\limits_{i=1}^{j} \exp\left[-\frac{(X-X_i)^T(X-X_i)}{2\delta^2}\right]} \tag{5}$$

where j is the number of training samples, δ is the smoothing factor, X is the network input variable, and X_i is the learning sample corresponding to the ith neuron. The weights factor for each observation Y_i is the squared Euclid distance between the corresponding sample X_i and the input variable X. The smoothness factor δ has a significant effect on the model. In this study, 5-fold cross-validation was used to identify the value of δ.

2.3.5. Combined Models, Sample Segmentation, and Accuracy Assessment

The principal components extracted from PLSR were used as input variables for the BPNN and GRNN models to overcome the overfitting problem of artificial neural network models. The purpose of extracting the principal components is to reduce the dimensionality of the spectral data, thus reducing the complexity of the BPNN and GRNN model. Therefore, the number of principal components should not be too large. The number of input variables can influence the structure and performance of the BPNN and GRNN models. We found that for most of the fractional differential spectra, it is able to represent more than 60% of the variability when the number of principal components is six. To better compare the performance of different fractional differential spectra, we set the number of extracted principal components as six. The combined PLSR+BPNN and PLSR+GRNN models were used to compare their simulated performance against PLSR, BPNN, and GRNN.

The field samples were randomly split into two datasets using the randperm function in MATLAB R2020a, with the training dataset accounting for 3/4 of the validation datasets and 1/4 of the total samples (Table 2).

Table 2. Descriptive statistics for data sets.

Samples	Number	Mean	Standard Deviation	Coefficient of Variation (%)
Total samples	301	17.97	6.05	33.68
Training sets	225	17.93	6.23	34.75
Validation sets	76	18.08	5.53	30.57

The accuracy of the model was assessed using the coefficient of determination (R^2), root mean squared error (RMSE), and the ratio of performance to deviation (RPD) [15]. An RPD greater than or equal to 2 indicates that the model has excellent predictability, while less than 2 and greater than or equal to 1.4 indicates that the model can make a rough estimate of the sample. An RPD of less than 1.4 indicates that the model cannot predict the sample [43].

3. Results

3.1. Leaf N/P Ratio, Fractional Differentiation of Reflectance, and Their Correlation

The mean values of leaf TN, TP, and N/P ratio are 18.51 mg/g, 1.17 mg/g, and 17.97, respectively (Figure 2). The TN content of our study is slightly lower than global and continent levels, with values 20.1 mg/g [44] and 20.20 mg/g [45], respectively. This TP content is significantly lower than the global average of 1.80 mg/g [44] or 1.99 mg/g [46]. The leaf N/P ratio from our study is similar to the results of Yang et al. [47]. The maximum and minimum value of the N/P ratio is 1.34 and 36.94, respectively, with a variation coefficient of 33.68%.

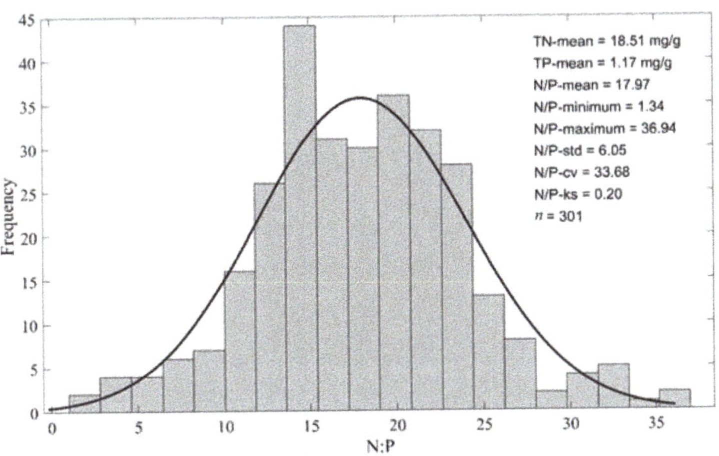

Figure 2. The leaf N/P ratio frequency distribution. TN-mean, TP-mean, and N/P-mean are the mean values of total nitrogen, total phosphorus, and N/P ratio, respectively. N/P-minimum, N/P-maximum, N/P-std, N/P-cv, and N/P-ks are the N/P ratio minimum, maximum, standard deviation, coefficient of variation, and one-sample Kolmogorov–Smirnov test, respectively. *n* is the number of samples.

Figure 3 shows the variation of the differential spectral reflectance from 0 to 3 orders. The shape of spectral curves of different orders is smoothly transitional. Compared to integer differentiation (FD (0.0), FD (1.0), FD (2.0), and FD (3.0)), fractional differentiation methods produce more detailed information about the spectrum, and these data subsequently allow for more complex leaf N/P ratio inversion training methods.

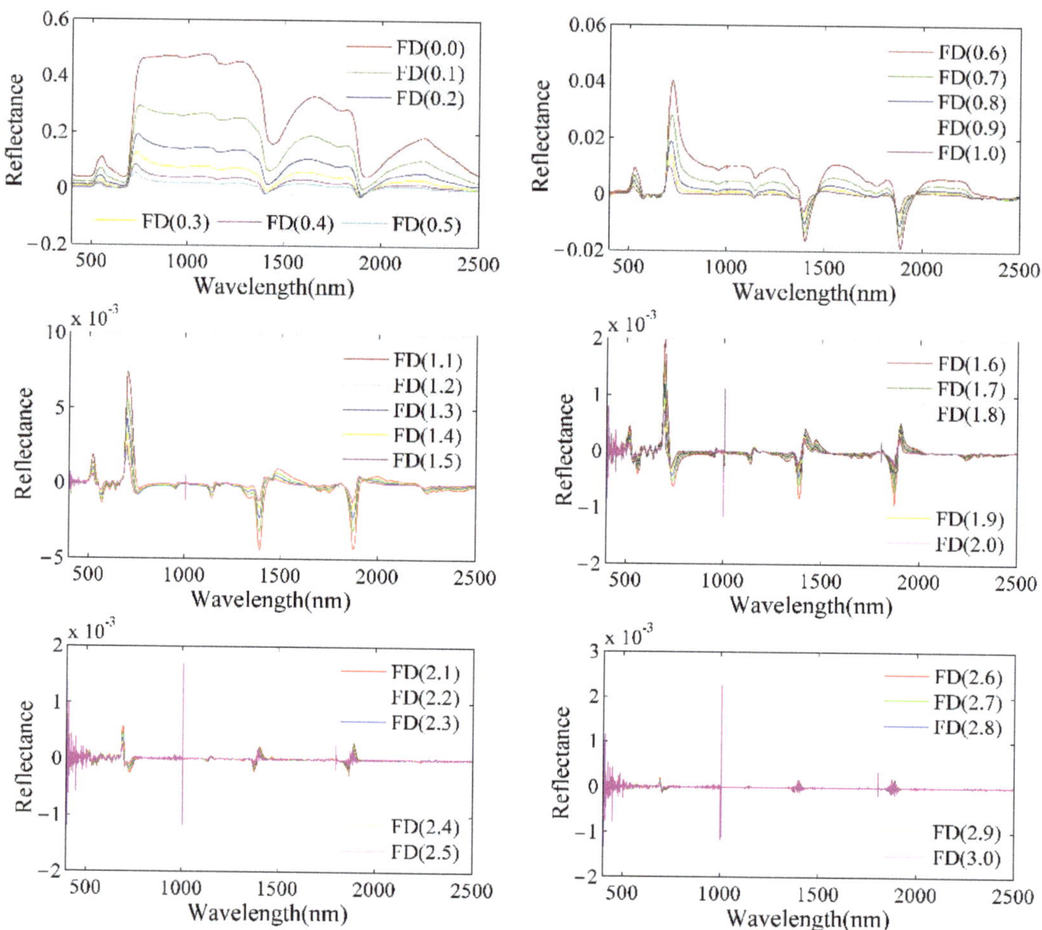

Figure 3. Fractional differential curves of plant leaf spectral reflectance (average of all the collected samples). If the curves overlap, the former will be overwritten by the latter.

The leaf N/P ratio of plants displays a significant correlation with the fractional differential spectra for wavelengths ranging from 400–730 nm (Figure 4). From FD (0) to FD (3), the maximum absolute value of the correlation coefficient displays a unimodal distribution as the fractional differentiation increases from FD (0) to FD (3) with a peak value of 0.44 for FD (1.6).

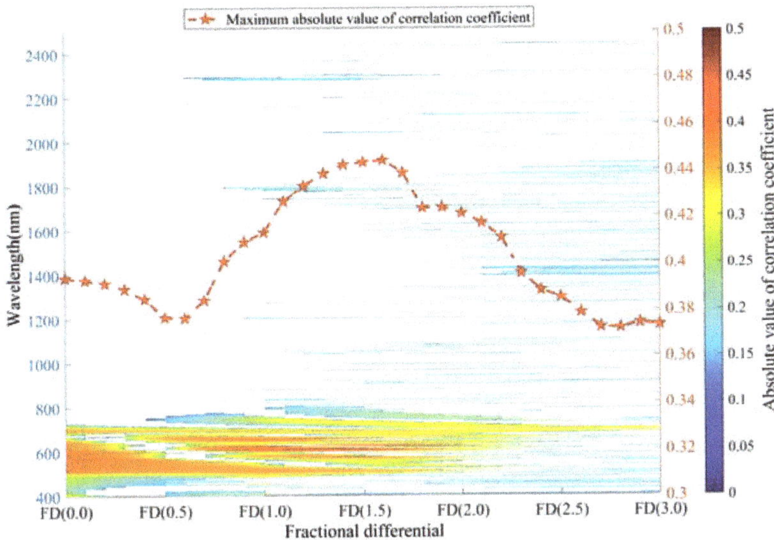

Figure 4. Distribution for the absolute values of correlation coefficients between fractional differentiation spectra and N/P ratio (vertical axis on the left side) and the maximum absolute value of each order fractional differentiation correlation coefficient (vertical axis on the right side). The color bar indicates the magnitude of the correlation coefficient. The white areas of the graph show where the spectral bands did not pass the 0.05 significance test.

3.2. Performance of a Single Model Using Fractional Differentiation of Reflectance

The accuracy of the PLSR model for predicting leaf N/P ratio of plants continuously increases with increasing fractional order for training sets but displays a unimodal distribution for validation sets (Figure 5a), yielding the highest R^2 values of 0.66 for FD (2.1). The prediction capability of the PLSR model gradually improves with the increase in fractional order from FD (0.6) to FD (2.1). However, the prediction capability of the PLSR model displayed a higher accuracy in training sets than for validation sets, especially after FD (2.1). This finding suggests that overfitting of the PLSR model is an issue for using this method to predict the leaf N/P ratio of plants.

The accuracy of the BPNN model in predicting the leaf N/P ratio of plants increases with increasing orders of fractional differentiation. The highest R^2 value for this method is 0.48 for validation sets and 0.92 for training sets around FD (1.1), with values remaining stable across higher fractional orders (Figure 5b). However, the overfitting problem is still present in the BPNN model when predicting the leaf N/P ratio of plants, as shown by the large differences in R^2 between the training and validation sets.

The value of R^2 of the GRNN model displays a general increasing trend as the fractional differentiation orders increase between FD (0) and FD (1.2) for training sets and then remain stable (Figure 5c). The R^2 value for the validation set increases with increasing fractional order from FD (0.0) to FD (1.7). However, similar to the BPNN model, the overfitting problem still exists in the GRNN model, as shown by the differences in the values of R^2 between training and validation sets.

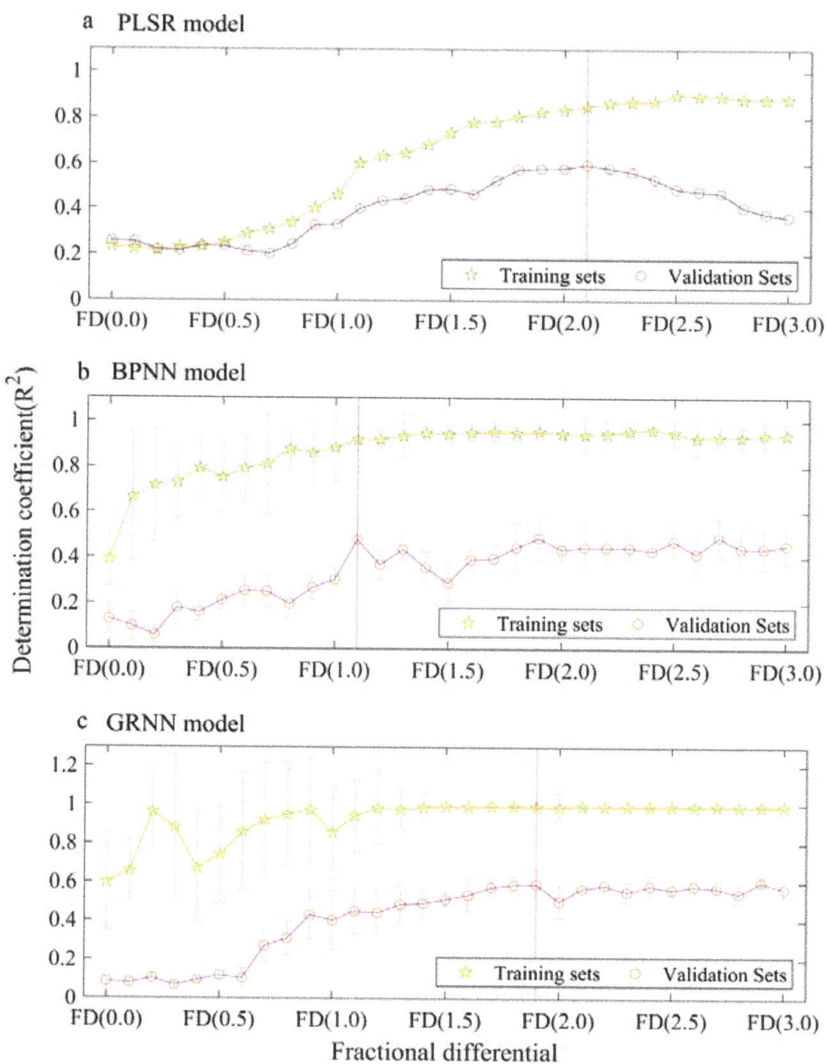

Figure 5. Trends of single model ((**a**) is the PLSR model, (**b**) is the BPNN model, and (**c**) is the GRNN model) methods from zero to third order of fractional differentiation R^2. The error bars represent two standard errors of each estimate, where available. The red vertical line indicates the position of the optimal fractional differentiation.

3.3. Performance of Combined Models Using Fractional Differentiation of Reflectance

To overcome the overfitting problem of models in predicting the N/P ratio of plant leaves, a combined model using PLSR+BPNN methods was applied. This PLSR+BPNN model used the principal components extracted from the PLSR model as input variables. The overfitting issue appears to be well-controlled by this method, as shown by the minor differences in R^2 between the training and validation sets (Figure 6a). The PLSR+BPNN model displays the best performance in predicting the leaf N/P ratio of plants when fractional differentiation was set to FD (2.3), yielding R^2, RMSE, and RPD values of 0.90, 1.94, and 3.21, respectively, for training sets and 0.79, 2.71, and 2.04 for validation sets (Figure 7).

The combined PLSR+GRNN model also uses principal components extracted from the PLSR model as input variables. When the fractional differential is larger than 1.7, the differences in R^2 values between the training and validation sets are minor (Figure 6b), suggesting that the overfitting is well-controlled compared to using the PLSR or GRNN models individually. The PLSR+GRNN performed well when the fractional differentiation was set to FD (2.6), with R^2, RMSE, and RPD values of 0.91, 1.98, and 3.15, respectively, for the training sets, and 0.81, 2.46, and 2.25, respectively, for validation sets (Figure 7).

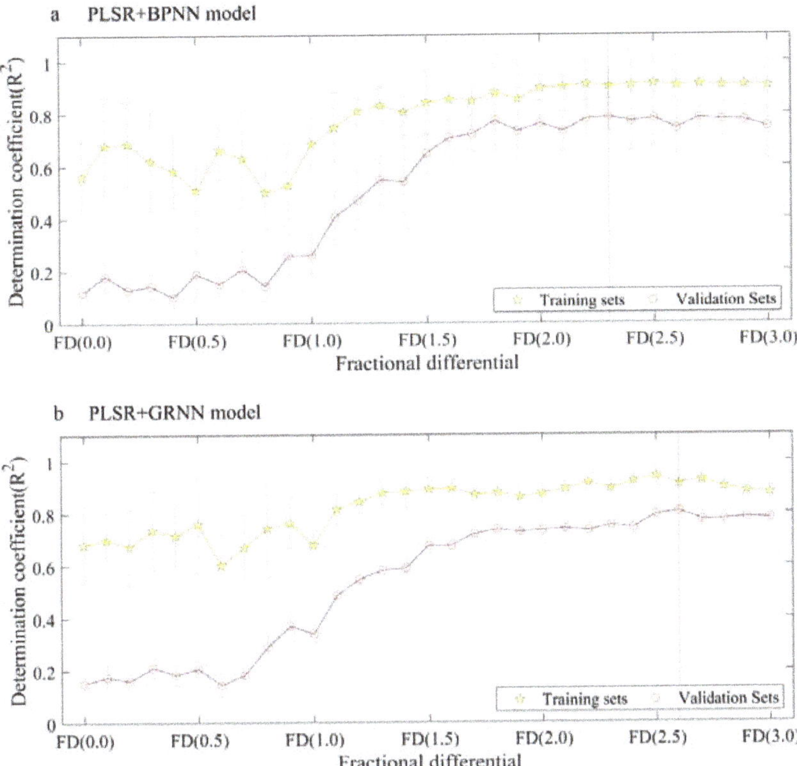

Figure 6. Trends of the differentiation coefficient (R^2) for combined model ((**a**) is the PLSR+BPNN model and (**b**) is the PLSR+GRNN model) methods. The error bars represent two standard errors of each estimate. The red vertical line indicates the position of the optimal fractional differentiation.

3.4. Model Comparison and Optimal Model Selection

In this study, five models, namely PLSR, BPNN, GRNN, PLSR+BPNN, and PLSR+GRNN, combined with fractional differentiation techniques, were used to predict the leaf N/P ratio of plants in the karst area of Guangxi Province. The optimal fractional differentiation prediction results of each model are shown in Figure 7. The prediction accuracy of these five models can be ranked in descending order as GRNN, BPNN, PLSR+GRNN, PLSR+BPNN, and PLSR according to the coefficient of determination (R^2) of the training sets, and PLSR+GRNN, PLSR+BPNN, PLSR, GRNN, and BPNN according to the coefficient of determination (R^2) of the validation sets.

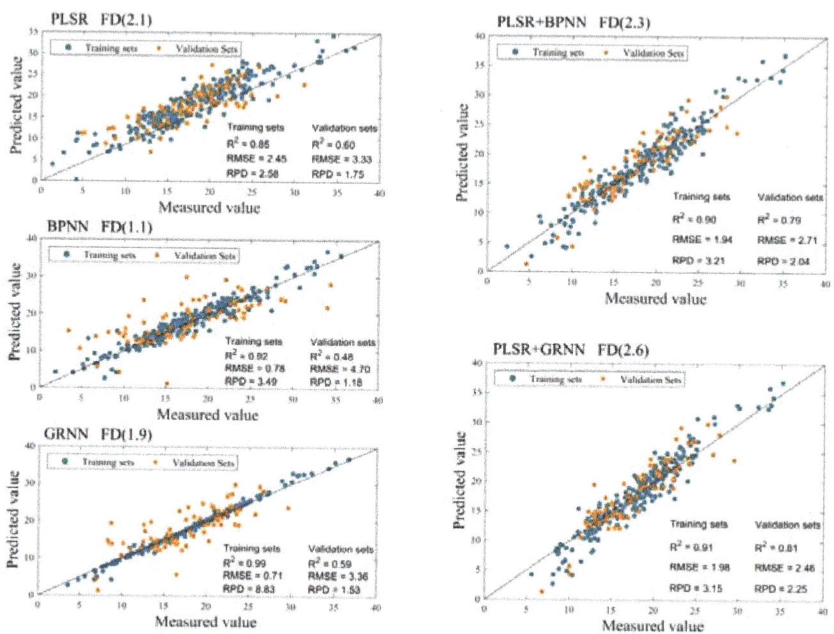

Figure 7. Leaf N/P ratio prediction accuracy of optimal fractional differentiation for each model ($p < 0.01$).

The PLSR+BPNN and PLSR+GRNN are excellent models for predicting the leaf N/P ratio of plants in karst area, as they display high prediction accuracy and successfully control overfitting. The PLSR+GRNN model is slightly better than the PLSR+BPNN and is selected as the optimal model in this study.

3.5. Advantages of Fractional Differentiation

Fractional differentiation of spectra can improve the performance of models in predicting the N/P ratio of plant leaves. The optimal differentiation for the five models used in this study is fractional differentiation rather than integer differentiation (Table 3). For example, the best fractional differentiation of the PLSR model is FD (2.1), with an RPD of 2.45 for the training sets and an RPD of 1.57 for the validation sets. The PLSR model with a fractional differentiation of 2.1 produced more accurate and robust values than zero, first, second, and third orders differentiation. Additionally, the optimal fractional differentiation of the PLSR+BPNN model and the PLSR+GRNN model is FD (2.3) and FD (2.6), respectively, which both produce better values than zero, first, second, and third orders differentiation. These results suggest that the fractional differential transform plays a positive role in predicting the leaf N/P ratio of plants.

Table 3. Accuracy of optimal fractional differentiation versus integer differentiation.

Model	Orders	Training Sets R^2	Training Sets p	Training Sets RMSE	Training Sets RPD	Validation Sets R^2	Validation Sets p	Validation Sets RMSE	Validation Sets RPD
	FD (0.0)	0.23	0.00	5.51	1.15	0.26	0.00	4.51	1.16
	FD (1.0)	0.46	0.00	4.62	1.37	0.33	0.00	4.37	1.19
PLSR	FD (2.0)	0.84	0.00	2.55	2.47	0.58	0.00	3.40	1.53
	FD (3.0)	0.88	0.00	2.17	2.91	0.37	0.00	4.32	1.20
	FD (2.1)	0.85	0.00	2.45	2.58	0.60	0.00	3.33	1.57
	FD (0.0)	0.39	0.00	5.96	1.05	0.13	0.00	8.12	0.68
	FD (1.0)	0.88	0.00	2.25	2.77	0.30	0.00	5.33	1.04
BPNN	FD (2.0)	0.95	0.00	1.59	3.91	0.44	0.00	5.03	1.10
	FD (3.0)	0.94	0.00	1.63	3.83	0.46	0.00	4.32	1.28
	FD (1.1)	0.92	0.00	1.78	3.49	0.48	0.00	4.70	1.18

Table 3. *Cont.*

Model	Orders	Training Sets R^2	Training Sets p	Training Sets RMSE	Training Sets RPD	Validation Sets R^2	Validation Sets p	Validation Sets RMSE	Validation Sets RPD
GRNN	FD (0.0)	0.60	0.00	4.43	1.41	0.09	0.01	5.38	1.03
	FD (1.0)	0.86	0.00	3.27	1.91	0.40	0.00	4.37	1.26
	FD (2.0)	0.99	0.00	0.86	7.20	0.50	0.00	4.33	1.28
	FD (3.0)	0.99	0.00	0.83	7.54	0.57	0.00	3.69	1.50
	FD (1.9)	0.99	0.00	0.71	8.83	0.59	0.00	3.61	1.53
PLSR+BPNN	FD (0.0)	0.56	0.00	4.13	1.51	0.12	0.00	5.67	0.98
	FD (1.0)	0.68	0.00	3.54	1.76	0.26	0.00	5.08	1.09
	FD (2.0)	0.90	0.00	2.00	3.11	0.76	0.00	2.78	1.99
	FD (3.0)	0.90	0.00	2.03	3.07	0.75	0.00	3.01	1.84
	FD (2.3)	0.90	0.00	1.94	3.21	0.79	0.00	2.71	2.04
PLSR+GRNN	FD (0.0)	0.68	0.00	3.95	1.58	0.80	0.00	5.08	1.09
	FD (1.0)	0.68	0.00	3.87	1.61	0.80	0.00	4.49	1.23
	FD (2.0)	0.87	0.00	2.44	2.56	0.80	0.00	2.86	1.93
	FD (3.0)	0.88	0.00	2.36	2.64	0.80	0.00	2.67	2.07
	FD (2.6)	0.91	0.00	1.98	3.15	0.81	0.00	2.46	2.25

4. Discussion

4.1. Distribution of Sensitive Wavelengths

The spectral reflectance from 400–730 nm, and especially 520–650 nm, are sensitive wavelengths for predicting the leaf N/P ratio of plants. This result is consistent with previous studies, such as Cui et al. [8], who found that near 650 nm was the best wavelength for estimating the N/P ratio. Hansen and Schjoerring [48] also showed that spectral reflectance near 530 nm and 720 nm are important wavelengths for estimating the N concentration of wheat. In addition, Xu et al. [49] found that spectral reflectance of 540–560 nm and 760–780 nm are sensitive wavelengths for the C/N ratio of wheat and barley leaves. There are some differences in sensitive wavelengths related to the leaf N/P ratio between the karst and non-karst plants, which may be caused by adaptions to survive in karst environments. For example, plants growing in karst areas have decreased stomatal conductance, thickened palisade tissue, and increased keratinization due to thin soil layers and relatively low air humidity [50]. These physiological differences can impact the radiative transfer processes of plant leaves, leading to changes in the sensitive wavebands.

4.2. Control Overfitting

We reduced the noise in the spectral reflectance by fractional differentiation and improved the representativeness of the feature variables for the model input. Moreover, we applied the PLSR model to extract principal components to reduce the dimensionality of the spectral data. The reduced-dimensional spectral data are used as input variables of the BPNN and GRNN models. In this way, we proposed two composite models, PLSR+BPNN and PLSR+GRNN. The prediction performance of PLSR+BPNN and PLSR+GRNN models will be better than simple models such as spectral index and ordinary linear regression. The simple model is more adapted to a single plant species [14,51,52] and very sensitive to databases [16]. In contrast, our model is more adaptable to complex environments as it is set up in a mixed forest database. On the other hand, this method decreases the complexity of the model while ensuring minimal loss of spectral information and reducing the occurrence of overfitting. These results are consistent with previous studies [53] that show overfitting issues could be overcome by using non-negative principal component analysis (NPCA) to extract principal components as input variables for machine learning. Although combining PCA or PLSR models with machine learning methods does not help improve the model prediction capability, it effectively minimizes the overfitting problem of machine learning methods.

4.3. Application for Mixed Forest

Empirical models tend to be site-, time-, and species-specific and are therefore unsuitable for large-scale analyses [54]. Previous studies investigating the inversion of plant biochemical parameters are also biased towards specific regions, such as wetlands [24] or grasslands [28], or species such as wheat [55] or rice [17]. Although high accuracy can be obtained from the inversion of biochemical parameters for a single species [15], plants are more likely to co-exist in a mixed form in the natural environment, and studies on only one species are not applicable to diverse ecosystems. Studies across various species need to be conducted before these methods can be applied to mixed forest environments. However, it is crucial to improve the model accuracy and robustness before the inversion of biochemical parameters of mixed species. We applied five models to predict the leaf N/P ratio of plants in a karst area. Among them, the PLSR+BPNN and PLSR+GRNN models have the best performance, with high prediction accuracy and robustness. The performance of both the PLSR+BPNN and PLSR+GRNN models was better than the models used by Cheng et al. [16] in terms of the coefficient of determination (R^2). This improved performance is mainly due to the full consideration of both linear and nonlinear relationships between leaf biochemical parameters and field spectroradiometer data in our study.

Sample composition also impacts the model performance. For example, there was a significant difference in the performance of the BPNN model between our study and Cui et al. [8] when comparing the coefficient of determination for validation tests, and no overfitting was observed in the results of Cui et al. [8]. These differences in performance were due to the large variability in sample composition, with leaf N/P ratio of plants in this study ranging from 1.34 to 36.94 compared to that of phragmites communis N/P ratio ranging from 6.7 to 15.9 in the Cui et al. study [8].

5. Conclusions

Estimating the N/P ratio of plant leaves using field spectroradiometer data is challenging. We estimated the variation of leaf N/P ratio of plants in a karst area of Guangxi Province using five models, namely PLSR, BPNN, GRNN, PLSR+BPNN, and PLSR+GRNN. The sensitivity wavelengths of the leaf N/P ratio of plants are mainly in the range of 400–730 nm. Applying a single model (such as PLSR, or BPNN, and or GRNN) can estimate the leaf N/P ratio of plants, but all methods produce significant overfitting of the data. In contrast, the combined models of PLSR+BPNN and PLSR+GRNN can avoid the overfitting problem in predicting the leaf N/P ratio of plants, and have high accuracy prediction capabilities. In addition, using fractional differentiation methods can effectively improve the prediction capability of the model in estimating the leaf N/P ratio across a variety of plant species. This study provides a valuable scientific basis for long-term dynamic monitoring of plant biochemical parameters using field spectroradiometer data.

Author Contributions: Conceptualization, W.H., J.W. and Y.Y.; Methodology, W.H. and Y.L.; Software, W.H.; Validation, J.W., Y.Y. and Y.L.; Formal Analysis, W.H. and Y.L.; Investigation, W.H., L.Y., L.N. and Y.L.; Resources, Y.Y.; Data Curation, L.N.; Writing—Original Draft Preparation, W.H. and L.Y.; Writing—Review and Editing, J.W., Y.Y., D.G. and Y.L.; Visualization, W.H.; Supervision, Y.L.; Project Administration, Y.Y.; Funding Acquisition, W.H. and Y.L. All authors have read and agreed to the published version of the manuscript.

Funding: This research was supported by the National Natural Science Foundation of Guangxi (2019GXNSFBA245036), the National Natural Science Foundation of China (32060369), the Basic Scientific Research Fund of Guangxi Academy of Science (2019YJJ1009), and the Basic Ability Improvement Scientific Research Fund of Young and Middle-aged Teachers in Guangxi Universities (2020KY58008).

Institutional Review Board Statement: Not applicable.

Informed Consent Statement: Not applicable.

Data Availability Statement: Not applicable.

Acknowledgments: We thank the Nonggang National Nature Reserve and the Huixian Karst Ecology and Water Ecology Research Station for supporting the field experiments. We also thank Dongxing Li, Ting Chen, Qiumei Teng, Weiting Hu, Jingbao Chen, Xiangxiang Chen, Lijie Qin, Ning Li and Jianmiao Long for helping with field data collection and the N and P element extraction experiments.

Conflicts of Interest: The authors declare no conflict of interest.

References

1. Vrede, T.; Dobberfuhl, D.R.; Kooijman, S.A.L.M.; Elser, J.J. Fundamental connections among organism C: N: P stoichiometry, macromolecular composition, and growth. *Ecology* **2004**, *85*, 1217–1229. [CrossRef]
2. Elser, J.J.; Sterner, R.W.; Gorokhova, E.; Fagan, W.F.; Markow, T.A.; Cotner, J.B.; Harrison, J.F.; Hobbie, S.E.; Odell, G.M.; Weider, L.W. Biological stoichiometry from genes to ecosystems. *Ecol. Lett.* **2000**, *3*, 540–550. [CrossRef]
3. Liu, J.; Gou, X.; Zhang, F.; Bian, R.; Yin, D. Spatial patterns in the C: N: P stoichiometry in Qinghai spruce and the soil across the Qilian Mountains, China. *Catena* **2020**, *196*, 104814. [CrossRef]
4. Elser, J.J.; Fagan, W.F.; Kerkhoff, A.J.; Swenson, N.G.; Enquist, B.J. Biological stoichiometry of plant production: Metabolism, scaling and ecological response to global change. *New Phytol.* **2010**, *186*, 593–608. [CrossRef]
5. Batterman, S.A.; Wurzburger, N.; Hedin, L.O.; Austin, A. Nitrogen and phosphorus interact to control tropical symbiotic N_2 fixation: A test inInga punctata. *J. Ecol.* **2013**, *101*, 1400–1408. [CrossRef]
6. Koerselman, W.; Meuleman, A.M.F. The Vegetation N: P Ratio: A New Tool to Detect the Nature of Nutrient Limitation. *J. Appl. Ecol.* **1996**, *33*, 1441–1450. [CrossRef]
7. Gusewell, S. Variation in nitrogen and phosphorus concentrations of wetland plants. Perspect. *Plant Ecol. Evol. Syst.* **2002**, *5*, 37–61. [CrossRef]
8. Cui, L.; Dou, Z.; Liu, Z.; Zuo, X.; Lei, Y.; Li, J.; Zhao, X.; Zhai, X.; Pan, X.; Li, W. Hyperspectral inversion of phragmites communis carbon, nitrogen, and phosphorus stoichiometry using three models. *Remote Sens.* **2020**, *12*, 1998. [CrossRef]
9. Rei, S.; Quan, W. Towards a Universal Hyperspectral Index to Assess Chlorophyll Content in Deciduous Forests. *Remote Sens.* **2017**, *9*, 191.
10. Zhao, H.S.; Zhu, X.C.; Li, C.; Wei, Y.; Zhao, G.X.; Jiang, Y.M. Improving the accuracy of the hyperspectral model for apple canopy water content prediction using the equidistant sampling method. *Sci. Rep.* **2017**, *7*, 11192. [CrossRef] [PubMed]
11. Yamashita, H.; Sonobe, R.; Hirono, Y.; Morita, A.; Ikka, T. Dissection of hyperspectral reflectance to estimate nitrogen and chlorophyll contents in tea leaves based on machine learning algorithms. *Sci. Rep.* **2020**, *10*, 17360. [CrossRef]
12. Barnaby, J.Y.; Huggins, T.D.; Lee, H.; McClung, A.M.; Pinson, S.R.M.; Oh, M.; Bauchan, G.R.; Tarpley, L.; Lee, K.; Kim, M.S.; et al. Vis/NIR hyperspectral imaging distinguishes sub-population, production environment, and physicochemical grain properties in rice. *Sci. Rep.* **2020**, *10*, 9284. [CrossRef]
13. El-Hendawy, S.; Al-Suhaibani, N.; Alotaibi, M.; Hassan, W.; Elsayed, S.; Tahir, M.U.; Mohamed, A.I.; Schmidhalter, U. Estimating growth and photosynthetic properties of wheat grown in simulated saline field conditions using hyperspectral reflectance sensing and multivariate analysis. *Sci. Rep.* **2019**, *9*, 16473. [CrossRef]
14. Meacham-Hensold, K.; Montes, C.M.; Wu, J.; Guan, K.; Fu, P.; Ainsworth, E.A.; Pederson, T.; Moore, C.E.; Brown, K.L.; Raines, C.; et al. High-throughput field phenotyping using hyperspectral reflectance and partial least squares regression (PLSR) reveals genetic modifications to photosynthetic capacity. *Remote Sens. Environ.* **2019**, *231*, 111176. [CrossRef] [PubMed]
15. Sonobe, R.; Hirono, Y.; Oi, A. Non-destructive detection of tea leaf chlorophyll content using hyperspectral reflectance and machine learning algorithms. *Plants* **2020**, *9*, 368. [CrossRef]
16. Cheng, T.; Rivard, B.; Sánchez-Azofeifa, A.G.; Féret, J.-B.; Jacquemoud, S.; Ustin, S.L. Deriving leaf mass per area (LMA) from foliar reflectance across a variety of plant species using continuous wavelet analysis. *ISPRS J. Photogramm. Remote Sens.* **2014**, *87*, 28–38. [CrossRef]
17. Inoue, Y.; Sakaiya, E.; Zhu, Y.; Takahashi, W. Diagnostic mapping of canopy nitrogen content in rice based on hyperspectral measurements. *Remote Sens. Environ.* **2012**, *126*, 210–221. [CrossRef]
18. Doktor, D.; Lausch, A.; Spengler, D.; Thurner, M. Extraction of plant physiological status from hyperspectral signatures using machine learning methods. *Remote Sens.* **2014**, *6*, 12247–12274. [CrossRef]
19. Oh, K.; Chung, Y.C.; Kim, K.W.; Kim, W.S.; Oh, I.S. Classification and visualization of Alzheimer's disease using volumetric convolutional neural network and transfer learning. *Sci. Rep.* **2019**, *9*, 18150. [CrossRef]
20. Gansfort, B.; Traunspurger, W. Environmental factors and river network position allow prediction of benthic community assemblies: A model of nematode metacommunities. *Sci. Rep.* **2019**, *9*, 14716. [CrossRef]
21. Sengupta, D.; Bandyopadhyay, S.; Sinha, D. A scoring scheme for online feature selection: Simulating model performance without retraining. *IEEE Trans. Neural Netw. Learn. Syst.* **2017**, *28*, 405–414. [CrossRef]
22. Cheng, Q.; Zhou, H.; Cheng, J. The fisher-markov selector: Fast selecting maximally separable feature subset for multiclass classification with applications to high-dimensional data. *IEEE Trans. Pattern Anal. Mach. Intell.* **2010**, *33*, 1217–1233. [CrossRef]
23. Barzegar, R.; Moghaddam, A.A.; Deo, R.; Fijani, E.; Tziritis, E. Mapping groundwater contamination risk of multiple aquifers using multi-model ensemble of machine learning algorithms. *Sci. Total Environ.* **2018**, *621*, 697–712. [CrossRef]
24. Dou, Z.; Li, Y.; Cui, L.; Pan, X.; Ma, Q.; Huang, Y.; Lei, Y.; Li, J.; Zhao, X.; Li, W. Hyperspectral inversion of Suaeda salsa bi omass under different types of human activity in Liaohe Estuary wetland in north-eastern China. *MAR Freshw. Res.* **2020**, *71*, 482–492. [CrossRef]
25. Guo, C.; Guo, X. Estimating leaf chlorophyll and nitrogen content of wetland emergent plants using hyperspectral data in the visible domain. *Spectrosc. Lett.* **2015**, *49*, 180–187. [CrossRef]

26. Peuelas, J.; Gamon, J.A.; Fredeen, A.L.; Merino, J.; Field, C.B. Reflectance indices associated with physiological changes in nitrogen- and water-limited sunflower leaves. *Remote Sens. Environ.* **1994**, *48*, 135–146. [CrossRef]

27. Benkhettou, N.; Brito da Cruz, A.M.C.; Torres, D.F.M. A fractional calculus on arbitrary time scales: Fractional differentiation and fractional integration. *Sig. Process.* **2015**, *107*, 230–237. [CrossRef]

28. Gao, J.; Liu, J.; Liang, T.; Hou, M.; Ge, J.; Feng, Q.; Wu, C.; Li, W. Mapping the forage nitrogen-phosphorus ratio based on Sentinel-2 MSI data and a random forest algorithm in an alpine grassland ecosystem of the Tibetan Plateau. *Remote Sens.* **2020**, *12*, 2929. [CrossRef]

29. Yuan, D. World correlation of karst ecosystem: Objectives and implementation plan. *China Adv. Earth Sci.* **2001**, *4*, 461–466 (In Chinese)

30. Luo, Z.; Tang, S.; Jiang, Z.; Chen, J.; Fang, H.; Li, C. Conservation of terrestrial vertebrates in a global hotspot of karst area in southwestern China. *Sci. Rep.* **2016**, *6*, 25717. [CrossRef]

31. Orme, C.D.; Davies, R.G.; Burgess, M.; Eigenbrod, F.; Pickup, N.; Olson, V.A.; Webster, A.J.; Ding, T.S.; Rasmussen, P.C.; Ridgely, R.S.; et al. Global hotspots of species richness are not congruent with endemism or threat. *Nature* **2005**, *436*, 1016–1019. [CrossRef]

32. Zeng, F.; Peng, W.; Song, T.; Wang, K.; Wu, H.; Song, X.; Zeng, Z. Changes in vegetation after 22 years' natural restoration in the karst disturbed area in northwestern Guangxi, China. *Acta Ecol. Sin.* **2007**, *27*, 5110–5119. [CrossRef]

33. SU, Z.; LI, X. The types of natural vegetation in karst region of Guangxi and its classified system. *Guihaia* **2003**, *23*, 289–293.

34. Shimadzu, S.; Seo, M.; Terashima, I.; Yamori, W. Whole irradiated plant leaves showed faster photosynthetic induction than individually irradiated leaves via improved stomatal opening. *Front. Plant Sci.* **2019**, *10*, 1512. [CrossRef] [PubMed]

35. Richard, H.L.; Donald, L.S. Soil Science of America and American Society of Agronomy. In *Methods of Soil Analysis. Part 3 Chemical Methods*; Soil Science Society of America: Madison, WI, USA, 1996; pp. 1085–1121.

36. Bao, S.D. *Soil and Agricultural Chemistry Analysis*; Agriculture Publication: Beijing, China, 2000; pp. 71–79.

37. Kharintsev, S.S.; Salakhov, M. A simple method to extract spectral parameters using fractional derivative spectrometry. *Spectrochim. Acta A Mol. Biomol. Spectrosc.* **2004**, *60*, 2125–2133. [CrossRef] [PubMed]

38. Han, Z.; Li, S.; Liu, H. Composite learning sliding mode synchronization of chaotic fractional-order neural networks. *J. Adv. Res.* **2020**, *25*, 87–96. [CrossRef] [PubMed]

39. Rumelhart, D.E.; Hinton, G.E.; Williams, R.J. Learning representations by back-propagating errors. *Nature* **1986**, *323*, 533–536. [CrossRef]

40. Stathakis, D. How many hidden layers and nodes? *Int. J. Remote Sens.* **2009**, *30*, 2133–2147. [CrossRef]

41. Specht, D.F. A general regression neural network. *IEEE Trans. Neural Netw.* **1991**, *2*, 568–576. [CrossRef]

42. Lotfinejad, M.; Hafezi, R.; Khanali, M.; Hosseini, S.; Mehrpooya, M.; Shamshirband, S. A comparative assessment of predicting daily solar radiation using bat neural network (BNN), generalized regression neural network (GRNN), and neuro-fuzzy (NF) system: A case study. *Energies* **2018**, *11*, 1188. [CrossRef]

43. Chang, C.-W.; Laird, D.A.; Mausbach, M.J.; Hurburgh, C.R. Near-infrared reflectance spectroscopy–principal components regression analyses of soil properties. *Soil Sci. Soc. Am. J.* **2001**, *65*, 480–490. [CrossRef]

44. Reich, P.B.; Oleksyn, J. Global patterns of plant leaf N and P in relation to temperature and latitude. *Proc. Natl. Acad. Sci. USA* **2004**, *101*, 11001–11006. [CrossRef]

45. Han, W.; Fang, J.; Guo, D.; Zhang, Y. Leaf nitrogen and phosphorus stoichiometry across 753 terrestrial plant species in China. *New Phytol.* **2005**, *168*, 377–385. [CrossRef] [PubMed]

46. Elser, J.J.; Fagan, W.F.; Denno, R.F.; Dobberfuhl, D.R.; Folarin, A.; Huberty, A.; Interlandi, S.; Kilham, S.S.; McCauley, E.; Schulz, K.L. Nutritional constraints in terrestrial and freshwater food webs. *Nature* **2000**, *408*, 578–580. [CrossRef] [PubMed]

47. Yang, H.; Li, Q.; Tu, C.; Cao, J. Carbon, nitrogen and phosphorus stoichiometry of typical plants in karst area of Maocun, Guilin. *Guangxi Zhiwu/Guihaia* **2015**, *35*, 493–555.

48. Hansen, P.M.; Schjoerring, J.K. Reflectance measurement of canopy biomass and nitrogen status in wheat crops using normalized difference vegetation indices and partial least squares regression. *Remote Sens. Environ.* **2003**, *86*, 542–553. [CrossRef]

49. Xu, X.; Yang, G.; Yang, X.; Li, Z.; Feng, H.; Xu, B.; Zhao, X. Monitoring ratio of carbon to nitrogen (C/N) in wheat and barley leaves by using spectral slope features with branch-and-bound algorithm. *Sci. Rep.* **2018**, *8*, 10034. [CrossRef]

50. Li, R.; Wang, S.; Du Xuelian, Y.G. Relationship among leaf anatomical characters and foliar$\delta\sim$ (13) C values of six woody species for karst rocky desertification areas. *China Sci. Silvae Sin.* **2008**, *44*, 29–34.

51. Liu, X.; Guanter, L.; Liu, L.; Damm, A.; Malenovský, Z.; Rascher, U.; Peng, D.; Du, S.; Gastellu-Etchegorry, J.-P. Downscaling of solar-induced chlorophyll fluorescence from canopy level to photosystem level using a random forest model. *Remote Sens. Environ.* **2018**, *231*, 110772. [CrossRef]

52. Jay, S.; Maupas, F.; Bendoula, R.; Gorretta, N. Retrieving LAI, chlorophyll and nitrogen contents in sugar beet crops from multi-angular optical remote sensing: Comparison of vegetation indices and PROSAIL inversion for field phenotyping. *Field Crop. Res.* **2017**, *210*, 33–46. [CrossRef]

53. Han, X. Nonnegative principal component analysis for cancer molecular pattern discovery. *IEEE/ACM Trans. Comput. Biol. Bioinform.* **2010**, *7*, 537–549. [PubMed]

24. Zhu, Z.; Bi, J.; Pan, Y.; Ganguly, S.; Anav, A.; Xu, L.; Samanta, A.; Piao, S.; Nemani, R.; Myneni, R. Global data sets of vegetation leaf area index (LAI) 3g and fraction of photosynthetically active radiation (FPAR) 3g derived from global inventory modeling and mapping studies (GIMMS) normalized difference vegetation index (NDVI 3g) for the period 1981 to 2011. *Remote Sens.* **2013**, *5*, 927–948.

25. Camino, C.; Gonzalez-Dugo, V.; Hernandez, P.; Zarco-Tejada, P.J. Radiative transfer Vcmax estimation from hyperspectral imagery and SIF retrievals to assess photosynthetic performance in rainfed and irrigated plant phenotyping trials. *Remote Sens. Environ.* **2019**, *231*, 111186. [CrossRef]

MDPI

St. Alban-Anlage 66

4052 Basel

Switzerland

Tel. +41 61 683 77 34

Fax +41 61 302 89 18

www.mdpi.com

Remote Sensing Editorial Office

E-mail: remotesensing@mdpi.com

www.mdpi.com/journal/remotesensing